Social History of Africa
CUTTING DOWN TREES

SOCIAL HISTORY OF AFRICA

Series Editors:
Allen Isaacman and Jean Hay

CUTTING DOWN TREES
GENDER, NUTRITION, AND AGRICULTURAL CHANGE IN THE NORTHERN PROVINCE OF ZAMBIA
1890–1990

Henrietta L. Moore
London School of Economics

Megan Vaughan
Nuffield College, Oxford

HEINEMANN

Portsmouth, NH

JAMES CURREY

London

UNIVERSITY OF
ZAMBIA PRESS

Lusaka

Heinemann
A division of Reed
Publishing (USA) Inc.
361 Hanover Street
Portsmouth, NH 03801-3959

James Currey Ltd
54b Thornhill Square, Islington
London N1 1BE

University of Zambia Press
P.O. Box 32379
Lusaka 10101

ISBN 0-435-08088-1 (Heinemann cloth)
ISBN 0-435-08090-3 (Heinemann paper)
ISBN 0-85255-662-4 (James Currey cloth)
ISBN 0-85255-612-8 (James Currey paper)

First published 1994.

Library of Congress Cataloging-in-Publication Data
Moore, Henrietta L.
    Cutting down trees : gender, nutrition, and agricultural change in
the Northern Province of Zambia, 1890–1990 / Henrietta L. Moore,
Megan Vaughan.
        p.   cm.
    Includes bibliographical references (p.   ) and index.
    ISBN 0-435-08088-1 (cloth). — ISBN 0-435-08090-3 (paper)
    1. Bemba (African people)—Agriculture.   2. Bemba (African
people)—Economic conditions.   3. Northern Province (Zambia)—
Economic conditions.   4. Food supply—Zambia—Northern Province.
5. Bemba (African people)—Nutrition.   I. Vaughan, Megan.
II. Title.
DT3058.B46M66   1993
330.96894—dc20   93-5650
                                                                    CIP

British Library Cataloguing in Publication Data
Moore, Henrietta L.
    Cutting Down Trees: Agricultural Change,
    Anthropology and History Among the Bemba
    of Northern Zambia, 1890–1990.—(Social
    History of Africa Series)
    I. Title.   II. Vaughan, Megan   III. Series
    968.94

Cover design by Jenny Greenleaf.
Printed in the United States of America on acid-free paper.
97 96 95 94 93 BB 1 2 3 4 5 6

IN MEMORIAM
Audrey Isabelle Richards
and
Ann Tweedie-Waggott

CONTENTS

ACKNOWLEDGMENTS

This book is one outcome of a collaborative research project, set up in 1986, between the University of Zambia and the African Studies Centre, University of Cambridge. We would like to acknowledge, in particular, the enormous contribution made to this project by Dr. Chipasha Luchembe, formerly of the Department of History, University of Zambia. It was he who nurtured our interest in the Northern Province and guided our understanding of the history and anthropology of the region.

While in Zambia we were affiliated with both the Department of History and the Institute of African Studies. We are grateful to the director of the Institute and to Ilse Mwanza for their help, support and encouragement. Permission to conduct research was granted by the Government of Zambia, and the research was funded by the British Academy, the Smuts Research Fund (University of Cambridge), the Hayter Fund (University of Oxford) and the London School of Economics and Political Science. We are grateful to all these bodies for their support.

A number of other institutions and individuals in Zambia greatly facilitated the research. These include the director and staff of the National Archives of Zambia, Dr. Bonnie Keller, Dr. Hugh Macmillan and Viv and Chaloka Beyani. While conducting fieldwork in the Northern Province, we benefited enormously from the hospitality and local knowledge of Nakulu Kasonde and Theodoro Luchembe. We owe a special debt to the White Fathers of Chilubula Mission, and especially Father Dan Sherry, whose friendship has meant a great deal to us. We would also like to thank the Sisters of Child Jesus and the staff of Chilubula Mission Hospital and St. Mary's Secondary School. The sisters took us in on many occasions and made us welcome despite our ceaseless questions. The Brothers of the Sacred Heart at Malole provided us with hospitality and help. We are also grateful to the late Archbishop of Kasama for permission to consult parish and diocesan records. Also in Kasama, we benefited from discussions with the staff of the Adaptive Research Planning Team, especially Dr. Richard Bolt. We are very grateful for their help, as well as for assistance from the Regional Planning Team and the Integrated Rural Development Project (SMC) at Mpika. We would like especially to thank Liz and Mike Shone for their hospitality and friendship.

We have received much help over the many years it has taken to complete this research project from the African Studies Centre, Cambridge; and we would like to thank the director and staff of the centre, and most especially Paula Munro, for their constant encouragement and support. We are also extremely grateful to the White Fathers for permission to consult their archives in Rome and for the help and assistance of the former archivist, Père Lamey. We acknowledge with thanks the help of Dr. Angela Raspin, archivist at the London School of Economics; the staff of Rhodes House Library, Oxford, and Chris Wright of the Royal Anthropological Institute, London. We are indebted to the following individuals and institutions for permission to consult, quote from and republish material: Mr. Stanley Tweedie-Waggott (for permission to consult and quote from the papers of the late Ann Tweedie); Rhodes House Library (for permission to quote from the Melland papers); the White Fathers (for permission to quote from their archives); the Royal Anthropological Institute (for permission to reprint photographs from the Richards archive); the London School of Economics (for permission to consult the Richards archive); to the editors of *Africa* and to Dr. Peter Stromgaard for permission to reprint the figure on page   ; and to Manchester University Press for permission to reprint maps on pages .

Finally, we would like to thank Marilyn Strathern and Luise White for reading earlier drafts of this work and for their insightful comments. Our greatest debt, however, is to the people in the Chilubula area who bore the brunt of our research efforts. We owe an equal debt to our research assistants, Mr. Boniface Sambo, Mr. Henry Musonda, Mr. Francis Mubanga, Mr. Abraham Chalota Sichilya, and Mr. Joseph Nkumbula. Their concerned interest in the causes and consequences of rural poverty; their professionalism, hard work and dedication, and their willingness to interpret for us the "local knowledge" of communities in the area have made this book possible.

Introduction
In And Out of Context:
The Problems of
a Re-Study

When we began research in the Northern Province in Zambia in 1986, we intended to conduct a re-study of the work of anthropologist Audrey Richards. Published in 1939, Richards' book *Land, Labour and Diet: An Economic Study of the Bemba Tribe* (Richards 1939) spoke to many concerns of the 1980s. Her attention to gender relations, to the burden of women's workload in a rural economy, and her attempt to measure well-being in terms of diet, all found echoes in the more recent literature on the problems of "development" in rural Africa in general, and in the Northern Province of Zambia in particular. Her painstakingly detailed descriptions of what African rural producers actually do, and what they know about soils and vegetation, trees and bees, and mushrooms and caterpillars also found echoes in the academic and "development" worlds, in the literature on "indigenous knowledge" and in the literature on the ecological history of Africa. Richards had chosen to study the Bemba-speaking people, famous for their "slash-and-burn" system of agriculture (citemene), a system whose decline many an expert had predicted, but which was very evidently still being practiced (and still being studied) in the 1980s. The combination of Richards' published work on the Bemba and her unpublished papers (housed in the archives of the London School of Economics) seemed to provide a perfect baseline for a longitudinal study of changes in agricultural production, diet, and gender relations in a rural African society. Richards' work also had other attractions. She not only documented in detail how rural people made a living, what they produced, and what they consumed, but she also documented the social relations of production and consumption in this society. These social relations, it seemed, could not be seen as being in any way secondary to what people did; rather they were part of the very substance of life, imbricated in every meal

and in every day of agricultural labor. For Richards, an account of a system of production and consumption was also (and necessarily) an account of the kinship and
political systems of the Bemba people, as well as an account of a symbolic system.
The power and persuasiveness of *Land, Labour and Diet* lies, in part, in her demonstration of the inseparability of the material and symbolic worlds and of their
mutually constituted nature. We eventually came to have many criticisms of this
very aspect of Richards' account and approach. However, her concern with demonstrating that a system of production could only be understood if it were viewed
as constituting, and being constituted by, a set of social relations and a belief system was a welcome antidote to much of the contemporary literature on rural production systems in Africa, which often seems to assume that the economics of
production can be modeled with little or no knowledge of the relations between the
individuals that make production possible.

Locating the Study and Providing a Context

There were many reasons, then, why conducting a re-study of Richards' work was
attractive to us. It was not long, however, before the apparently straightforward
notion of a re-study came to appear less simple, and we had to revise our plans.
There were both intellectual and mundane reasons for such revisions, and sometimes the intellectual and the mundane were closely connected. Because much of
the population of the Northern Province (or at least of the central plateau of the
Northern Province) continues to shift village sites, albeit less frequently than in the
past, this posed an obvious problem for anyone aiming to re-study the villages previously studied by Richards. It was not just that these villages might have moved
elsewhere and would have to be found, but rather that the composition of the village in the Northern Province was so fluid and unstable that it might not be obvious which of many possible settlements one would consider to be the descendant
of one of the villages identified in the 1930s.[1] The mobility of the population of the
Northern Province, the scattered nature of settlement, and the frequency of fission, began to make the rationale for a re-study of a specific village (or a few villages) less clear. It was not just the mobility of the people of the Northern Province
that concerned us. We also had to contend with the mobility, and often indeterminate location, of the anthropologist during the original study. As we pored over
Richards' papers in the London School of Economics, the absence of any maps that
might have supplemented her rather vague descriptions of where she, and the villages she studied, actually were, began to pose a serious problem. It seemed to us
that Richards' sense of location was more political than spatial. In many ways she
had not conducted the classic anthropologist's village study because she had
clearly moved over very large areas of this sparsely populated and remote area,
between villages that were sometimes more than 100 miles apart. Despite the fact
that, in mere practical terms, this approach must have involved a great deal of effort, Richards pays very little attention to it in her account. Unless one reads her
book very closely, it is possible to imagine that the villages she studied were within
walking or bicycling distance from each other, although this was far from true.

In many ways, we viewed Richards' coverage of a large geographical area as an
advantage because it would enable us to take a less local, more regional per-

spective. However, once in the Northern Province, it soon became apparent that her mobility had been of a particular kind. She had certainly moved over large areas, but her points of reference had been political rather than geographical, and these were almost entirely determined by her interest, not in production and consumption, but in the Bemba political system. Richards had moved between one politically significant Bemba settlement and another. Certainly there were contrasts to be drawn between these settlements, but these contrasts were insignificant as compared with those that might have been drawn had she studied the many non-Bemba and the more nominally Bemba areas that exist all over the plateau. Her version of the Bemba system of production, and particularly of its social and symbolic significance, might, we later concluded, have been rather different if her reference points had not been those of the Bemba chiefly system. The student of history or anthropology on reading Richards' text might assume that the area she studied was inhabited only by the Bemba, and that Bemba territory was synonymous with both Bemba identity and with its boundaries. These assumptions would be false, and one of the issues we seek to examine in this book is the extent to which Bemba ethnicity was constructed both practically and symbolically not only through the actions of Bemba-speaking people, but also through the process of being represented as an object of knowledge. (See Chapters 1 and 2.)

Further, it was clear that there existed on the margins of the central plateau, to the west, east and north, areas with very different ecologies and different systems of production. These largely non-Bemba areas were occasionally mentioned in Richards' account to draw a contrast between ethnicities but, in the longer history of this region, their importance has been far greater than this. To some degree these areas, which from a Bemba political point of view were occupied by marginal and insignificant peoples, came to form the most productive niches within the colonial economy of the province. (See Chapters 4 and 6.)

We decided, therefore, that our own work could never be a direct follow-up of Richards', but would have to begin by putting into context her observations both in time and space. This we have attempted to do throughout this book. Despite the fact that the emphasis of the research summarized here (including our own field research) lies heavily in the direction of the Bemba-speaking communities of the plateau, we have tried to contextualize these observations by moving constantly between different levels of analysis, from the workings of the Bemba household to the regional economy and back. The "Bemba bias" remains, however, and to that extent the original conception of a re-study has continued to have a determining influence on the outcome of our own work. Although we attempt throughout the book to keep a provincial-level perspective, for the most part our work focuses on the districts covered by Richards' research in the Kasama, Mpika, and Chinsali districts. Even within this largely Bemba-dominated area, however, we strove to emphasize the degree of variation in social and economic organization (as well as in microecology) which exists and which has exerted such a powerful influence on the local and regional history. This means that we have had to retain a constantly shifting perspective in each chapter, and to attempt, even as we discuss the specifics of our own work or those of other researchers, to recontextualize the details in a larger frame. We recognize the perils inherent in this method, but we are committed to trying to demonstrate the effects of larger-scale processes on household and local level relations. The Bemba have always been part of a larger picture and we do not wish to lose sight of this fact.

In trying to paint this larger picture, it became apparent to us that Richards' work not only had to be contextualized within a region, but that it was also necessary to place it in historical perspective. Like many other colonial anthropologists, Richards clearly saw the tasks of her research to be the documentation and elucidation of the workings of a "traditional" society and economy, and the predicting of the effects on this society and economy of incorporation into a colonial state. In the case of the Northern Province, it was the development of the migrant labor system which appeared to be the most disruptive force at work, and Richards paid close attention to this in her study. To predict the changes which would occur in Bemba society as a result of labor migration, however, she needed to paint a picture of Bemba society prior to the impact of colonialism and capitalism. At times Richards appeared to be painting such a picture through her observations of the workings of Bemba society and politics in the 1930s, implying that what she observed was a "traditional" society still intact, the workings of which could be assumed to have remained relatively unchanged since the distant past. At other times she acknowledged that important changes had already taken place and relied, for her account of "tradition," on the stories and memories of elders and chiefs. For a number of reasons we came to the conclusion that it was impossible to treat Richards' account of Bemba society unproblemmatically as a baseline against which to measure the changes wrought by colonialism and capitalism. It was clear from a reading of the colonial archives that the first thirty years of colonial and missionary presence in this area had brought significant disruption, a point made most forcefully by Henry Meebelo in his pioneering study of the Bemba-speaking people's reaction to colonialism, in which he stresses the importance of understanding the local responses to, and the perceptions of, the colonial state (Meebelo 1971). This is not to imply that Bemba society was totally transformed by colonialism and capitalism, for it clearly was not, but there is no question that significant changes did take place.

One important but rather intangible area in which the colonial presence had made itself felt was in the peoples' own representations of themselves and their history. Bemba history, we argue, was being actively constructed in this period, in the context of a colonial state in which such accounts of history were of enormous political and material significance. A great deal of important Bemba precolonial history remains unknown to us, but what we can say is that the accounts of a "traditional" system given to Richards and handed on to us by her must always be viewed as the accounts of a chiefly elite constructed in the context of a colonial state. We explore this argument in more detail in Chapter 1. It is even more crucial to recognize the degree to which Richards's account was constructed in dialogue with her Bemba informants, in their sure knowledge of the importance of custom and history in defining their relations with the colonial state. We must be wary, therefore, when depicting Richards's account, as we must, as that of an outsider and an anthropologist with scientific aspirations, that we do not neglect the active role which her Bemba informants played in the shaping of her account. This argument is elaborated further below.

A rather different aspect of our original plan to re-study Richards' research area was our interest in changes in food supply and nutrition over time. It seemed to us that, in comparison to most parts of rural Africa, the data available for the Northern Province on changes in patterns of production and consumption over time were unusually rich. There are very few historical reconstructions of African agricultural systems which include any assessment of the effects of interventions in those systems on food security. This is largely because there is usually insufficient data to enable such an analysis to be made. We recognized before we began our research that the existing material on the Northern Province, combining as it does anthropological, ecological, agricultural, and economic data, was probably unique. Not only had Richards and her colleague Lorna Gore-Browne collected household-level data on dietary composition and intake in the 1930s, but we also had access to the unpublished raw data on food consumption and sharing collected by Ann Tweedie in the same region in the late 1950s. Furthermore, this picture of household-level patterns of consumption could be brought up to date with the material collected and analyzed by researchers working on the Integrated Rural Development Programme in Mpika, whose interest in the effects of agricultural change on nutrition within the household mirrored Richards' interests and our own. We were also able to make use of the detailed work of the Adaptive Research Planning Team in Kasama and their very sophisticated analysis of changing crop patterns and agricultural livelihoods. We discuss this data in Chapters 7 and 8, and compare it with information we collected ourselves. We were also able to draw on colonial archives, missionary archives, hospital records, economic data on agricultural production, and a variety of other sources, including recent research by Zambian scholars. In making use of this data, we had to draw on a wide range of academic disciplines and develop a passing acquaintance with many different theories and discourses. We also had to rely on the hard work of many other researchers. We conducted field research intermittently in the Northern Province of Zambia over the period from 1986 to 1990. We collected data using participant-observation, oral history, survey, and semistructured interview methods. We concentrated on investigating the links between food production and nutrition and the changing nature of agricultural production and the sexual division of labor. We were particularly interested in investigating two ideas, which were put forward by Richards and other researchers, that the area is food-short both because of the nature of the slash-and-burn (citemene) agricultural system and because of male labor migration. The data we present in the following chapters are based on a nutrition survey of twenty households, sixty migrant histories (thirty women and thirty men), thirty interviews with male farmers, and thirty interviews with female farmers, as well as on interviews with agricultural and health officials, on economic data and archival materials. We also make use in Chapter 3 of unpublished data collected by Audrey Richards in the 1930s and Ann Tweedie in the late 1950s on kinship and household consumption patterns.

In bringing together a wide range of data from disparate sources, we were not only seeking to improve our empirical understanding of agriculture and food supply in the Northern Province, we were also attempting to develop new methodological practices for the writing of African historiography and anthropology. Combining mission archives with agricultural research data and information from the Integrated Rural Development Programmes reveals, not surprisingly, that all are forms of interested representation. The goal was not only to examine how knowledge is constructed in different contexts in different periods, but also to explore the ways in which seemingly discrete realms of knowledge feed on and over-determine each other at particular times. We argue that the particular set of slash-and-burn agricultural practices known as citemene provides a constant reference point throughout the period under study for the people of the Northern Province and for the scholars, experts, and outsiders who have studied, written about, and intervened in this area. It is only by looking at citemene, over a one-hundred-year period, both as a set of concrete practices and as a metaphor, that we have been able to understand how it is that interventions in the agricultural system have been continually shaped by engagement with powerful representations of the citemene system. These representations were produced by local people and outsiders, by farmers and agronomists, and by informants and anthropologists. This mode of knowledge production was simultaneously a mode of intervention. Thus, Richards's account, like all the other accounts, was not only a representation of the Bemba agricultural system, but also an intervention in that system. By providing our own account of agriculture in the Northern Province, we have not simply sought to re-study Richards' data, but rather we have attempted to set her research alongside other accounts. This point is discussed further in the following text and in Chapters 1 and 6. However, the particular story we have tried to tell is one about the interpenetration of different accounts, and their consequences, over a long period of time. Telling this story has caused us to juxtapose different sorts of data that are rarely brought together in any systematic way. The result is a particular method of contextualization which seeks to demonstrate degrees of autonomy and interdependence between different discursive frames and strategic practices. One positive product of this method has been to remind those of us who needed reminding that Audrey Richards' account of agriculture in the province is only one among many.

In order to further our particular method of contextualization, we had to ambitiously aim to reconstruct, in as much detail as possible, the changes in the agricultural systems of the Northern Province over the century from the 1890s to the 1990s. Our reconstruction of these changes is, we hope, as complete as it can be, but it must always be recognized that there will be gaps in any attempt to provide a detailed reconstruction of an agricultural system over a one-hundred year period.[2]

In addition to documenting changes in the agricultural systems of the province, we had also hoped to say something about changes in nutritional standards over the period from 1890 to 1990. However, the role of nutrition in our finished account, is somewhat different from that which we had originally envisioned. It has not proved possible to reconstruct in any detail or with any certainty the precise changes which have taken place in the composition of local diets over this period. Richards and Gore-Browne collected their data when nutritional science was in its infancy, and there are many gaps and inconsistencies in their material; we discuss these problems in Chapters 3 and 7.[3] Because any such data are extremely rare, however, we have made the most of it and feel certain that it can be used to indicate the composition, if not the volume, of intake in selected villages in the 1930s. To use this material and Tweedie's later observations to speak of changes in the adequacy or inadequacy of the local diet would, however, be more difficult. Not only are the data probably insufficiently accurate for this purpose, but changes in the accepted wisdom of nutritional science complicates this task. Scientists now

stress the multiple causes of malnutrition and undernutrition and the close interactions between infection and malnutrition. Furthermore, the whole question of
defining necessary or desirable levels of intake of various nutrients has been
greatly complicated by the discovery of individual adaptive mechanisms. In the
1930s, the methods of nutritional science were crude although nutritionists spoke
with some confidence on the subject of nutritional requirements. In the 1990s,
more detailed and sophisticated research has brought with it a great deal more
caution.[4] The fact that we do not know, with any certainty, what appropriate nutritional measures would be for this area, nor how intake has changed over time,
does not, of course, imply that malnutrition and undernutrition are not serious
problems here, for they manifestly are. It does mean, however, that our focus of
interest in nutrition is less on quantification and more on what the material can tell
us about the social production of malnutrition or, to put it more positively, the elements of social organization which prevent malnutrition from being a more extensive problem than it is. This was also one of Richards' central concerns, as well
as an issue addressed by Tweedie's research.

We have used Richards' and Tweedie's raw data (rather than their analyses) to
say something about the importance of women's social networks and sharing
mechanisms in determining the fate of household-level nutrition, and we have
compared their material with our own. (See Chapters 3 and 7.) Without minimiz-
ing the changes which have taken place since the 1930s, nor the daily problems and
worries which women face in provisioning their households, we have nevertheless
argued that there has not been the generalized breakdown in sharing mechanisms,
which Richards and others predicted for this area. In the parts of this book which
deal explicitly with nutrition (Chapters 3 and 7) as well as in other chapters, we
argue that the idiom of "breakdown" and the analysis which accompanies it, obscure a great deal more than they illuminate.

Our interest in the relationship between regional and local processes, between
household consumption and the wider economy, and between labor migration and
the changing nature of the sexual division of labor meant paradoxically that a village study was not appropriate to the problems we wished to investigate. Colonial
Zambia had been the forging ground of village studies in anthropology, but to have
focused on a single village, given the high levels of residential mobility and social
differentiation observable in the province, would have obscured many of the processes we wanted to investigate. Single villages in the Northern Province are not
large, and no single village was likely to contain sufficient numbers of successful
maize farmers or of female-headed households to answer the questions we wished
to ask. Although we required village- and household-level data, we could not focus on a single village. Many of the questions we wanted to ask about labor migration were not best approached through studying a single village, because it
was the relationship of returning migrants to a wider area and to a regional economy which turned out to be most significant, and not simply their connections
with a single village. This issue is explored in Chapter 6. Likewise, the investigation of food security and nutritional adequacy, although requiring village- and
household-level data, could not be understood in its proper context by studying a
single village. We were thus forced to move away from what some would consider
to be a traditional anthropological approach, and had to rethink the methodological
parameters of an anthropological and historical investigation which needed to

reconcile the most intimate details of household negotiation with the dictates of a regional economy transformed by colonialism and capitalism. We chose to focus on an area which incorporated several villages as a way of gaining both household- and village-level data, while still maintaining a broader focus. Chapters 6, 7, and 8 make use of the field data we collected.

Writing and Context

The practical and methodological problems we encountered while conducting this study were paralleled, and indeed connected to, equally important conceptual and intellectual difficulties concerned with disciplines and their representations. The practice of writing anthropology and history has changed a great deal in the recent past, but it is sometimes easy to forget that these changes have come about as a result of a sustained challenge from both inside and outside academia, and not simply as a consequence of shifts in theoretical frameworks. There has been much discussion within the discipline of anthropology about the crumbling authority of the anthropologist as author.[5] This self-reflexive criticism has been inspired both by the influence of poststructuralist and postmodernist theory in the discipline, and by a growing awareness of the methodological and political necessities of a contemporary anthropology which must seek to make sense of the mutual impact of diverse worlds (Clifford and Marcus 1986; Marcus and Fischer 1986; Clifford 1988).

Anthropologists, of course, never had a mandate to represent others, but certainly they acted for a time as though they did. When Audrey Richards went to Africa to study the Bemba people, there was no question of her authority being questioned, at least textually. Richards' informants may have had views about her representations of themselves, but if they did they have not come down to us. There are many self-evident reasons for this. However, one suggested reason why this might be so is that Richards' expressed views of the nature of the Bemba polity and of Bemba identity happened to coincide rather closely, as mentioned above, with those of the Paramount Chief, the members of the royal clan, and senior Bemba men in general. We discuss this in detail in the first chapter of the book because it is so crucial to what follows. However, in making this statement we do not intend to suggest that Richards was duped by her informants or that she was inattentive to the views of women, or that her ethnography is defective. It is our intention to give substance to the complex situation of mutual dialogue, comprehension, and miscomprehension, which is the context of fieldwork, of ethnographic practice in general, and of the colonial and postcolonial encounter. The Bemba-speaking people were not simply represented by administrators, missionaries, and anthropologists, but rather they were actively engaged in representing themselves. They, or rather some of their representatives, had an interest in telling a particular historical story, constructing a view of their society, creating a set of customs, establishing a collection of powerful symbols, and furnishing themselves with an ethnic identity and a set of social institutions so pervasive and so powerful that its canonization in Richards' account seems perfectly comprehensible. In a sense, the centrality of Richards' account to the discipline of anthropology is a measure of the success of their project.

This project was never, however, a complete one, either practically or intellectually. It was fractured and piecemeal, tentatively established in certain circumstances and places, only to be subverted and challenged in others. It was both conscious and unconscious, worked out eventfully and fitfully, rather than conceived and executed. This does not mean that the project did not involve strategy, inventiveness, guile, and determination. Like the political actions of the peasant farmers described by James Scott (1989), it was rarely explicitly orchestrated. If the chiefs were its instigators, they were frequently overtaken by it, only to find their political strategies turned on their heads. The shifting terrain of meaning and action on which it was constructed meant that it was neither historically contingent nor consciously planned.

There is, of course, an enormous gulf—one mostly of power—between those who write and produce textual representations and those who do not, even if the latter are involved in the wider process of representation and creation of meaning. We do not intend to gloss over this difference, nor ignore its effects on local representations and interpretations of the typologizing and totalizing effects of Western intellectual thought. However, anthropologists, even those who are committed to a postmodernist critical re-evaluation of the discipline and of the practices of its practitioners, too often seem to forget the role that those who are studied have in the construction of texts. The practicalities and the consequences of textualization can never be controlled, and this is symbolized by the fact that most anthropologists write their texts greatly removed in space and time from the mutually constituted circumstances of fieldwork. But, precisely because anthropological texts are the product of a complex dialogical process, they can never be completely realized as the straightforward product of cultural domination and disciplinary convention. Whatever the anthropologist as author may intend, such texts always exceed the intentions of their authors. Even though anthropologists may control the production of the text and set the agenda of inquiry, they are never entirely free agents. The constraints that govern their work and the parameters of meaning set down are not just those of disciplinary convention, style, and genre. The representation of the Bemba people which emerges in *Land, Labour and Diet*, for example, is neither a pure invention of the anthropologist's mind, nor a direct recording of the experience of the Bemba people. It is both a record and an interpretation, and in each of these senses it is the product, not only of the anthropologist, but also of her informants. Despite the anthropologist's efforts to provide one dominant set of meanings for the lives and the events she describes, it is still possible to read this text as an indeterminate one, full of the contradictions, contestations, and uncertainties which existed in everyday life. Richards' representations, in other words, do have a foothold in the complexities of the real world, but more important they have as one of their sets of referents those practices, meanings, and values that the Bemba-speaking people held to be of central importance.

The politics of residence and food, symbolized by and made concrete in the citemene system, were of key strategic and cultural importance to the people of the Northern Plateau before Richards arrived. It was for this reason that citemene was the domain of contestation between Bemba chiefs and their people as well as between missionaries, colonial officials, and the local people from the late nineteenth century onward. It is true that Richards arrived in the area with an interest in questions of nutrition and food, but it is equally true that she had intended to study the

Bemba political system, and that her final choice of research area was determined by the consistent Bemba preoccupation with citemene, an obsession shared by the missionaries and the colonial officials. (See Chapter 1.)[6] The current criticisms by anthropologists and others of the invention of societies and cultures through the process of textualization have an important role to play both inside and outside the discipline, but to claim that these processes float free of all the significant others involved in their production is a strange kind of inverted arrogance.

Richards, like all anthropologists, was profoundly influenced by the contexts in which she worked. One context was that provided by Malinowski and the London School of Economics. This framed her interest in the relationship between ritual beliefs and relations of production and her insistence on treating the Bemba polity and the practice of citemene agriculture as bounded systems. This emphasis on system gave rise to an understandable anxiety, which was shared with many other anthropologists at the time, about the causes and consequences of what was taken to be evidence of social breakdown. The second context was one provided by the metropolitan government's interest in nutrition in the colonies, the work of the colonial nutrition committee, on which she sat with Raymond Firth, and the developing science of nutrition. A third context was that of colonial Northern Rhodesia, and the social, political, and intellectual circles into which Richards was drawn.[7]

We found when we began our work that we could not ignore these contexts. Richards drew not just on intellectual approaches and social mores prevalent at the time (so obvious as to hardly bear mentioning) but she also drew on a wide variety of discourses that influenced the texture of her interpretations, the substance of her representations, and the form of her descriptive language, without ever necessarily directly determining these things. Thus, as we began to read colonial archives, the writings of colonial ecologists and agriculturalists, the notes of the colonial committee on nutrition, the writings of certain Bemba officials, the reported speech of mine workers and farmers, court cases, parish records, and a host of other sources, we began to hear things. And what we heard were cadences, rhythms, and forms of language that encapsulated metaphors and sometimes stood metonymically for things referred to elsewhere, and it was not always easy to know exactly where that elsewhere was. At moments, all was explicit, and we could with certainty say that one thing was directly influenced by another, or that Richards was drawing on her functionalist training, for example. But, this was not always straightforward because the different discourses of anthropology, administration, ecology, and so on, all of which informed representations of the Northern Province in the 1930s and 1940s, interpenetrated with and mimicked each other. Sometimes this copying was intentional, sometimes it came about as a result of a common philosophical heritage (and no doubt a common schooling) but at other times it was far more textual and contingent: the result of many conversations, of remarks overheard, of correspondence read, of a shared set of practical experiences, and a knowledge of a physical and social landscape, which was only barely acknowledged.

Tracing these discourses, their shared metaphors, and mutually resonant natures was not an easy task. What was even more daunting was the realization that the life histories we collected from women and labor migrants, the descriptions of agricultural tasks, and the changing nature of life and the landscape over the last sixty years as perceived by our informants both influenced and were influenced by the same set of discourses. We deliberately use the word influence because it is not possible to describe a set of determinate causes or determinations. We struggled to make sense of this fact, and slowly we came to realize how mutually determining were the accounts, experiences, and representations of colonizer and colonized, anthropologist and informants, ecologists and farmers, labor recruiters and mineworkers, and missionaries and converts. There is no doubt that those with the economic and political power, the education, and the control of textual production were more powerful than those without. We are not arguing for a naive liberalism or an eradication of asymmetry, exploitation, and domination. However, we are not willing to write the people of the Northern Province out of their own history under the guise of making a critique of the dominant forms of colonial texts and colonial power.

As we conducted our own work, we became acutely aware of how influenced we were by our own contexts of various kinds, and of how our understanding of what was going on in the Northern Province in the 1980s was influenced by the people we worked with. This experience of contextualization was not simply an intellectual one, but one that was forged through practical engagement with an agricultural system, a declining economy, and a kind of practical economy of partially shared meanings and symbols. Every anthropologist and historian knows that one comes to understand the politics, agriculture, and ecology of an area through a practical engagement which both grounds and goes beyond intellectual appreciation, but knowing this does not always allow us to see exactly where our knowledge comes from and what it inheres in.

This point was brought home to us in an unexpected way. We began to realise that just as colonial interventions in the agricultural system of the Northern Province had been profoundly influenced by the texts of Colin Trapnell, Audrey Richards, and others (Chapter 1), so too were more modern interventions. This influence was not necessarily direct, either in the past or in the present; rather it was the result of an evolving and shared history. Particular representations and understandings of the agricultural system inevitably guided the interventions made in that system, as well as the solutions proposed to its problems. Indeed, the question of what were perceived as problems was determined in an interplay between representation and intervention. We discuss in Chapter 6 how the view of the Northern Province as a labor reserve has distorted our understanding of the nature of social and economic change. We discovered that the parameters for this distorted understanding were laid down by Richards and various officials in the 1930s, and when we started work in the mid 1980s, they had changed very little, if at all. In Chapter 2, we examine how the discourses of colonial ecologists characterized the agricultural systems of the province in a particular way, not only determining their apparent form, but also proposing what courses of intervention would be appropriate. Contemporary research on agricultural systems in the province has worked in much the same way, and is heavily influenced by the work of colonial ecologists and anthropologists. The relevant texts are rarely mentioned in contemporary government reports, but their language, metaphors, problems, and solutions are all there working in these modern texts as present absences.

The problems and solutions of the agricultural systems of the province were never, and are not now, proposed simply by experts and outsiders. The people of the province who had to respond to the economic and social changes wrought by colonialism, nationalism, and a variety of policy and development initiatives since have always taken an active part in the determination both of problems and solutions. They spoke to colonial administrators and anthropologists, as they speak now to contemporary officials and development consultants, arguing their cases, representing the nature of their activities, characterizing the problems of the economy, describing the nature of their social systems, resisting other interpretations, picking up rhetoric they know to be effective, and welding it into narratives of their own making. They drag recalcitrant researchers of all kinds out to far removed plots to indicate the wilting stems of plants they think inappropriate for their needs, they divert critics away from their citemene activities, and they answer directly and firmly when asked what they think the problems are. They know that officials and agricultural assistants have been away on training courses, sometimes in Europe, and they listen as the rhetoric of problem and solution is returned to them. They adopt and adapt this rhetoric themselves, making the outsiders believe more than ever that the problems they have identified and the solutions they have proposed are genuinely inherent in a given state of affairs.

Of course, not everyone participates equally in this process. Local communities are themselves riven by social differentiation and distinctions of various kinds. The rhetoric of problem and solution so useful to relations with researchers, consultants, and officials can also be put to good use in local contexts, helping to commandeer resources and to control other people. We discuss examples of this in Chapter 8 with regard to access to land. Such attempts at control are often partial and vulnerable to failure, as well as to unforseen consequences, but they are only one of the many ways in which the discourses of ecology, agriculture, government, and anthropology are brought into a mutually constitutive relation with local practices and local politics.

The people of Northern Province did not, to any great extent, produce written texts, nor were they able to control the production, distribution, and/or survival of their texts. In our text, we have made use of written sources in the form of letters, narrative accounts, and records of meetings when they have been available. In some sense it is possible to argue that these documents, along with the interviews and oral history we collected during the course of our research, represent "African voices." It has become popular in recent African historiography to stress the importance of incorporating accounts by local people into the text, and to cease privileging the accounts of outsiders, experts, and scholars. The same trend is evident in anthropological writing. However, these corrective measures, although absolutely essential and long overdue, are not enough. We suggest that one reason why they are not enough is because of the emphasis on language, which the focus on "African voices," local accounts, informants' beliefs, etc., implies. This is not to suggest that where we have direct access to such sources, they should not be privileged. However, in many instances, local people's representations of and interventions in a particular situation are not necessarily evident simply in speech or linguistic form. Much of what we describe in this book as local representations or conceptualizations are evident as practices rather than as discourse. Colonial officials recognized this all too well. For example, individuals or groups who did not want to conform to various residence requirements simply removed themselves elsewhere. Paradoxically, in the absence of direct local "voices," meaningful and strategic practices of this kind survive well in mission and colonial archives where they are often described obsessively. In the same way, development documents and agricultural research papers often contain not only descriptions of the strategies people employ in terms of their agricultural activities, but recognizable traces of locally constructed representations of those activities, representations that were often intended as interventions, and which would later be magically re-presented as the conclusions of the researchers and thus incorporated into the solutions posed to particular problems, as we argued above. The same is true of anthropological accounts, as we noted earlier, including that of Audrey Richards. This process of presentation and re-presentation is part of the active construction of history and agency, in a situation where no one's representations are entirely free of those of engaged others. Thus, concrete practices are as much evidence of agency and self-presentation as are "voices." In arguing this point, it is not necessary to suggest that such practices are always strategic, in the sense of being consciously formulated. What is necessary is that we assume that actions and representations are indissolubly linked. We have tried throughout this book to draw out and demonstrate the importance of considering practical activities both as representations and interventions. We have not tried to suggest, nor would we wish to, that this kind of approach replaces personal testimony. Where we have interview and oral history data, we have used it, particularly in Chapter 6. What does seem worthwhile, however, is to stress the importance of the mutual interpenetration of coexistent practices and representations. If we do not do this we are in danger of denying local people a significant domain of action, as well as consistently excluding them from the texts produced by scholars, officials, and experts on the grounds that they did not write them themselves. John and Jean Comaroff have made similar arguments, pointing out that missionary writings are not free of Tswana representations because both were bound by their mutual engagement in a struggle over material and symbolic resources (Comaroff and Comaroff 1991: Introduction).

As we have stated, we have tried throughout this book to contextualize and discuss the various discourses, both locally and externally produced, that have been involved in the construction of the Northern Province as an object of knowledge and as a subject of intervention. This has meant placing Richards' work in its historical context, but it has also meant engaging with the writings of colonial officials, ecologists, and others as a series of representations. To explore the degree to which these representations are representations, we have often had to displace them from their context, from the naturalized parameters of meaning within which they make sense, and to read against the grain of their intentions. We believe that our account of the Northern Province is a more appropriate one precisely because it reads so many others inside and outside their contexts, and places different interpretations and sets of meanings alongside each other. We have nevertheless had to contend with the problem that besets any researcher influenced by social constructionism in its "weaker " form, which is that we must simultaneously treat accounts as though they were both factual and simultaneously "constructed."

In coming to terms with the way colonial ecologists wrote about the citemene systems and in exploring how their constructions authorized various interventions in those systems, we have had to recognize also that these accounts are descriptions of a set of practices that existed. In arguing, for example, that the colonial ecologist Colin Trapnell provided a specific account of the citemene systems, we do not wish to suggest that people did not cut down trees or grow millet. In suggesting that Audrey Richards painted a very particular picture of Bemba kinship and patterns of consumption, we do not intend to imply that social relations did not exist or that the Bemba people were not short of food at certain times of the year. We have to recognize that we simultaneously use accounts of all kinds, whether from the past or the present, both as representations and as data. There is no escape from the unease which this dualism produces. All accounts, including our own, are constructed accounts, but they are also accounts of something. This should not be construed as supporting an argument about residual materialism of the kind which suggests that at bottom and in the final analysis facts are facts. It is rather part of a more general plea for a recognition of the fact that in a constituted world, we must recognize the limits of a position that would suggest that we or anybody else had "invented" the lives and histories that form the substance of this book.

Narrative Contradictions

In recognizing the mutually constitutive nature of discourses in the Northern Province, we do not wish to imply that all these discourses are in agreement or that they are all equally powerful. We have often been able to deconstruct various discursive and narrative moves by reading people's practices against their accounts, as stated above. We have also deconstructed certain texts and examined the way in which contradictions, tensions, and ambiguities have worked within them to produce a line of argument constantly threatened by their return. Richards' texts have been subjected to a particularly close critical reading and re-evaluation in light of a wide range of alternative interpretations and sets of information. The result of this is that we have not in our own account produced a single, smooth narrative of agricultural change in the Northern Province of Zambia. Such a narrative does not exist nor has it ever done so, and we have often allowed the contradictions in the material to remain apparent.

There are many different interpretations of the facts presented here, and although we sometimes clearly favor one interpretation over another, at times we have hesitated to guide our readers too firmly in any one direction. We have argued at length about whether we would ever have recognized this indeterminacy had we not inhabited a post-Foucauldian academic context. Certainly our account also owes much to the successes of Marxist theory in history and anthropology, and to what has come to be called practice or praxis theory in anthropology,[8] but our commitment to a method that allows for a multiplicity of interpretations to coexist and that highlights, rather than obscures, the contradictions and tensions inherent in the history of agriculture in the province has had other, perhaps surprising, consequences. Working against narrative closure and the silencing of multiple voices has resulted in a reliance on description and the building up of layers and layers of information in the text. As a result, we have written what is in many ways an empiricist text underscored by a great deal of labor in the form of old-fashioned data collection.[9] Whether Foucault ever realized the embarrassing predicament in which his theories would cast the jobbing social scientist is not clear. But, both the predicament and the paradox remain, and we are committed to them because there are a number of coexistent interpretations in this text, and we wanted to create a space in which the disagreements and convergences between them could be heard.

Having submitted the narrative strategies of others to detailed scrutiny, we also had to decide what form our own representations would take. Part of this decision, as discussed earlier, involved making various methodological choices and commitments. We also, however, had to make various choices about the narrative structure of the book, and about disciplinary conventions, including questions of style and genre. We do not claim to have solved all the problems. The book aims to be thematic, it makes use of chronology and narrative where these aid our interpretation and abandons them where they do not. The result is that we have not attempted to produce a chronological account and neither have we attempted to produce a single, closed narrative. We have not smoothed over contradictions where they have occurred in the data, nor have we tried to avoid them or hide from them. We have left them for the reader precisely because they are an integral part of our own account. In the same way, we have not followed anthropological convention in terms of presenting an account of the area and of social institutions at the beginning of the book, and then placing the elaborated and personal data later, on top of the foundations, so to speak. We have tried to subvert this disciplinary convention and make that subversion work for us. Kinship is not mentioned until Chapter 3 and is not fully discussed until the last chapters of the book. The nature of the household both in the province at large and in the area in which we conducted field research is only brought in where it is relevant. We have sought to make our way of writing embody the account we wish to provide, and as such we have tried, while still being utterly dependent on them, as we must be, not to privilege anthropology and history unduly.

NOTES

1. This problem was predicted by Dr. John Iliffe in the early stages of the project. His skepticism turned out to be well founded.

2. In our case, this is particularly noticeable for the 1970s, from which, despite of important work by Bwalya (1979); Bratton (1980); Bates (1976), and Dodge (1977) on resettlement schemes, marketing, agricultural policy and local level political structures, there is a dearth of household-level consumption data and/or community-level information on agricultural production and the changing nature of the sexual division of labor, with the exception of Pottier's re-study of the Mambwe, which was conducted in the late 1970s (Pottier 1988). Our own interview and oral history data failed to plug this gap. This is particularly regrettable in light of the collapse of copper prices after 1973 and the decline in male employment. See also recent work by Chipasha Luchembe (1992) and Mwelwa Musambachime (1992) in the volume on Zambian colonial history edited by Chipungu (1992).

3. This fact was pointed out to us by Mrs. Gore-Brown herself who cautioned against drawing any firm conclusions from the data which she and Richards had collected.

4. For a discussion of this issue see Pacey and Payne (1985, especially Chs. 3 and 5) and Dasgupta and Ray (1990).

5. In fact, there is very little evidence to suggest that the authority of the anthropologist as author is crumbling. Ironically, poststructuralist and postmodernist theories have probably shored up this authority by legitimizing a new generation of politically and theoretically sensitive individuals (see Moore, 1993b).

6. Richards's earlier work had been published as *Hunger and Work in a Savage Tribe* (1932). But, note that her interest in questions of nutrition in the colonies and her involvement in the Colonial Nutrition Committee were the result of her fieldwork in Northern Rhodesia (Zambia) in 1931 and 1933. (See Ch. 3.)

7. For an account of these different contexts and their effects on Richards's personal and intellectual life, see Gladstone (1985; 1986; 1987; 1992) and Malinowska-Wayne, (1985).

8. See Ortner (1984) for an overview of the development of theory in anthropology; see also Comaroff and Comaroff (1991; 1992) for an illuminating discussion of the relationship between anthropology and history, which both describes and takes advantage of these theoretical developments.

9. Kapferer (1988) made this point nicely when reviewing the effects of postmodernist theory and Foucauldian theory on anthropological writing.

1

The Colonial Construction of Knowledge: History and Anthropology

Our focus is on a system of production, and the history we have to tell, like the system of production itself, operates simultaneously at a number of different levels. By the early 1930s when Audrey Richards arrived to conduct her fieldwork and put the Bemba on the ethnographic map, the people of the northern plateau of Northern Rhodesia had experienced a range of colonial interventions, some of which are described in this chapter. Richards was particularly anxious to document the fate of the Bemba agricultural system in these changed circumstances and to quantify the effects of these changes on standards of nutrition, an issue that we address in more detail in Chapter 3. However, as an anthropologist, Richards was also concerned with demonstrating that the material conditions under which a "primitive" people lived could not be separated from their political and symbolic systems. The power and persuasiveness of Richards' book, *Land, Labour and Diet* (Richards 1939) lay and still lies in the painstaking demonstration of the inseparability of the material consequences and the cultural meanings of a system of production and consumption. It seemed to Richards that the identity of the Bemba as an ethnic group, their citemene production system, and their chiefly political system were inextricably bound together, and she presented numerous examples of this interrelatedness throughout her account.

It is not our intention here to deny the reality of the situation which Richards described, but by contextualizing her work in a wider historical frame we have inevitably arrived at a somewhat different analysis of the relationship between politics, production, and identity on the northern plateau. Briefly, our account emphasizes variability, flexibility, and to some extent the lack of "fit" between representation and reality. Rather than viewing Bemba identity as inherited unproblemmatically from the precolonial past, we prefer to see it as being actively constructed in the colonial period, and to see Richards as one of many actors involved consciously or unconsciously in this construction. In this chapter, we de-

1

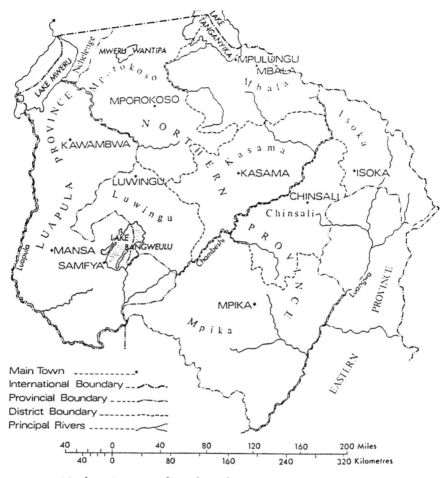

Northern Province of Zambia: administrative boundaries, 1970

Map 1.1 Source: Meebelo, 1977.

scribe the construction of "the Bemba" as an object of knowledge and show how and why it is that this process of construction should have been so intimately tied up with the question of agriculture (citemene) and agricultural production. We argue that in the late nineteenth and early twentieth centuries, the Bemba-speaking peoples of the Northern Province were actively engaged in forging a social and political identity, or rather identities, in a situation where the interests of chiefs and commoners were far from identical. These differing interests were practically and symbolically represented by continuing negotiations over residence and food production. At the heart, therefore, of these negotiations was the shifting cultivation system of citemene. Where were villages and gardens to be sited? Could chiefs control the labor of commoners or indeed of their sons-in-law? The system of production was clearly bound to the system of reproduction and what was at stake in both was the question of chiefly power. It came as no surprise then that missionaries and colonial officials should also wish to control residence and food production. It mattered not that questions of taxation and conversion had replaced issues of chiefly power. The problem of ruling was essentially the same. It also came as no surprise that the battle ground should once again be citemene. How would it be possible to tax and convert a population who would not stay still? How could missionaries and colonial officials gain access to the food supplies and labor they needed? The control of people and of agricultural production was as central to the missionary and colonial project as it was to that of the Bemba chiefs. The system of citemene as described by Richards was the product of these protracted and contested negotiations, negotiations that not only affected the meanings and values ascribed to the citemene systems, but which also changed the actual practice and form of citemene.

Chiefly Power and Bemba Identity

By the 1930s, the Bemba-speaking people were known in colonial and anthropological accounts for two things. First, they were known for their complex chieftaincy system, centered on a Paramount Chief (Chitimukulu). Second, they were known for their system of production (citemene), which was viewed as wasteful and unsustainable. These two aspects of the Bemba peoples' way of life were, even before Richards had worked in Northern Rhodesia, seen as connected historically. The connection hinged, we would argue, on a notion of masculine identity. In this account, Bemba history was seen as the history of a "warrior" tribe. Bemba men, it seemed, had raided and dominated neighboring groups through violence, and their chiefs had dominated their subjects through fear. Given this history, early colonial officials and missionaries were unsurprised that Bemba men showed little interest in agricultural pursuits with the marked exception of their apparent preoccupation with cutting down trees for the citemene gardens. No longer wielding the spear, they now wielded the axe.

Throughout the colonial period this association remained firmly embedded in colonial documentation of the area, reappearing prominently at the time of nationalist political activity in this area, when Bemba masculinity was once again under scrutiny (Chapter 5). Such a lasting and powerful association cannot simply be viewed as the product of the anthropological imagination or the colonial "official mind," however. This was a reality constructed actively by Africans with access to the ears of outsiders, and with an investment in a certain account of Bemba history, and particularly the history of the nineteenth century.

Let us begin by separating the two elements of this account of Bemba identity: a particular political system and a particular production system, for it is only by separating them that we may come to see how they became so interrelated in contemporary accounts. The Bemba political system evoked a mixture of horror and admiration on the part of the early European observers, missionaries, and the British South Africa Company (BSAC) officials who set up the early colonial administration of the area. Much time was spent by those who came either to administer or evangelize in puzzling through the intricacies of the chiefly system and uncovering its rules of inheritance. From the 1930s onward, this task was largely taken over by professional anthropologists, and the example of the Bemba chiefly system spawned a small political anthropological industry of its own.[1] Though colonial officials, missionaries, and anthropologists still sometimes admitted to bafflement, as time went on "the system" appeared generally more comprehensible, and its rules of descent were clearer. Knowledge of how the Bemba chiefly system operated became a mark of more general claims to "know," and the subject of some competition between colonial officials, anthropologists, and others. European observers frequently took sides, for example, during inheritance disputes, backing their favorite candidate with extraordinarily complex historical and anthropological arguments. What all of this professional and amateur anthropological analysis tended to obscure, however, was the fact that this political system was itself relatively new when the first European intrusions took place in this area; that it was a system which had been built on a very effective publicity machine which operated through violence, but which had always been unstable at the margins; that the power of its elaborate system of chiefly control was more apparent than real.

The account of the Bemba political system which had come into being by the late 1930s can be summarized with reference to two of Richards' articles: her piece in the volume *African Political Systems* in 1940 (Richards, 1940a) and her later piece "The Bemba of North-Eastern Rhodesia" (Richards 1951). In these accounts Richards first stressed, the broad similarities between the Bemba and other 'Bantu' political systems. What appeared to characterize them all was that authority was based on descent: the chief was a more powerful version of a family head, and political centralization resulted from the domination of one family line or clan over others (Richards, 1940:83). Despite these similarities, however, it seemed to her that the Bemba were "to all intents and purposes a homogenous group," forming a "quite distinct political unit" from the Bisa, Lala, Lunda, and other neighboring groups with similar traditions of origin" (Richards 1940a:86). Richards, like others, saw a key institution of the social and political structure of the Bemba as chieftainship, centered around the Paramount Chief (Chitimukulu) (Richards 1951:168): "I call chieftainship the dominant institution among the Bemba," she wrote "because the belief in his power, both political and religious, is the main source of tribal cohesion throughout this scarcely populated area" (Richards 1951:168).

The Paramount Chief was drawn from the royal clan (one clan among some thirty other clans), known as the Benang'andu, and traced his descent in a matrilineal line. The Chitimukulu ruled over his own territory in the center of Bemba country (Lubemba), but also acted as overlord to a number of territorial chiefs who succeeded to fixed titles and were also drawn from the same Benang'andu line. Among these chieftainships were those of Mwamba, Nkolemfumu, and Nkula. These royal chiefs with their own territories were, Richards noted, arranged in an order of precedence "according to nearness to the centre of the country— Lubemba—and the antiquity of the office" (Richards 1940a:92). Consequently, on the death of a Chitimukulu, all the incumbents of major chieftaincies also moved up one rung on the ladder, shifting site and taking their followers with them.[2] Under these major chiefs were other subchiefs, also drawn from the royal clan, as well as sisters and uterine nieces of the paramount, who were counted as chieftainesses. The heirs to all these chieftaincies were theoretically the brothers or uterine nephews, and they were given village headmanships until a suitable chieftaincy vacancy arose for them. In addition, sons and grandsons of chiefs could, under certain circumstances, succeed to special chieftaincies (Richards 1951:169). A further important feature of the Bemba political system was the tribal council, or Bakabilo, which consisted of thirty or forty hereditary officials, many of royal descent, whose job it was to act as guardians of the chief's relics, to oversee religious matters, to perform tribal ceremonies, and to act as regents on the chief's death.

Richards saw kinship underpinning this structure and providing coherence. Succession to office was, she argued, "based on descent in nearly every case," and for this reason it was "essential to study the dogma of descent by which these powers are believed to be transferred from one generation to another, and the legal rules of succession by which status and office are passed from one man to another" (Richards 1940a:96). Richards' account of the descent system underpinning the political system of the Bemba can be interpreted in several different ways. Although at times she stressed that the Bemba were a matrilineal people operating clear matrilineal rules of succession (Richards 1951:174; 1940a:87), at other times she stressed bilaterality. A Bemba man, she explained, not only belonged firmly to his matrilineal clan and to a smaller matrilineal descent group, or 'house', but he also belonged to a bilaterally based group, his "ulupwa." This was a "body of kinsmen with whom the Bemba cooperate actively in daily life" and it cut across the matrilineally based groups: "The balance between the powers of the maternal and paternal relatives is a very even one in Bemba society," Richards wrote, "in spite of the legal emphasis on the matrilineal side. . ." (Richards 1940a:88).

Alongside this tension in Richards' account between matrilineality and bilaterality was another, related tension. While emphasising the power of the chiefs and the territorial nature of chieftaincy, she also described residence patterns as extremely fluid. The size of a local residence grouping or village (mushi) depended on the ability of a headman to attract distant relatives to him, from both the matrilineal and patrilineal side. But the village was an impermanent unit. This impermanence was clearly related to the citemene cultivation system but also, as she explained, to the existence of the ulupwa because: "A Bemba is a member of an ulupwa and may move as he pleases to live with any of the relatives composing it, and he is the subject of a chief and may obtain permission to live in any part of the latter's territory, but his ties to a given locality are not particularly strong" (Richards 1940a:90). This residential instability not only proved to be a countervailing force for the political structure that Richards described, but it was also a matter of great concern to successive colonial administrators, as we shall see.

In her account of the Bemba political system then, Richards presented evidence for the importance of matrilineal ideology and a system of chieftaincy based on the matrilineal "dogma of descent." At the same time she presented evidence that cut across this picture, for it seemed often enough that residence patterns hardly reflected matrilineal ideology (although they could always be said to reflect a kinship ideology of some description), and neither did chiefly territorial boundaries and chiefly hierarchies seem to prevent extensive mobility and instability of residential groups. In explaining this apparent contradiction, Richards came to place enormous emphasis on the role of chiefly ritual authority in binding the whole system together, and it was here that she emphasized the link between the political system and the production system, that is, between the Chitimukuluship and citemene. This link enabled her to solve the puzzle of Bemba chiefly authority: how could a system be so centralized and yet so decentralized at the same time, and how, in such a sparsely populated area, did Bemba chiefs maintain their authority? It could not easily be argued that they had done so primarily through the control of trade and accumulation of wealth, because although this was certainly a factor in Bemba nineteenth century expansion, Bemba chiefs had never been centrally involved in the regulation of long-distance trade.[3] Richards argued rather that they had maintained their authority through the exercise of important ritual powers over the land and over fertility which had bound the farthest-flung Bemba household to the Paramount chief in Lubemba, which, in the past, this ritual authority would have been backed up by other sanctions. The Benang'andu chiefs had bound large numbers of men to themselves as warriors, who would have raided surrounding districts, exacted tribute, and held new lands for them. A chief's power had been measured by the number of followers he could gather around him, and the amount of tribute and tribute labor he could command. In the course of the European occupation, Richards argued, this aspect of chiefly power had diminished, but the ritual authority of the Chitimukulu and his territorial chiefs remained strong, and this was the key to any understanding of the Bemba economic and social system.

Richards' Bemba informants stressed the belief that the Benang'andu chiefs descended from an original tribal ancestress, and that when they succeeded to a chieftaincy they inherited directly the spirit of the dead chief or mupashi (Richards 1951:169). This spirit was believed not only to act as a guardian of the chief himself, but also to enter the wombs of pregnant women in the dead chief's territory and to be born as guardian spirits to their children. Thus, owing to his inheritance of a mupashi, the person of the chief was endowed with immense powers over the well-being of his entire people and territory. As a result, the chief's health and welfare, and the conduct of his sexual life were of the greatest significance to all of his subjects and were hedged around with taboos. In *Land, Labour and Diet*, Richards described in detail the chiefly rites that she saw as having the clearest influence on Bemba agricultural activities. These rites, she argued, had a double nature in that they both enlisted the help of ancestral spirits as well as stressing the political authority associated with the ownership and cultivation of the land (Richards 1939:362). Among these rites was the tree-cutting ceremony (or ukutema rite), which Richards argued had been particularly important in the past in binding chiefly authority to ordinary agricultural activity. In one of her more colorful passages, Richards described the ukutema ceremonies performed at Chitimukulu's and Mwamba's gardens in 1933/1934. In this passage, the link between ritual, chiefly authority, and the construction of Bemba masculinity seems particularly clear:

> *The young men seize their axes, and rush wooping up the trees, squabbling as to who should take the highest trunk. They dare each other to incredible feats and fling taunts at each other as they climb. Each falling branch is greeted with a special triumph cry. I collected about forty different ukutema cries at the cutting of Citimukulu's and Mwamba's gardens in 1933–4. These are formalized boastings like the praise-songs commonly shouted before a Bantu chief. The cutter likens*

himself to an animal who climbs high, or to a fierce chief who muti-
lates his subjects, cutting off their limbs like the branches of this tree.
A squirrel might be afraid of such a tree, he says, but not he!
(Richards 1939:291).

Other important ceremonies included the firing of the land at the beginning of
the agricultural season, and tribute-giving after the harvest. A central feature of all
of the ceremonies Richards described was that they gained much of their power
through a ritual association with the sexuality of each Bemba subject, but particularly of the chiefs. All ceremonial was hedged around with sexual taboos and pu-
rification rites, the fertility and prosperity of the land depending crucially on their
proper execution (Richards 1939:364–5). Writing in the past tense, because she felt
that many of these rites were no longer being followed as asiduously as in the past,
Richards concluded that:

> . . . *the magico-religious rites of the Bemba were certainly an inte-*
> *gral part of their whole economic system. The ceremonies gave an im-*
> *petus to their ordinary agricultural activities, and enabled the chief to*
> *organize their efforts more effectively, and thus support the whole ma-*
> *chinery of tribal government. The performance of the ritual con-*
> *stantly emphasised the supernatural powers of the chief, both in his*
> *own person and by his right of access to the tribal spirits. On this*
> *deeply-seated belief in the attributes of their mfumu, economic incen-*
> *tives and the use of the land ultimately depended. It is this faith which*
> *has been weakened nowadays with pronounced effects on the whole*
> *morale of the tribe, as far as agriculture is concerned.*
> (Richards 1939:380).

Thus, through her analysis of ritual, Richards was apparently able to solve the
puzzle of how Bemba chiefs exercised their power. In constructing this analysis,
Richards referred directly and indirectly to a past in which the power of Bemba
chiefs over their subjects had been much greater, when their ritual power had been
backed up by their military might and their direct physical surveillance of the large
numbers of people who lived within their stockades. In Richards' view, the problems facing the Bemba people in the twentieth century arose in no small part from
the decline of this chiefly authority that had acted, in the past, as a central force in
ensuring the economic welfare of the whole "tribe."

However, there are other possible interpretations of Bemba history, and these
different interpretations imply at the very least a shift of emphasis in any analysis
of the relationship between political authority, production, and Bemba identity.
Writing after Richards, and with the benefit of access to new historical work, other
anthropologists have placed a somewhat different slant on the evidence for the
workings of the Bemba political system. Richard Werbner argued in a 1967 article
that Richards had overemphasised the extent of centralization within the Bemba
chiefly system. Through an analysis of the history of "civil strife" among members
of the Bemba royal line, Werbner concluded that the Bemba political system was
based on a "complex and variable" distribution of power that depended on the absence of a fixed hierarchy of royals and notables controlled from the center (Werb-

ner 1967:23). By shifting attention away from the matrilineal succession of Benang'andu chiefs to other important offices within the Bemba system, including the "lords of the marches" who controlled the borders of the nineteenth century state, the Bakabilo, and the Queen Mother, Werbner was able to argue that it was a complex set of crosscutting alliances among diverse elements that allowed some degree of centralization around the Chitimukulu to persist: ". . . if any feature has persisted from the nineteenth century kingdom . . . it is the flexibility and varia-tion in relations between administrators which has continued to be an essential and defining feature of the Bemba political system" (Werbner 1967:26; see also Rob-erts 1973:Ch. 9). In another reanalysis of the Bemba political system written in 1975, Epstein concentrated on the history of Bemba military expansion in the nineteenth century and its relationship to chieftaincy. Bemba chiefs, he argued, were highly successful military commanders, and in the course of the nineteenth century this military command had become increasingly centralized. To some extent, he ar-gued, the Bemba chiefs were victims of their own successes because once the last major military opposition in the form of the Ngoni had been removed, there was no longer an enemy "to confront them that required the mobilisation of the entire people," and so there was increasing scope for local autonomy and independence within the kingdom. "What must be noted here," wrote Epstein, "is that a stable system with power unequivocally established at the centre was never achieved" (Epstein 1975:204).

Some decades earlier, W. V. (Vernon) Brelsford, a colonial officer in the North-ern Province and an avid student of the Bemba, had argued something very sim-ilar. Richards had exaggerated, he implied, the ritual powers of Bemba chiefs and their priests. These, he argued, had always been limited to the areas over which the chiefs had direct control: "the spiritual association with land is broken when political power over it is lost" (Brelsford 1944:3). Although himself a great system-atizer of knowledge on the Bemba, Brelsford seemed to be arguing in this piece that such systematization distorted reality: "It is a mistake to generalize, as is so often done, and assert that the native is hide bound by custom and non-adaptive. It is more correct to generalize in the opposite direction and say that in every phase of his physical and mental life is he showing his adaptiveness to modern conditions" (Brelsford 1944:4).

A similar note of caution was to be found in the painstaking reconstruction of Bemba history carried out by Andrew Roberts in the 1970s (Roberts 1973). By tak-ing a longer historical perspective and a more regional view, Roberts was able to argue that "The undoubted military achievements of the Bemba have prompted several writers to overemphasise the 'centralised' nature of Bemba government, and this tendency has probably been reinforced by nostalgic exaggeration on the part of informants around the paramount's capital" (Roberts 1973:305; see also Rob-erts 1981). Indeed, what emerges most strikingly from Roberts' reconstruction is not only how recent was the coming into being of the Bemba state as an entity, but also how brief was its hegemony. The Bemba had emerged as a distinct group un-der a Chitimukulu in the late eighteenth century, and in the course of the nine-teenth century they had expanded to create a more or less unified polity that spread over most of the present day Northern Province. However, this unity did not survive the death of Chitimukulu Chitapankwa in 1893. Thus, when the first

European traders, administrators, and missionaries arrived on the scene, the polity was already suffering a decline.

It is not possible here to explore in depth the complex history of the rise and decline of the Bemba state, described in detail in the work of Andrew Roberts. Two things do seem clear, however. First, the degree to which Bemba chiefs were able, even at the height of their power, to control the day-to-day activities and the residence patterns of their subjects seems open to question; and second, the period in which they may have exerted such control was relatively brief. We know, from historical material and from the fascinating life story of Bwembya collected by Richards, that within their late nineteenth century stockades, major Bemba chiefs had exerted frightening powers over their immediate subjects, including the power to control their sexuality, and the power to mutilate and slaughter (Richards 1936). However, it is not clear how much power the chiefs ever exerted over those who lived in the furthest corners of this vast Bemba territory. According to Bwembya, much seemed to depend on a messenger system, in which trusted runners would convey the wishes of the chief to his more distant kinsmen and subjects, but this seems to imply a rather minimalistic kind of control. Indeed, recent work by Hugo Hinfelaar on religious change amongst Bemba women throws further doubt on the whole question of Bemba chiefly control (Hinfelaar 1989). Hinfelaar argues that the hegemony of the Benang'andu chiefs was never complete, and that their claim to ritual control over ordinary Bemba commoners was fragile and contested even before the entry of Christian missionaries and British administrators. Whereas Richards had very largely focused her research on the centers of royal chiefly power, Hinfelaar concentrated on the physical and political margins of the Bemba kingdom. Whereas Richards concentrated her attention on the relationship between chiefly power and masculinity, Hinfelaar, by focusing on the beliefs and practices of commoner women, was able to see less centralization and more subversion.

What is lacking in Richards' account of the Bemba polity is any notion of contestation, other than in her rather formalized discussions of chiefly succession disputes. Official colonial documentation of this also suffered from an oversystematization informed by anthropological analysis. When called upon to adjudicate a chiefly dispute, for example, colonial officials would first establish the "rules" and would then apply them.[4] Whether the "rules" of matrilineal descent and of chiefly authority had ever been applied (or even coherently constructed) in the past was not a question that they asked. We do not wish to argue that Richards or anyone else "invented" Bemba political tradition, because this would clearly be an oversimplification, but Richards, among others, was involved in the important task of the codification of custom, a codification that came to have clear political consequences in the era of Indirect Rule.[5] For this reason, although Richards frequently described complexity and instability in the Bemba "system," it always remained for her a "system." Anthropologists were only one set of actors in this construction of the Bemba polity. As Hinfelaar's work makes clear, Christian missionaries and their adherents were, in some ways, more significant in this regard because their close contact with ordinary people and their efforts to translate a Christian message quite literally into Bemba speech had far-reaching effects. In particular, the Catholic missionaries constructed their own ethnographic models of the Bemba that focused heavily on chiefly ritual powers, ignoring a range of alternative (mostly female and nonchiefly) sources of important ritual authority. These were then to re-emerge in the late colonial period within the Lumpa church of Alice Lenshina and in the Ba-Emilio movement.[6]

If anthropological accounts and colonial practice in this area tended to over-emphasize the degree to which the Bemba state had been a state in the past, what of the association between Bemba identity and the citemene system that became such an enduring theme of the colonial period? It seems to us quite likely that the ritual link between chiefly authority, Bemba masculinity, and the citemene system, which Richards' informants stressed, was not a vestige of the past but a more recent construction. In order to substantiate this view we need to look a little more closely at the nineteenth century history of this area.

Citemene and Masculinity

It is very difficult to find any evidence for an association between Bemba identity, political authority, and the citemene system in the nineteenth century. This obviously may be partly accounted for by the general lack of historical evidence for Bemba agricultural methods before the beginning of the twentieth century.[7] It seems likely from what we know of the soils and vegetation of the northeastern plateau that some form of citemene cultivation had been practiced there for centuries. However, the "Bemba of Chitimukulu" who migrated into the area from the late eighteenth century are remembered in the oral traditions less as agriculturalists and more as raiders. "Bemba" was a political identity in the nineteenth century, and Bemba masculine identity at least was probably tied up more with the wielding of the spear than with the wielding of an axe. Indeed, if Hinfelaar is correct, it seems likely that it was Bemba women who had exercised ritual authority over agricultural production and fertility, an authority that was challenged but never entirely usurped by the Benang'andu chiefs in the nineteenth century (Hinfelaar 1989: Ch. 2). If we follow Hinfelaar in seeing the establishment of chiefly ritual authority over the land and production as a contested process in the nineteenth century, then we can better understand the situation that Richards described for the early twentieth century when other forms of chiefly authority had suffered a massive decline.

With the gradual elimination of the slave trade in the 1890s, officials of the BSAC had slowly established a skeletal kind of rule over Bemba chiefly territory.[8] At the same time, members of the Catholic missionary order of the White Fathers were moving into the Bemba heartland from their base in Mambwe country to the north. It is clear from the accounts of these missionaries that the Bemba chiefs' ability to centralize and control large populations was already being threatened by famine by the mid-1890s. In 1895, for example, they reported that what had once been the large and impressive village of the Bemba chief Makasa, was depopulated by famine and Bemba people were seen seeking refuge in Nyamwanga villages (WFC: Nyassa 2, 1895, No. 65:196).[9] Throughout the late 1890s and the early 1900s, population dispersal continued. Perhaps this was because of the pressure on the agricultural system inevitably brought about by the defensive concentration of population, but it was also undoubtedly a result of the loosening of control of the chiefs. In 1899, Captain Close, British representative to the Anglo-German boundary commission in the north remarked that villages on the plateau were very small, often made up of a mixture of "tribes," and that this was indicative of a loss of power on the part of the chiefs (BS1/31: F.O. to Sec. BSAC 3.5.1899). On the southernmost borders of the declining Bemba kingdom the situation was much the same. Walking south from Chilubula to Chilonga and into Bisa country the White Fathers counted a total of only 300 people in nine days, but saw numbers of ruined Bisa villages that had been abandoned because of Bemba and Ngoni raids. These Bisa villagers had fled to the treeless shores of Lake Bangweulu, and were busy evolving a new system of production there. It is in the reported account of a conversation between the Bisa Chief Kopa and the White Fathers on this occasion that we get our first piece of evidence for the "ethnicization" of agricultural practice. The Bisa, explained Chief Kopa, "cultivent par terre, tandis que les Babemba, leurs voisins, cultivent en l'air."[10] Cultivation "en l'air," explained the White Fathers, referred to the Bemba practice of lopping the branches off the trees and then burning them.

A Bemba chief further elaborated on this point, telling the White Fathers that God (Lesa) had ordained that they, the Bemba, should "grimper dans les arbres comme les singes et crever de faim;" whereas the Bisa had been ordered to "cultiver la terre afin d'avoir beaucoup de nourriture" (WFC: Nyassa 2, 1901, No. 90:158).[11] In 1899, the missionaries described citemene as it was practiced around the newly established Chilubula mission (WFC: Nyassa 2, 1990, No 86:266).

It was the dispersal of population that very quickly brought the people of the plateau into conflict with the newly established British South Africa Company administration. Many ordinary people seized this first opportunity to escape from the direct control of the Benang'andu chiefs, leaving their stockaded settlements and setting up small villages (Gouldsbury and Sheane 1911; Roberts 1973:284–5; Meebelo 1971). Having contributed to the breakdown of chiefly control, the new administration very quickly began to regret its demise. For the BSAC (and later the colonial administrations of the northeastern plateau) the basic and recurring problem was that of raising taxes sufficient to pay the costs of administration without jeopardizing the area's food supply. The collection of tax demanded that the population be known and controllable. But most Bemba seemed to prefer to live in small, scattered and relatively impermanent settlements and their system of production encouraged this. Therefore, the first decade of BSAC rule, saw a series of haphazard and sometimes contradictory interventions designed to address this puzzle. The people of the plateau, having only just escaped the control of one set of chiefs, were quick to resist the attempts of this new set of rulers to regulate their settlement patterns and police their relationship with the forest. In what evolved as a kind of guerilla war between cultivators and administrators, being Bemba became inextricably bound up with the male right to cut trees. In this process, the citemene system acquired new and powerful political meanings.

Although the early BSAC administration was skeletal, one should not underestimate either the disruptive effects of the new trading economy that it introduced to replace the raiding economy of the nineteenth century, or the degree of competition that arose between the administration and Bemba chiefs over the control of the population. In an area of such sparse population as the northern plateau, the extraction of labor for porterage could be highly significant. In 1898, for example, Captain Close had estimated that there were about 300 men working as carriers for the Boundary Commission, over 1000 for a Belgian expedition, and another 1000 for the African Lakes Company and the BSAC. Given that, in his estimation, there were only two "grown men" per square mile all over the plateau, this "represented a depopulation of the men over more than 1000 square miles, or about one seventh of all available men on the plateau" (BS1/31: F.O. to Sec BSAC 3.5.1899). This was possibly an exaggeration, but it is clear that the turn of the century saw the northern plateau buzzing with a degree of trading activity that has perhaps never been fully replicated since. Caravans of porters carrying goods for the administration, or for the trading companies, traversed the area continually, facilitating the spread of smallpox as they did so (WFD: Vol. I 24.2. 1902). In these years, as Gouldsbury and Sheane later rather wistfully recalled, "loads were plentiful and wealth circulated briskly" (Gouldsbury and Sheane 1911:45). There were more Europeans of various sorts in the area during the period 1900–1902 than there were to be at the outbreak of World War I. Many were involved in the construction of the telegraph, others were transient hunters and traders. All required and recruited labour, in turn facilitating the collection of tax, which began in 1901. Once the construction of the telegraph was complete, however, so the level of commercial activity receded: "one by one the Europeans, their various tasks completed, withdrew to other spheres, until only the missionaries and administrative officials remained" (Gouldsbury and Sheane 1911:45). One of these officials was a young man named Frank Melland, who was employed in 1901 by the BSAC to take charge of their Mpika station. Melland's responsibilities included raising taxes and moving loads from one administrative station to another. In theory, the employment of porters to move the loads would result in the payment of taxes, but in practice things were rarely as simple as this. Frequently Melland would either find himself with 1000 loads to move at a time of year (March/April, for example) when noone wished to work because they were too busy in their gardens, or alternatively, with only half the taxes collected and no work to offer. With only one or two messengers at his disposal, his ability to recruit labor and to collect taxes depended heavily on the cooperation of those chiefs who were paid a rebate for this purpose. Not infrequently, these chiefs refused to cooperate, viewing Melland's actions as directly competitive with their own already diminishing ability to command labor (Melland Diaries Vol. III 1904: 7.3.04; 8.6.02; 8.11.04).

The problems faced by Melland were typical of this area in the decades to come, as we shall see in later chapters. He could not afford to forcibly recruit labor if this were to jeopardize the production of food in the area, and because he was also responsible for providing rations to his laborers, there was a tricky balance to be struck. This was further complicated by the attitude of the chiefs who rightly saw his demands on labor as rivaling their own, and also by the very patchy and erratic nature of the new trading economy in this area. Months would go by in which no work was available locally, and so no taxes could be paid. Labor migration from this area was consequently an early phenonenon. In 1904, for example, Melland found himself seriously behind in tax collection, so he sent 150 men to look for work at Fort Jameson and at the Bwana M'kubwa mine.

Recruitment of labor and payment of taxes were closely related to the other major intervention of the early BSAC administration, which was its attempts to control citemene cultivation and settlement patterns. No sooner had the population dispersed from chiefly stockades, the BSAC began trying to concentrate it again. It is difficult to unravel the original motives of administrators' attempts to control settlement and cultivation patterns from beneath the elaborate discourse on citemene and the "Bemba mentality." By the interwar period, as we shall see in Chapter 2, a scientific account had been elaborated that provided an ecological rationale for interventions in the citemene system. In this early period, however, there was no such elaborate justification, only passing references to the "wasteful" nature of the system. The primary motives for intervention were almost certainly the need to collect taxes in this sparsely populated area, and the larger problem of maintaining control. It was, in fact the building of seasonal dwellings (mitanda) in the citemene fields, rather than the practice of citemene itself, which was the original target. In 1903, the third year of taxation, BSAC administrators toured villages near the White Fathers' Chilubula Mission and burned houses in an attempt to discourage the building of mitanda (WFD: 24–25, 5.1903). The building of mitanda, although it came to symbolize many other things, was a practice that tended to reduce the frequency with which villages had to be shifted, but for the BSAC it represented uncontrollability. It seemed to them impossible to know where anyone was, because although individuals might live in named villages, for part of the year they might well be resident somewhere unidentifiable in the forest.

The BSAC administration made its first concerted attempt to control this practice over a wide area in 1906, calling together a meeting of chiefs at each of their administrative centers. At these meetings it was announced that henceforth both the citemene system and the building of mitanda were abolished. What was actually meant by the abolition of citemene was an enforced modification of the system; the main concerns of the administration were that people should live in "proper" villages (defined as twenty huts or more), that they should not spend half the year living in mitanda, and that their chiefs and headmen should be able to enforce the payment of taxes. Assuming that the dispersal of population that had occurred in the previous few years was a direct result of the agricultural system, they set out to modify this system. Citemene fields were to be confined to the immediate environs of the village, and no mitanda were to be permitted. There was to be no cutting of citemene on the banks of streams, nor in the forest, but only in what the administration referred to as "open country." At the same time, villagers were instructed to "hoe energetically" beds of cassava, sweet potatoes, groundnuts, and beans in permanent gardens: "If scarcity comes," announced the administrator, "it is possible to eat cassava." If all of these instructions were followed, the people would be rewarded with permission to kill game (KDH 1/1, Vol. 2:126; Meebelo, 1971:105).

The extent to which these regulations could be enforced by the skeleton administration must remain a matter for doubt. The severe food shortage that occurred in 1907 was more likely to have been precipitated by the widespread locust invasion of 1906 than by the restriction on cutting (KSD 4/1, Vol. 1). However, the administration did report some success in their attempts to force people into large villages. In the area of the Bisa Chief Kopa, in the Mpika district, it was reported that the population had begun to center on only five villages, while in the Bemba Chief Luchembe's area, sixty-three villages had been amalgamated into nineteen (KSD 4/1, vol 2:27). The significance of the policy, however, lay not so much in any effect it might have had on cultivation patterns, but rather in the widespread hostility it engendered. This hostility was sufficient to cause real, if brief, alarm among the BSAC administration. In Mpika, Frank Melland reported in 1908 that the attitude of the Bemba had given rise to "a certain amount of anxiety." They had changed, he wrote "from the cheerful attitude they had always borne" (KSD 4/1, Vol. 2:27; Meebelo 1971:107).

Fear of rebellion, anxieties over food shortage, and the near impossibility of enforcement, led the administration to reallow the clearing of citemene fields, but only if a number of other conditions were adhered to. First, mitanda were not to be allowed, but only rough shelters (sakwe) in which the male head of household was to be allowed to sleep (but not his wife or children) at times of the cutting of branches, gathering of branches and fencing of the field. Women and children would only be allowed to sleep in the sakwe during the harvest period. No live-stock were to be kept in the citemene fields, neither were grain-bins to be erected there. Finally, cultivation would not be allowed in citemene gardens that were "re-mote from the villages" (KSD 4/1, Vol. 2: Indaba at Chilonga Feb 1909). At Chilonga, according to the White Fathers, the people seized eagerly on the oppor-tunity to cut again (WFC: Nyassa No. 3, 1909:399). But with citemene reallowed, a long struggle ensued between the people and the administration over the building of mitanda. On this point the BSAC officials were determined to hold out, and will-ing to punish any infringement. In 1909, the White Fathers at Chilubula reported that a BSAC magistrate on tour had found some people from the village of Makubi living in mitanda built like "les vraies maisons," together with their goats and chickens. As a punishment they were forced to repair roads and bridges in the area (WFD: Chilubula: 11.5.1909).

Throughout the years up to the outbreak of World War I, such incidents were repeatedly reported by the White Fathers, who occasionally interceded on the part of the people living around their mission stations. In an incident that was typical of many occurring in the more accessible parts of the then Awemba district, in 1913 the collector from Kasama touring near Chilubula, ordered the destruction of over sixty granaries that he had found constructed near mitanda. This destruction was carried out, despite the White Fathers' somewhat alarmist suggestion that the Bemba might interpret his action as a prelude to war: "La nourriture," wrote the Father Superior to the collector, "est chose sacrée, la gaspiller c'est exaspérer les gens: pour eux, autrefois, abattre les greniers c'etait le cri d'alarme pour guerre" (WFD: Chilubula: 6.9.1913).[12] "La nourriture" was not the only thing that was sa-cred to the Bemba people, however. It is not possible to know with any certainty how people interpreted the administration's actions with regard to citemene cul-tivation and mitanda-dwelling, but we know enough to be able to say that more than food was at stake. Control over the ritual meanings of the forest was some-thing that the Benang'andu chiefs had struggled to centralize in the course of the nineteenth century. This control, the exercise of which was intimately related to beliefs about the relationship between the dangers of sexuality and the dangers of the forest, was probably never fully achieved, but it was an important area of po-litical contestation between Bemba chiefs and their subjects (Hinfelaar 1989: Ch. 2). This contestation can only have been heightened by the entry of a third party: the BSAC administration. We can only imagine, for example, how the following reg-ulation concerning who might sleep in mitanda was interpreted. Messengers were instructed to investigate families working in their citemene fields. If the man was staying in the mutanda and his wife was at the village, or vice versa, than all was well. If, however, he found that both man and wife were staying in the mutanda, then the man was to be arrested and taken to Kasama to be fined. It seemed that the administration, in addition to policing the forest, was also entering the symbolically charged area of sexual relations: it was preventing husbands and wives from sleeping together. The people were annoyed (WFD: Chilubula: 31.7.1912).

Whether on account of the administration's symbolic interventions or not, this was not a period of plenty on the plateau. Serious food shortages occurred throughout the period up to the outbreak of World War I. Shortage was widespread in 1909. Two White Fathers who had walked the route from the Luangwa valley in the east to Chilubula reported serious famine all along the route (WFD: Chilubula: Jan. (n.d) 1909). In 1911, the administration reported that the last three months of the year had seen "serious famine" throughout the Mpika subdivision. In 1912, the crops of much of this subdivision failed as a result of drought. (KSD 4/1, Vol. 1:15). In January of 1913, the White Fathers, who had been touring from Chilubula, reported that they had found many people in the villages (rather than in mitanda) all were suffering from hunger (WFD: Chilubula:17.1.1913; 7.2.1913). Serious as these food shortages were, they do not appear to have developed into widespread famine, and the reasons for their occurrence are likely to have been more complex than can be accounted for by the interventions of the BSAC administration. This was, however, a period of vulnerability. Between the establishment of the BSAC rule and the outbreak of the World War I, there had been just over a decade in which the people of this region had attempted to establish or re-establish viable systems of agricultural production, but the conditions under which they did so contained a large element of unpredictability. Not only did the administration attempt haphazardly to alter patterns of settlement and conditions of production, but Bemba chiefs also fought for recognition and control over their evermore dispersed subject population. The battle over mitanda and citemene was not only a battle over a material way of life, but also a battle over symbolic control, and over the meaning of being Bemba. It was during this period, rather than in the distant past, that the association between cutting the trees and Bemba masculinity, which Richards was to describe so vividly in the 1930s, came into being. The citemene fields cut out of the forest became an important symbol of male Bemba autonomy, even as these same men became drawn into the evolving labor migrant economy.

New Forms of Production and Control

Alongside the battle over citemene, other important changes were taking place, including the beginnings of labor migration. By 1904 it was clear that there was insufficient employment available on the plateau to facilitate the payment of hut tax, and men began making their way to Southern Rhodesia, Nyasaland, and the Congo to find work. The Rhodesian Native Labour Bureau found Bemba men to be willing recruits. The White Fathers looked on disapprovingly as increasing numbers of their converts had their "heads turned" by the prospect of high wages (WFC: Nyassa 3, 1907, No. 136). Bishop Dupont hoped that the appallingly high death rates in the mines of the South, the dangers of independent migration (including the likelihood at this time of catching smallpox or sleeping sickness), would soon calm "nos Babemba si avides d'aventures et de voyages" (WFC: Nyassa 3, 1907–08:242-3).[13] But the migration of labor continued to increase. In

1912, "large numbers" of converts were recruited for the mines of Transvaal and for Katanga. The priests attributed that year's food shortage to the disruptive effect of labor recruitment of various kinds, combined with the administration's sporadic demands for food to feed that labor: "aujourd'hui c'est un messager qui vient inscrire les hommes pour le travail, demain c'en est un autre avec ordre d'enrolement pour les mines, un troisième arrive pour faire ramasser des charges d'eleusine et de farine."[14] According to the priests, the old people were baffled by the unpredicability and apparent irrationality of the new regime: "Bwana nous n'y comprenons plus rien, d'un côte l'on nous dit de cultiver et d'apporter la nourriture au Boma, de l'autre l'on nous enleve nos enfants, les seuls capables de cultiver, que veux-tu que nous fassions?" (WFC: Bangweolo 1, 1912–13:613).[15]

The White Fathers, we must remember, had set themselves up in a Bemba chiefly heartland, one of their number briefly becoming a Bemba chief himself.[16] There were other areas of the plateau, however, that remained almost completely unaffected by the interventions of the early administration. The political importance accorded the Bemba chiefs by the BSAC administrators had also had the consequence that it was the central Bemba areas of the plateau that experienced the most direct interventions. More marginal Bemba communities, along with groups who identified themselves as Bisa or Lungu, were less likely to ever encounter a tax-collector or messenger. In the vast and inaccessible areas of the Bangweulu swamps, for example, and in the Chambeshi marshes, some communities were left almost entirely alone. The marginalization of these areas from the evolving colonial economy and polity was to have long term consequences for it was the ecological niches occupied by those who had been driven out of the central plateau by the Bemba chiefs in the nineteenth century that turned out to be the most productive areas of all (see Chapter 4). Because of the attention paid to those professing to be the real Bemba, however, these other communities and production systems were largely ignored. The northern plateau came to mean "the Bemba" in the colonial period to a degree that had probably never been achieved even in the most expansionary phase of the kingdom in the nineteenth century. At another level, however, there was increasing diversity in the modes of being and experiences of the peoples of this area, a diversity which, as we shall see in Chapter 2, was both identified and obscured by colonial ecological science.

The outbreak of the World War I brought new levels of disruption, entailing as it did a massively increased extraction of both labor and food supplies from this area. The Isoka, Abercorn, and Chinsali districts were declared "war-zones," but other parts of the then Awemba Province were also deeply affected. For example, a supply route was set up through Lake Bangweulu, and in 1915, 1500 canoes were requisitioned for this work (Northern Rhodesia, 1916:37). During 1917, 79,000 men were employed as carriers and 28,000 women and children employed to carry food over short distances (Northern Rhodesia, 1918). The year 1913 had been one of famine in some parts of the plateau; the White Fathers at Kayambi reporting a number of deaths directly attributable to famine having taken place between December 1912 and May 1913 (WFC: Bangweolo 1913:616). In 1914, however, people were reported to be responding with "bonne grace" to the administration's demand for food for the troops, which was purchased in exchange for cloth. Each village was required to contribute millet flour, cassava roots, and dried fish, which were transported on the heads of men, women, and children to the nearest boma, and sometimes all the way to Abercorn in the north (WFC: Bangweolo 1, 1914:313). Whether the serious food shortage reported further south, near Chilonga, in 1914, was connected with this requisitioning, is impossible to say (WFC: Bangweolo 1, 1914: 355). In 1915, however, there was another, and possibly more widespread shortage of food, this time nearer the front in the Abercorn district. The Fathers at Kayambi reported many deaths from starvation and speculated on the cause of what they called a famine, blaming it on the previous year's extraction of supplies by the administration. The next harvest looked good but, because of the lack of labor and the weakness of the people, many fields had remained unfenced and thus vulnerable to the depredations of wild animals (WFC: Bangweolo 1, 1915:14).

The northern border districts were clearly more adversely affected than the central plateau areas. Old African rivalries were revived by this inter-European war, and a new space opened up in which Bemba chiefs could reassert their hegemony (Fields 1985:134–136). The Paramount Chief, Chitimukulu, had apparently greeted the outbreak of war with enthusiasm, announcing that this was his opportunity to defeat the Mambwe people once and for all, and Chief Mwamba declared himself ready to go to the front (WFD: Chilubula: 15.9.1914).

Not only did the war result in the temporary and often brutal extraction of male labor for porterage but it also contributed to the development of a more permanent labor migrant economy. This trend was already evident before the war but accelerated during and after it. Having returned from conscription as porters for the war, many men turned around immediately and set off for the mines of Katanga or the railways of Southern Rhodesia "in quest of gold," as the White Fathers put it (WFC: Bangweolo 1, 1915–16:169). A man employed locally in the northern districts as a porter or construction worker would find it hard to earn his 5/-tax and the 10/- - 15/- he needed to buy his wife a cloth (WFC: Bangweolo 1, 1916–17:129). Conversely, a migrant to Katanga or the south could reportedly earn sums large enough to allow him to return as "le petit monsieur" (WFC: Bangweolo 1, 1919–20:252–3). Postwar inflation further increased the attractions of labor migration. In 1918–19 prices for "native goods" had risen by over 300% with calico now costing 1/- per yard rather than 4d and hoes 5/- (Northern Rhodesia, 1919:62). The depressed market for the products of Northern Rhodesia made employment within the territory increasingly difficult to obtain, and at the same time the fall in the value of the Belgian franc meant that employment in Katanga was less remunerative than before. More and more men from the "northern tribes" made their way south to Southern Rhodesia. In 1920–21 the Rhodesian Native Labour Bureau recruited 13,361 men, mostly from the north (Northern Rhodesia, 1921:72).

The brutality of labor coercion, among other factors, was instrumental in fueling the Watchtower revival in 1917 (Fields 1985:Ch. 4; Meebelo 1971:Ch. 4), and discontent over high prices kept the movement smoldering after the war. This movement, directed in part at least against the institutionalized role of Bemba chiefs under BSAC rule, provided an opportunity for those chiefs to demonstrate their loyalty to the administration and to further distinguish themselves from surrounding ethnic groups. In the last months of the war and in its immediate aftermath, the administration was anxious to appease the Bemba chiefs, and it was in this context that the issues of citemene and mitanda were once more featured on the political agenda. At a meeting between the Visiting Commissioner (Hugh Marshall) with chiefs and headmen, held in Kasama in 1918, the Commissioner asked if the men who had returned from the war were satisfied. The Paramount Chief Chitimukulu took this opportunity to say that, although many were "quite satisfied," some were complaining that they were "not allowed to live near their gardens to collect the crops, drive off the game, and make new gardens." In response, the Commissioner made the following hesitant concession:

> You will remember why the Administration wished the people to abandon their forest huts (mitanda): it was the chiefs' own desire because they were getting out of touch with their people. You also know that bigger villages were built and then the Administration agreed to allow you to kill certain kinds of game without licences. I have talked this over with Mr. Croad [Magistrate] and he agrees that some people may live in their gardens, this is because the men have worked well during the war and because we wish to keep up the food supply . . . You must understand that this is a special permission for this year and the coming season and all huts in the villages must be kept in repair . . . All chiefs and headmen must know exactly where their people are living and be ready to call upon them should they be required for war work.
>
> (BS3/418: Record of meeting of Chiefs and Headmen, Kasama 5.6.18).

By 1919, however, the policy of allowing mitanda under certain controlled conditions was once again modified, the administration having decided to allow the building of smaller villages, but to prohibit anyone from sleeping in mitanda (WFD: Chilubula 3.6.1.1919; 1.6.1919). The guerilla war over citemene and mitanda had resumed.

Despite repeated prohibitions on mitanda dwelling, the practice continued. Each April, the White Fathers reported, "le grand calme" descended on the villages surrounding their missions as people dispersed into the mitanda. And alongside this seasonal dispersal the continued fragmentation of villages (despite the "ten taxpayer" rule) continued, and was as pronounced in the Bisa areas around Lake Bangweolo as it was of the Bemba area of Chilubula (WFC: Bangweolo 1, Chilonga 1920–21:280). Throughout the 1920s, officials engaged in sporadic attempts to discourage mitanda, but the colonial state, unlike the early BSAC administration, was not inclined to jeopardize the food supply of the area by disrupting cultivation and settlement patterns energetically. By this time, the identification between the Bemba and citemene in official discourse was complete, and every new recruit to the colonial service in the province reported on it. But practical concerns dictated a more laissez-faire attitude to what continued to be viewed as a problem. Even a small colonial administration required considerable amounts of food to feed its workers, and this continued to pose difficulties in the 1920s as it had for Melland at the turn of the century. Meanwhile, the incidence of labor migration continued to increase, giving rise to a new concern over the viability of the citemene system in the absence of large numbers of men. The 1930 Report on Native Affairs listed a number of possible consequences of the "exodus to work" including a fall in the birth rate, the "breaking up of family life," the "breaking up of village life and tribal control" and insufficient cultivation and the consequent shortage of food (Northern Rhodesia, 1931:23).

It was in the same year that Audrey Richards arrived in Kasama to begin her fieldwork. By this time, the Bemba-speaking people had experienced nearly three decades of intervention from administrators and missionaries in the course of which Bemba identity had become bound up, not only with a chiefly system, but also with a mode of production—citemene. Richards' work further reinforced this construction of Bemba ethnicity through her emphasis on the inextricable connections between citemene as a system of production and a system of political and symbolic reproduction. As we will see in Chapter 3, Richards' engagement with the new science of nutrition enabled her to demonstrate what she saw as the profound dangers to the material welfare of the Bemba of the continued growth in labor migration. Citemene came then to represent in her work (and in the writings of colonial officials) not only the important system of production that it was, but also the people who practiced it. Richards did not invent the relationship between being Bemba and practicing citemene; this, as we have already seen, had been constructed by Bemba-speaking people themselves in the course of their encounter with early colonial officials. What the discourse of anthropology did, however, was to provide metaphors for this relationship that systematized and fixed it symbolically.

Anthropology was not the only discipline to have this effect. Richards' work drew on and contributed to another important body of colonial knowledge—colonial ecological science. The next chapter examines the models of citemene that have been produced by these scientists since the 1930s, and the interventions that have followed from these constructions.

2

The Colonial Construction of Knowledge: Ecology and Agriculture

From the 1930s onward, a new generation of agricultural and ecological scientists began conducting research on the practice of citemene systems of cultivation in Northern Rhodesia, specifically regarding their viability and their adaptability. However, this research was not conducted in an intellectual or political void. As we have seen in Chapter 1, by the 1930s the citemene system (or, more properly, systems) had acquired a number of distinct and powerful social and political meanings, both for their practitioners and for colonial administrators. The anthropologist Audrey Richards was to construct these meanings as traditional and timeless, and was to give particular prominence to the association between Bemba chiefly authority, Bemba ethnicity, and the sexual division of labor within citemene. But, as we have already indicated, it may be more accurate to see these and other associations as having been given their real force by the very fragility of the Bemba political system and by the "guerilla warfare" that developed between Bemba commoners and colonial administrators over the right to dispersed residence and the building of mitanda.

However recent in construction these meanings and associations may have been, by the 1930s they were powerful indeed, and were given added force by a number of developments. The first was the rise of a more widespread colonial concern over the African environment and ecological degradation.[1] The second, and related factor, was the development of the new science of nutrition and its application to the colonies (Worboys 1988). In addition to these broader influences, there were more locally specific factors. The most important of these was the extension of the labor migrant economy in Northern Rhodesia and the increasing absence of men from the Northern Province in particular, as labor migrants. The Northern Province came to be known in this period as a labor migrant reserve, and one that caused concern to colonial administrators. Given the apparently precarious nature of the agricultural system in the area, and the fact that this system, as

20

Richards had demonstrated, depended on a marked gender division of labor, the question arose as to whether this area could possibly remain self-sufficient in food. The Bemba economic and social system was widely believed to be "breaking down." The convergence of these factors was to contribute to the creation of a powerful colonial discourse on the citemene system of the northern plateau.

Citemene had long been seen by observers as a "wasteful" method of cultivation, but now, with the development of ecological science, it seemed possible to measure the extent of this "wastefulness" and to determine the limits of the system's viability in relation both to population pressure and to labor availability. With regard to the latter, Richards' work emphasized that the sexual division of labor within citemene was not only socially, but ritually sanctioned, and that the absence of the male tree-cutters would render the system of production even more vulnerable than it already seemed. Meanwhile, industrial employers were drawing attention to the poor nutritional and health status of many rural African populations from whom their labor was drawn, and nutritional scientists were devising ways to assess the inadequacy of these rural African diets. Richards was to apply these new techniques to her own study of the Bemba agricultural system, as we shall see in Chapter 3. In this chapter, however, we concentrate on an examination of the ways in which ecological and agricultural science, from the 1930s to the late 1980s, has constructed knowledge about the citemene system. In our account we treat the research findings of scientists who have worked in this area in two different ways. First, we examine it as a contribution to what became a dominant colonial (and post-colonial) discourse on "the problem of citemene" and on the Bemba-speaking people. Second, we treat this research as primary historical evidence for the practice of and changes in, the citemene systems of cultivation. There is, inevitably, a tension produced by this dual approach, but it is a tension that we regard not only as unavoidable but also as potentially productive. By neither dismissing the results of colonial science as mere "colonial representations" nor taking the evidence of that science entirely at face value, we hope to be able to shed some light on the mutually constitutive relationship between material changes in the practice of an agricultural system, the often complex and detailed observations of scientists, and the creation of a colonial discourse.

What is perhaps most evident from these scientific accounts themselves is another tension: the tension between a wish to describe complexity, adaptability, and variability, and the wish to typologize, systematize, and make comprehensible. Those elements of the scientists' accounts that were taken up and assimilated into a more generalized administrative discourse on citemene were, perhaps inevitably, the typologizing and systematizing elements, but a close reading of these accounts produces a very different impression, and one which throws doubt on the very idea of a "system." In these accounts, agricultural practices, cropping patterns, and even such physical structures as soils seem constantly to be in danger of escaping from the typologies created to contain them. Our own discussion, reflecting that of the scientists, focuses not only on the construction of a scientific discourse, but also on the ingenuity and adaptability of communities frequently labeled as "backward" cultivators. As will become evident in later chapters, some appreciation of the complexity of the interactions between cultivators and their environments in this area is necessary for any analysis of the patterns of continuity and change that characterized the halting integration of this area into the market economy. It is also only with an appreciation of some of the technical debates about population pressure, carrying capacity, and the limits of adaptation in the citemene systems that we can begin to understand why the predictions of the colonial scientists, administrators, and anthropologists about the fate of the people of the Northern Province and of their agricultural systems have not come to pass, at least, not quite in the way expected.

North and South, Large and Small:
Types and Typologies of Citemene

Put at its most simple, citemene is a form of shifting-cultivation, in which the movement of fields and settlements is occasioned by the exhaustion of suitable woodland for cutting and burning. This is a form of cultivation practiced in most parts of the plateau that occupies northwestern and northeastern Zambia, wherever the rainfall exceeds 1,000 mm and the soils are heavily leached (Schultz, 1976:50). Research conducted from the 1930s onward into agricultural systems, soils, and vegetation, identified two different types of "traditional" citemene in the Northern Province: the "Large-circle" or Northern citemene system and the "Small-circle" or Southern citemene system (Trapnell 1953; Trapnell and Clothier 1957; Allan 1949, 1965; Peters 1950; Alder 1958). However, the scientists who constructed the north/south, large-circle/small-circle typology also recognized that a large number of variations existed within these two citemene systems, particularly on the geographical (and social) margins of the plateau. Most notably, in areas near Lake Bangweulu in the west, along the Tanzanian border of the north, and around Isoka in the east, citemene was, and is, combined with other economic strategies (such as fishing and cattle-keeping) and with non-citemene cultivation practices in ways that produce distinctive agricultural systems. These modes of livelihood, while incorporating to a greater or lesser extent the growing of finger millet and/or the cultivation of burnt gardens, also incorporated a great deal more.

This evident complexity, which has always been well recognized by agricultural researchers, resulted in the construction of a number of subsidiary typologies that aimed to capture and contain observable variations (Trapnell 1953; Schultz 1976; Reid et al. 1986). For example, in addition to the "Northern" and "Southern" citemene systems, the colonial ecological scientist, Colin Trapnell, identified both a "Western" citemene system in which cassava was grown as a joint staple with millet, and a "Lake Basin" system, in which cassava grown on mounds predominated, but where millet and bulrush millet were also grown on burnt gardens (Trapnell 1953:59–63, 65–70). He also described a "Northern Grassland" system, characteristic, it was said, of the Mambwe area along the Tanzanian border, where millet was sown on grass-mounds (Trapnell 1953:56–58). This typology has had a long life. In the 1980s, the Zambian Government's Adaptive Research Planning Team (ARPT) based at Kasama has recently identified five "farming systems" or "zones" in the Northern Province that correspond reasonably closely in some aspects to Trapnell's 1930s classification (Reid et al. 1986). These are (zone 1) the Lakes Depression, a cassava/fish system covering Kaputa and part of Mbala Districts, which Trapnell classified under his Lake Basin system; (zone 2) the Central Plateau, described as traditionally a citemene-based finger millet/bean system, which is synonymous with Trapnell's Northern and Southern citemene systems; (zone 3) the North Eastern Plateau, a highly productive maize/cassava/finger millet system with a tradition of cattle keeping, classified by Trapnell as the Northern Grassland or Mambwe system; (zone 4) the Chambeshi-Bangweulu Floodplains, a predominantly cassava/fish system, also covered by Trapnell's "Lake Basin System," and (zone 5) the Luangwa valley, described by Trapnell as the "Valley System." The ARPT classification is said to be based on ecological and socioeconomic considerations, as opposed to Trapnell's which was primarily based on cultivation methods and staple food crops. However, since cultivation methods and staple food crops are products of an interactive relationship between ecological and socioeconomic factors, there is considerable convergence between the two systems of classification, even though they were constructed fifty years apart.

The simultaneous emphasis on similarity and difference that is evidenced in these accounts is probably the feature of all typologies. In this case, the central and common feature that is held to bind these various sets of practices together is the cutting and burning of trees. It is the cutting of the trees and the burning of the branches into ashes which, in the minds of scientists, administrators and others, defines a set of practices as a form of citemene. This is the case despite the fact that researchers from Trapnell onwards have also clearly documented the fact that all of these "systems" involve a combination of shifting and semipermanent cultivation. People who practice citemene, both in the knowable past and in present-day Zambia, are also involved in the cultivation of semipermanent gardens around villages (Allan 1965:68–69; Schultz 1976:58; Trapnell 1953:54–55). Although these gardens are of various types and sizes, they are an integral and important part of the agricultural system, and the evidence suggests that they have been increasing in importance throughout the last century, most especially in the 1970s and 1980s (see Chapter 8). It is, therefore, important to bear in mind that the so-called citemene systems are actually made up of a number of agricultural strategies and practices, of which the citemene fields are only a part.

The difficulty of reducing such complexity to a "system" has sometimes been addressed by using observable variability as a basis for establishing an evolutionary significance. This provided the model of citemene with both an historical and an "ethnic" dimension, and it was through this model that the notion of the "breakdown" of a system was most frequently expressed. Some practices were thought to be older than others and some ethnic groups were thought to be more "advanced" than others; here, the natural scientists were clearly influenced, not only by their own observations, but also by anthropological models that associated different "tribal" peoples with different modes of livelihood.

For example, a theme in the literature from the 1930s to the present day is the relationship between hoe-cultivation and citemene practices, a relationship assumed by Allan, for example, to be diachronic:

> There is . . . some evidence, from aerial photographs in which the pock-marked appearance of land subjected to repeated cycles of small-circle citemene appears as a faint background to the land-use patterns of to-day, that the system was formerly more widely practised, and a number of otherwise normal hoe-cultivation systems show apparent evidence of former citemene practice, or include subsidiary gardens.

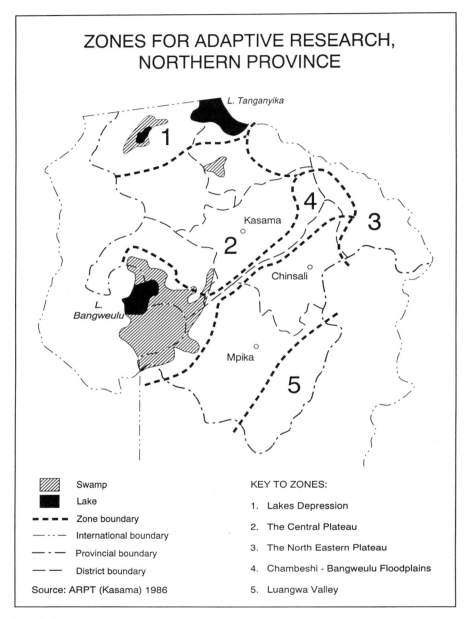

ZONES FOR ADAPTIVE RESEARCH, NORTHERN PROVINCE

Swamp
Lake
Zone boundary
International boundary
Provincial boundary
District boundary

Source: ARPT (Kasama) 1986

KEY TO ZONES:

1. Lakes Depression
2. The Central Plateau
3. The North Eastern Plateau
4. Chambeshi - Bangweulu Floodplains
5. Luangwa Valley

Map 2.1.

*Indeed, it would be a simple matter to present an apparently evolu-
tionary series ranging from small-circle through large-circle and
semi-citemene systems to normal hoe-cultivation"*
(Allan, 1965:69).

More recently, Peter Stromgaard has suggested that what came to be known
after ethnic groups as the "Lamba system" of block citemene, the "Lala system" of
small-circle citemene, and the "Mambwe system" of grass-mounded cultivation
(fundikila), are all evolutionary responses to the depletion of forest cover, and are
thus secondary in some sense to large-circle citemene (Stromgaard 1985a; 1988a;
1989). However, Colin Trapnell had always taken a more critical view of the idea of
an evolutionary sequence and he suggested that because citemene systems were
constantly evolving and changing in response to a variety of factors, socioeconomic
as well as ecological, they could not be fitted into a simple system of classification.
He further suggested that any single evolutionary sequence was likely to be difficult to establish, because many of the observable variations were clearly microre-
sponses to local conditions (Trapnell 1953:34,37,39–56).[2]

It is clear from the colonial historical record that much of the variability described by ecological and agricultural scientists was frequently lost as a result of the
anxiety of officials to define and control what was, in fact, an enormously complex
set of agricultural practices. As we shall see in later chapters, ethnic and evolutionary paradigms held an attractive simplicity for those charged with ensuring social order and food security in a sparsely populated area, large parts of which they
had never even visited.

But colonial scientists of citemene were charged with the task, not only of describing and classifying, but also of predicting. They and their postcolonial suc-
cessors were asked to calculate the productivity and the carrying capacity of what
appeared to many untrained European eyes as wasteful and unsustainable systems. They were asked to answer the question of whether citemene could survive
in the face of two developments: (1) the growth of population and (2) the absence
of men as labor migrants. In order to answer the question "Can citemene survive?"
they produced a wealth of detail on practices and productivity, the results of which
we discuss next.

Environmental Adaptations: Land Preparation
and Carrying Capacity

Aerial photographs of the Northern Province do indeed show in Allan's words a
"pock-marked" appearance, created by the repeated cycles of clearing, cultivation,
and regeneration that defines citemene systems. In the so-called "large-circle" system, citemene branches are looped from trees in a specific area of woodland. These
branches are left to dry and are then collected together in a large circle or oval,
which is burned in early November when the first rains come. In this "large-circle"
system, a citemene field usually covers an area of 0.5–1.0 hectares, but the area of
woodland cut will have been six to ten times that size. This type of citemene came,
in the colonial period, to be viewed as characteristic of the Bemba people, despite
the fact that it was and is practiced by several other groups on the plateau. The

word citemene is derived from the Bemba word kutema (to cut down trees) and among the Bemba, the term icitemene (pl. ifitemene) is applied to a new millet garden during preparation.[3] When the garden has been sown, it is known as an ubukula (pl. amakula), and in its second and further years of life, it is called an icifwani (pl. ififwani) until it reverts to bush (mpanga) (Richards, 1939:243–244). Thus, each citemene field has, in Bemba terms, its own life-cycle, during which its uses and meanings change.

In "small-circle" citemene systems, the trees are cut down at waist-height, as opposed to being pollarded as they are in the "large-circle" system. They are then collected together for burning. In this system, many small fields are laid out within the cleared area, and these fields will represent only 1/10–1/20 of the total area cut. In both the "large-circle" and "small-circle" systems, fields may be fenced, if necessary, to prevent pigs and other animals from destroying crops (Richards 1939:297–298).

Research shows that the "large-circle" and "small-circle" systems differ not only in the shape and size of their fields, and in the ratio of cultivated land to cleared area, but in the overall period of cultivation and in the rotation of crops. There are many local variants, but "large-circle" citemene generally involves the continuous use of the field over a period of four to six years, and an alternating and mixed cultivation of finger millet, cassava, maize, groundnuts, beans, and sorghum. These crops and others are also grown on the village gardens. According to accounts of this system, each farming household should ideally prepare a new citemene field every season, with the result that all the crops in the rotation cycle will be available to that household each year (Schultz 1976:56–58; Stromgaard 1984a; 1985b).[4] In the "small-circle" system, the citemene fields—in which finger millet is grown in the first year and groundnuts are grown in the second—are abandoned after three years. A number of crops, notably sorghum and cassava in the past, but more recently maize and cassava, are grown in semipermanent village gardens (Peters 1950:7–11; Schultz 1976:60–61). The "small-circle system," which came to be associated with the Lala people, has often been regarded by agriculturalists, and was certainly considered by colonial officials, to be more wasteful than the "large-circle" system (Trapnell 1953:106). This is because the use-life of the citemene fields is shorter in the "small-circle" system, the ratio of cut woodland to cultivated field is greater, and the complete felling of trees, as opposed to the lopping of the branches, is thought to retard the woodland recovery rates.[5]

There was a tendency among nonspecialist observers to view any system that involved the burning of trees and eventual abandonment of fields as in some sense wasteful. Even Audrey Richards, who clearly admired the skill and knowledge of the cultivators she observed (Richards 1939:289,307,319), felt compelled at other points in her book to describe citemene as ". . . one of the most primitive forms of bush cultivation" (Richards 1939:19). Richards echoed the prevailing mythology of the colonial period that held that citemene was backward and primitive, but this existed side-by-side with the view that this was a system well-adapted to local environmental conditions, as well as being flexible and responsive to change.

Agricultural research in the 1930s and 1940s made it clear that citemene systems could be seen to have a number of advantages in the context of the high rainfall and badly leached soils that were characteristic of the northern plateau. Allan, for example, pointed out that citemene methods gave better and much more reli-

'The burnt Citemene field'
Photo: Audrey Richards, 1930s

able results on the weak plateau soils than did hoe cultivation. Agricultural experiments conducted in the 1930s (and repeated many times since) found that citemene produced the most reliable yields for finger millet, and when one agricultural researcher tried "ordinary farm practices" with manure and fertilizer, it was clear that such methods involved much greater cost but gave no better results

'Tree cutting'
Photo: Audrey Richards, 1930s

than citemene. In fact, manuring had a negative rather than a positive effect because it had a selective influence on the weed *Eleusine Indica* and thus actually served to depress the millet yield (Allan 1965:72, cited Moffat 1932). Allan also cited the results from an experiment conducted at the Lunzuwa agricultural station in which millet grown on dug plots prepared according to the Mambwe grass-mound system was compared with that grown on citemene plots over five seasons, during which three millet crops were taken. As can be seen from Table 2.1, cite-mene yields were higher and more reliable than those obtained using what was generally felt to be the more "advanced" Mambwe grass-mound system.

Further research revealed a complex set of reasons accounting for the productivity of the citemene system. The fertilizing effects of ash were clearly important, but other things were happening when a citemene field was burned. Richards noted that the phosphate and potash status of the soil were enhanced by burning, and that freshly burnt soil also contained a high concentration of calcium that improved its physical condition (Richards 1939:289). Furthermore, burning appeared to decrease soil acidity, which meant that the added phosphates could be more eas-

'Women piling branches'
Photo: Audrey Richards, 1930s

TABLE 2.1. Millet Yields: Citemene and Dug
Plots Compared

Year	Dug kg/ha		Citemene kg/ha
1936	Failure		1,798
1937	—	Groundnuts	—
1938	489		1,638
1939	—	Beans	—
1940	650		1,333

Source: Allan 1965:73

ily maintained (Allan 1965:73). It seemed then, that it was the combination of ash and burn that accounted for the beneficial effects of the citemene system, and that the use of ash alone, or burn alone, did not produce the same favorable results. Allan also found that doubling the quantities of ash supplied and doubling the intensity of burn gave only small yield increases—suggesting to him that the quantity of timber habitually used by citemene cultivators was more or less optimal (Allan 1965:73). Another important benefit of the method was found to be connected to the rate of decomposition of organic matter. Researchers found that a large flush of nitrogen was released by the rapid decomposition of the humus at the beginning of the rains. The drier the soil was before wetting, the greater would be the flush of decomposition. This indicated that the drying of the soil through burning served to increase the availability of nitrogen in the growing season (Allan 1965:74; Mansfield 1973; Haug 1981:12–18).

It seemed, then, that citemene was a very effective means of conferring a high, if transient fertility on soils of very low intrinsic value (Allan 1965:74). In general, in the Northern Province soils of good physical and chemical status are rare, though pockets of more fertile soils do exist, and soil fertility can vary greatly within local areas and even within the same soil "type" (Kalima 1983; Brammer 1976; Richards 1939: 320, 323).[6] In the more "typical" acidic and heavily leached soils, the millet yields of citemene fields were, and are, hard to improve upon. A more recent researcher has gone as far as to say that ". . . the 'cut and burn' technique of citemene, as well as the mound cultivation where the grass is naturally composted and which is practised in the 'village gardens' of citemene systems . . . are adaptations which make these systems relatively independent of soil fertility" (Schultz 1976: 12–13).

The general thrust of arguments made by the majority of agricultural researchers from the 1930s to the present day, then, seems to be that citemene is a potentially adaptive system, well-suited to prevailing ecological conditions on the plateau. But the same researchers have been concerned, not only to assess the productivity of the system, but also to investigate its long-term viability. Such a system, it was argued, was highly productive for areas of low population density that allowed full regeneration of the forest, but could it be sustained without serious environmental degradation when densities increased?

This question, which was asked by successive colonial and post-colonial administrators concerned with the food security of the area and haunted by the spectre of environmental decline and social breakdown, led researchers to attempt to calculate the carrying capacity of the citemene farming systems. These attempts, though often far from successful, revealed further complex interrelationships between the agricultural practices of citemene, the physical properties of the soils, and the social and political determinants of population and residence patterns. No assessment of the sustainability of citemene systems could be complete without addressing all of these factors. Here the models and methods of ecological science did not always prove equal to the task.

One of difficulties encountered in attempting to calculate the carrying capacity of citemene systems was the very fact that people who practice these systems do not rely entirely on their citemene fields for their substance requirements. They also cultivate substantial, semipermanent village gardens, as discussed above, and many systems also incorporate certain subsistence strategies and productive techniques that permit the maintenance of much higher population densities than could be maintained by a citemene system alone (Allan 1965:107). The narrow focus of colonial attention on cutting and burning obscured the fact that this was only one part of the way in which communities of the northern plateau fed and supported themselves. Once these other strategies were entered into the calculation, wide variations in critical population densities were revealed. For example, the Lake Basin system has been calculated to have a critical population density twenty times, and the Northern Grassland system fourteen times, that of the Northern and Southern citemene systems (Mansfield et al. 1975, vol 2:10). Although citemene had come, by the 1930s, to represent the agriculture of the whole of the Northern Province, variation between parts of the province was probably as important as the common element of cutting and burning. This theme of variation is taken up in more detail next, and is also explored in later chapters.

A further, and revealing, problem encountered in the task of assessing carrying capacity was the difficulty of estimating the percentage of land in the region that was suitable for citemene production. Estimates made in the 1930s and 1940s varied from 50%–70% of the total land surface of the northern plateau considered to be suitable for citemene (Allan 1965:131; Trapnell 1953:98–99; Schultz 1976;133,139). These assessments undoubtedly reflected actual variation in the proportion of woodland to total land surface over the whole of the northern plateau, but they also demonstrated the inadvisability of basing calculations on average percentages. In addition, figures for carrying capacity based on estimates of land suitably for a particular method of cultivation depended on assumptions about the eventual utilization of all suitable land and on assumptions about ideal population distributions. In other words, they depended on assumptions about social and political factors, as well as on assessments of the "natural" resources.

As we saw in Chapter 1, population distributions in the Northern Province were not only skewed as a result of the disruptions of the nineteenth century and of the aggregation of persons that was necessitated by the nature of local political systems, but were subsequently affected by the attempts of BSAC and colonial administrators to control the movement of population and to prevent what they saw as excessive dispersal. In later decades, population distribution was further influenced by the building of roads and the provision of markets (Richards 1939: 277–278), as well as by the availability of important resources such as fish, and by the distribution of the tsetse fly. No calculation of carrying capacity, then, could assume that populations would be "naturally" and regularly dispersed over the available cultivable land.

Most calculations of carrying capacity produced for the Northern Province, though ostensibly concerned with the availability of land for citemene cultivation, tend to point toward the importance of the noncitemene elements of production systems in determining the numbers of people who can be supported on any given piece of land. A recent example can be found in the results of research by Mansfield et al. (1975:vol.2) who calculated the average critical population densities of the five "systems" of the Northern Province, taking into account the different levels of land available for citemene agriculture (see Table 2.2).

The marked variability in the critical population densities that they found clearly reflected differences in cropping patterns and in production techniques, which are only obscured by the focus on the cutting and burning aspect of cite-

TABLE 2.2. Average Critical Population Densities for Different Levels of Land Available for Citemene in Number of Persons Per SqKm.

	% Total area with suitable land for traditional agriculture.					
	100%	75%	60%	50%	40%	30%
Citemene System						
Southern	3.0	2.3	1.8	1.5	—	—
Northern and Western	3.9	3.0	2.4	2.0	—	—
Lake Basin	80	68	60	40	32	—
Northern Grassland	56	42	34	28	19	17

Source: Mansfield et al. 1975, Vol. 2:10.

mene. The cassava/fish-based economy of the Lake Basin system and the grass/manure techniques of the Northern Grassland system are much more intensive than are the Northern and Western citemene systems, and they permit much higher population densities. Standard assessments of the carrying capacity of citemene systems are not appropriate, therefore, for these systems because their levels of production do not primarily depend on the cutting and burning of trees.

For those systems in the Northern Province that are more dependent on citemene fields, one crucial factor in assessing carrying capacity is woodland regeneration rates, and over this there has been much difference of opinion among scientists. The period necessary for a given area of woodland to regenerate is difficult to assess. It has always been thought to be shorter in large-circle citemene systems as opposed to small-circle citemene systems because of the practice of pollarding the trees rather than felling them. However, regeneration rates also depend on the overall management of burning. The British South Africa Company (BSAC) and colonial administrators were quick to promote what they called "early burning," which involved burning grass and detritus early in the season before it had become too dry. If this was not done, then when the citemene stacks were set alight just before the start of the rains, there was a risk of fire spreading to the surrounding trees and bush which were tinder dry by this time in the season. The result could be scorching of woodland outside the citemene circle, thus damaging wide areas of bush and further retarding regrowth.[7]

As far as the citemene cultivators themselves are concerned, the important issue is the length of time necessary for the trees to regenerate to a size suitable for recutting. This does not mean that the woodland has to be fully regenerated, but if it is not, it is then necessary to cut a larger area in order to make the required ash garden, and this, of course, affects any calculation of the amount of land required for sustainable citemene production. Various commentators have remarked that the proportion of cut woodland to garden area varies from 5:1 to 10:1 depending on the quality of the woodland (Allan 1965:131). This means that the amount of woodland required for citemene can double under some circumstances. The crucial point is that enough woodland must be cut to produce a sufficient quality and quantity of ash. Peters' study of the Lala of the Serenje plateau showed a clear relation between short regeneration periods for woodland and increasing ratios of cut woodland to prepared fields (see Table 2.3).

TABLE 2.3.   Regeneration Periods for Trees Compared With Area of Woodland Cut for Citemene

Regeneration period (years) of trees cut for fitemene	9–12	13–16	17–20	20+
Burned areas as a % of the total area cut	4.41	6.26	7.31	9.57

Source: Peters 1950:37.

Peters estimated that under this small-circle citemene system, the total land requirement per head—including land necessary for subsidiary gardens—was around 42 hectares. He calculated further that the carrying capacity was approximately 2.2 persons to the sq km. These figures were based, however, on the assumption that the period required for full regeneration of the woodland would be thirty-five years (Peters 1950:68–69). This was only one of many calculations of regeneration rates, however. Trapnell argued (1953:98–100) that in small-circle citemene, twenty years was the period required for regrowth, while Allan (1965:110) calculated it as twenty-two to twenty-five, and Mansfield et al. (1975, Vol. 2:7) thought it to be in the region of twenty to thirty years. On the basis of a regeneration period of twenty years, Trapnell estimated that under a small-circle citemene system, approximately 185 hectares of woodland would be needed to maintain a "normal family" in perpetuity (Trapnell 1953:100). Trapnell did not allow, as Peters did in his calculation of the carrying capacity of small-circle citemene, for the amount of land necessary for the subsidiary gardens, nor did he specify the number of persons in a "normal family." However, if we take Peters' parameters and apply them to Trapnell's figures, we can assume that the subsidiary gardens take up a further 5% of the total land required, which would bring the figure up to 194.25 hectares; and we can further assume a mean family size of 6.06 persons, which gives a figure of 32 hectares per person, as against Peters's figure of 42 hectares per person (Peters 1950:68–69). The 24% difference in the calculations for the amount of land required per person cannot be accounted for solely by the 43% difference in the number of years allowed for woodland regeneration; indeed, without access to the raw data, there is no way of knowing what accounts for the differences between Peters's and Trapnell's figures. However, the magnitude of the differences does emphasize once again the difficulty of estimating carrying capacities for citemene systems.

More recent research has produced the figures shown in Table 2.4 for land requirements per head for the different types of citemene system. The results shown in Table 2.4, like earlier calculations, are based on three interrelated, but problematic figures: (1) the quality and extent of woodland cover, (2) the period of fallow thought to be necessary for woodland regeneration, and (3) the area of woodland estimated as required for citemene each year. All of these figures are difficult to arrive at with any certainty, in part because, even within the more citemene-dependent agricultural systems, very local variations in both physical properties such as soil depth and social factors such as cropping preferences, create a situation of such complexity as to render any calculation suspect. As Trapnell himself argued when he came up with a carrying capacity of 1.6 per sq km for small-circle and 2.5–3.7 sq km for large-circle citemene: "It should be made clear that these

TABLE 2.4. Weighted Overall Area in Hectares Required Per Head for Traditional Agricultural Systems

	% Total Area With Land Suitable for Traditional Agriculture				
	75%	60%	50%	40%	30%
Citemene System					
Southern	44.0	55.0	66.0	—	—
Northern and Western	33.6	42.1	50.6	—	—
Lake Basin	1.6	2.1	2.5	—	—
Northern Grassland	2.4	3.0	3.6	4.3	14.7

Source: Mansfield, et al 1975, Vol. 2:10.

carrying capacities are given for purposes of illustration only and that the exact carrying capacity of land under the various forms of the citemene systems can only be arrived at by fuller local investigation" (Trapnell 1953:100).

If we take an historical (as opposed to evolutionist) perspective on the whole question of the viability, sustainability, and carrying capacity of citemene systems, further complications arise. Population densities for the Northern Province as a whole remain low, but the period 1890–1990 has clearly seen a large increase in the population of the Northern Province of Zambia (see Table 2.5), although the exact magnitude of that increase is difficult to calculate because the population figures themselves are so unreliable.[8] In addition to this general increase in population is the fact, noted earlier, that some areas of the plateau seem to have been able to support population densities far in excess of what has been estimated to be the average carrying capacity. In other words, the ability of citemene systems to survive and to support increasing populations seems to have far exceeded the expectations of those who have studied them.

As we indicated earlier, this "puzzle" can only be solved by extending the field of vision beyond the practice of citemene itself. Increases in population densities in the Northern Province have been skewed, indicating bias in the distribution of urban centers, infrastructural provision, and market potential, as well as differences in land use factors, including cropping patterns.

The cassava/fish systems (practiced for example, in the Luwingu District) and the grass-mound/cattle systems (practiced in the Isoka and Mbala Districts) are clearly able to support higher population densities, in part through a greater reliance on semipermanent cultivation techniques, rather than reliance on the traditional citemene systems (represented by the Kasama, Mporokoso, Mpika, and Chinsali Districts). However, the relationship between production practices and sustainable levels of population is multicausal and complex. In some cases, population pressure appears to have given rise to agricultural intensification, but in other cases, it appears that changes in agricultural techniques and cropping patterns permitted increased population levels.[9] Settlement decisions, both in the present and the past, have been influenced as much by such things as political allegiance, the attractions of urban living, opportunities for education and wage employment, and infrastructural provision, as by land quality and woodland availability (see Chapters 5 and 8). Richards, for example, noted in the 1930s that

TABLE 2.5. Population Density Per Sq km Over Time

District	Population Density Per Sq km			
	1958	1969	1980	1990
Kasama	3.3	5.1	7.2	9.3
Mbala	3.1	5.8	6.2	7.5
Isoka	2.7	5.6	6.8	8.8
Mporokoso	1.7	2.9	3.4	4.5
Chinsali	1.5	3.7	4.3	5.6
Mpika	1.1	1.4	2.0	2.8
Luwingu[1]	2.3	7.3	5.9	7.5
Chilubi[2]	—	—	7.2	8.6
Kaputu	—	—	3.4	4.0

Source: Allan 1965; Mansfield et al. 1975: Vol.5; Government of Zambia,
CSO Census 1980; 1990
[1]Previously part of Luwingu District
[2]Previously part of Mporokoso District

semipermanent mound cultivation had been adopted in a village near Kasama because people were unwilling to move their village to a new site where they would be able to practice citemene. Their reasons were simply that they did not wish to lose the opportunities for wage labor and income generation provided by the proximity of their village to an urban center (Richards 1939:322–324). Cash, far from being the cause of breakdown in these systems, has often been an important factor in their survival, as we shall see in later chapters.

In a recent study, Carol Kerven and Patrick Sikana have described changes in cropping patterns and land use in the village of Nsokolo, in the Mambwe area, where increases in population have apparently led to a decline in woodland suitable for citemene (Kerven and Sikana 1988). As a result, some people moved to growing finger millet on permanent fields using grass-mounds (fundikila). This change involved a corresponding change in soil selection, from the lighter soils of the midslope preferred by farmers for finger millet production on citemene fields to the heavier, more clay soils of the hill crests. However, the poor yields of finger millet on unburnt fields led many people to switch from finger millet as a staple crop to cassava, and later to hybrid maize. The original pressure on forested land can be partly attributed to the growth of the local village as the Chief's center. At the present time, the main crop grown on permanent fields (using grass mounds) is beans. Resource-rich households also grow hybrid maize (as a staple and a cash crop) on grass-mound fields using fertilizer. Resource-poor households tend to grow cassava on permanent fields when they can no longer find trees for citemene.

In this example, it can be argued that population pressure and the subsequent decline of available woodland have given rise to the adoption of new staple crops, new farming methods, changes in soil preferences, and new settlement patterns. Processes like these can seen going on all over the Northern Province today and they were certainly observed, albeit in slightly different form, in the past (e.g., Trapnell 1953; Allan 1965; Stromgaard 1984b, 1985a, 1988a, 1989, 1990a, 1990b). However, despite the low carrying capacity of so-called traditional citemene systems, it would not be correct to see population pressure as simply causing these changes. The cultivation of hybrid maize on permanent fields, for example, cannot be seen simply as the direct result of population pressure. The causal mechanisms here have much more to do with government policy (on agriculture and settlement), the decline in employment prospects and the value of real wages, subsidies to the producer, and the provision of credit facilities (factors that are discussed in Chapter 8). All of these factors combine in ways that make the cultivation of maize attractive to farmers. Maize cultivation serves to encourage permanent settlement and the concentration of populations within easy reach of the roads: the inevitable result is the cutting of surrounding woodland. In addition, changes in household labor requirements for maize production discourage many families from cultivating citemene fields at any appreciable distance from the main settlement. In such circumstances, population pressure and changes in land use are the consequence and not the cause of a change towards maize production.

Generalizations about critical population densities, then, are always likely to be complicated by a variety of socioeconomic factors that have little to do with ecological causes in any sense. Ecological and agricultural scientists have often been aware of this. For example, Allan said of the Northern Grassland system of the Mambwe:

> It is difficult to say what the Critical Population Density of the original system may have been or how far general degeneration of this system has proceeded as a result of population increase. . . . the process of change has been greatly complicated by other factors such as the absence of men needed for the heavy work of mounding, the pressure of a growing cattle population, the introduction of the plough and the recent emergence of a class of 'individual settlers' who are 'farming' relatively large acreages"
>
> (Allan 1965:138)

Trapnell noted in the 1930s that microecological variations (in soil fertility, for example) within the same area meant that some systems were able to support populations in excess of their estimated carrying capacity, although he emphasized that it was variations in local agricultural practices that probably accounted for the different levels at which population pressure brings about environmental degradation (Trapnell 1953:96–98). These variations included, of course, such things as differences in the nature and extent of cutting, but much more importantly, they involved variability in the type of crops grown, in crop rotation practices, and in the extent and nature of the so-called subsidiary or village gardens. Such variations were brought about in response to changes in production and consumption patterns as a consequence of socioeconomic and political change. They did not take place independently of ecological conditions, nor were they determined by them. It was clear to Trapnell that, even disregarding the importance of the non-citemene elements of the agricultural system, it was possible to find important variation in what cultivators did with their citemene fields—variations that could have direct consequences for calculations of carrying capacity and sustainability. For this reason, the scientists of citemene turned their attention to an examination of cultivation practices, and it was here that the discussion of the social organization of production provided by Richards became so influential.

The Cultivation of Citemene Gardens

In colonial administrative accounts of the Northern Province, citemene was usually presented as synonymous with millet production. Yet the detailed research of colonial scientists had demonstrated that the citemene garden was best conceived as a complex horticultural system involving mixed cropping.

According to these accounts, in the first year of cultivation, after the garden had been burnt, the main millet crop (amale yakalamba) was sown in December or January, when the ground was soft from the heavy rains. However, before the rains began, or at the very beginning of the rains, gourds, pumpkins, small cucumbers (and recently increasing amounts of cassava) were planted around the edge of the garden. It was also common to broadcast perennial sorghum or kaffir corn (sonkwe) over the garden at wide intervals before the main millet crop was sown, or to mix a few such seeds in with the millet at the time of sowing.[10] The taller sorghum plants would then shoot up here and there within the main millet stand (Richards 1939:296–297; Trapnell 1953:45; Vedeld 1981:39–40). A special variety of quick growing millet (mwangwe) that ripens in April was also sometimes planted in small ash patches in the garden area or on a small part of the main garden in November (Richards 1939:296; Vedeld 1981:40).[11] Trapnell reported in the 1930s that small-circle citemene cultivators in the south of the Mpika district included cowpeas on the periphery of their gardens and a small stand of maize in the center (Trapnell 1953:41).

The general point made in these accounts is that mixed cropping in the main citemene garden was common to all systems. The actual nature of that mixed cropping has not changed dramatically since the 1930s (e.g., Vedeld 1981:38–40; Stromgaard 1985a:44–45; Table 2.6, this text), but there are no hard and fast rules to be elicited and no one "system" to be discerned because crop mixtures always appear to have been responsive to labor availability, household composition, consumption preferences, and climatic exigency. However, there is evidence to suggest that, at the present time, with commercialization and the introduction of maize cash cropping, crop diversity in the agricultural systems of the province is declining. This has serious consequences for child nutrition, household reproduction, and intra-household resource allocation, issues that are discussed in Chapters 7 and 8.

The use of mixed cropping in the citemene systems is further evidence of the adaptability and flexibility of these systems and helps account for their survival. Experiments in Zambia on mixed cropping have indicated that crops grown in mixed culture give considerably higher total yields than those grown in pure culture.[12] There are also other benefits to be derived from mixed cropping, including insurance against crop failure, the leveling of labor peaks, disease control, and the efficient use of physical resources by plants with different needs and characteristics. The citemene systems use these benefits to the full, and mixed cropping occurs both in the citemene fields and in the so-called subsidiary or village gardens (Table 2.6). This is one of the reasons why the recent move to maize monocropping, which has occurred in some parts of the Northern Province, has had such a profound affect on the sexual division of labor, and especially on women's labor time.

In addition to taking account of mixed cropping, however, any analysis of citemene must also address the fact that crops are also rotated on the same field.

Crops and garden types

Name	Ubukula	Ichikumba bukula	Chikumba	Akakumba	Chikuka	Umunkumba	Imputa / Ibala	Imputa / Chifwani	Chibela	Ibala lya musalu	Mukanda	Fiputu	Chifwani, flat cult	Latin	Bemba
Crops grown															
Millet	•	•	•		•				•				•	Eleusine coracane	male
Fish poison	•	•	•	•	•									Tephrosia vogelii	kobamushi
Cassava	•	•	•		•	•			•				•	Manihot esculenta	kalundwe
Groundnut	•												•	Arachis hypogea	mbalala
Pumpkin														Cucurbita pepo	chipushi
African sugar cane														Saccharum officinarum	ichisali
Yam														Dioscorea spp.	chilungwa
Onion	•				•	•								Allium cepa	kanyense
Calabash	•	•		•	•	•	•	•	•				•	Crescentia cujote	nsupa
Tomato											•			Solanum spp.	mutuntula
Cucumber		•	•	•	•	•	•	•	•					Cucumis sativus	chibimbi
Cow pea														Vigna unguiculata	ilanda
Sorghum	•	•	•	•	•	•	•	•	•	•			•	Sorghum vulgare	masaka
Sweet potato														Ipomoea batatas	ichumbu
Tobacco							•	•						Nicotiana tabacum	fwaka
Bean	•	•	•	•	•	•	•	•	•	•			•	Phaseolus vulgaris	chilemba
Banana														Musa sapientum	nkonde
Maize													•	Zea mays	nyanje
Cannabis	•	•	•	•	•	•								Cannabis sativa	dagga
Water melon														Citrullus vulgaris	ntanga
Mango											•	•		Mangifera indica	mwembe

Garden types		Latin	Bemba
Ubukula	Big size, first year ash garden	Grass turf cultivation	Chibela
Ichikumba bukula	Medium, first year ash garden	Vegetable garden	Ibala lya musalu
Chikumba	Small, first year ash garden	Refuse heap garden	Mukanda
Akakumba	Smaller, first year ash garden	Flat quadrangular mound esp. f. tobacco	Fiputu
Chikuka	Very small, first year ash garden	Small village garden	Akalibala
Umunkumba	Big trunk burned	Mounds in ibala or chifwani	Imputa
Ibala	Village garden	Any old ash garden	Chifwani

Source: Stromgaard 1985a

Table 2.6.

In the second and subsequent years of its life, the citemene garden is known by the term "icifwani". Both Richards and Trapnell documented cropping sequences for icifwani gardens in the large-circle citemene systems (Richards 1939:313–319; Trapnell 1953:46). A common sequence consisted of millet in the first year followed by groundnuts in the second, millet in the third year and beans in the fourth. There were (and are) a number of other possible sequences, including allowing the perennial sorghum planted with the millet in the first year to spring up again with the groundnuts in the second year and then to take possession of the garden in the third year. In some instances, beans intercropped with other crops were then grown in the garden on mounds (mputa) in subsequent years. It was unusual to find two years of successive millet crops, but this situation did occur if the soil was good enough or if the first crop had failed. In such cases, the garden was then planted to beans or sorghum on mounds in the third year. There were also sequences in which millet was followed by the annual sorghum or kaffir corn (amasaka) or by further cassava plantings.

The enormous variability in crop sequences led Trapnell to produce a complex classification system that contained at least four subsidiary systems or variants within the large-circle or Northern citemene system: the "simple" cropping method, with or without cassava, the annual kaffir corn or "masaka" variant, and the "developed" system (Trapnell 1953:46). The complexity of his classification demonstrates once again that as much variation existed within the categories on which the typologies of the citemene systems were based as between them. Trapnell did recognize, however, the importance of the local variations he was able to document. His work thus exhibits and sustains a tension between his desire to classify and to associate certain production systems with "tribal" identities and his knowledge of microlevel variation and the effects of socioeconomic factors. For example, of the variations in crop sequences, he says:

> The elaborate nature of the system lies in the extent to which these sequences have been developed. Various forms of them are found in every region in which it is practised and more than one may be employed by a single village or family
>
> (Trapnell 1953:46).

Audrey Richards also provided a number of examples of crop sequences in single icifwani gardens, but her most notable contribution lay in the observations she made of how households managed their numerous icifwani fields (Richards 1939:314–319). Richards showed that the cutting of a new citemene every year combined with crop rotation on each individual icifwani field meant that in any given year every household had a number of fields growing a variety of different crops. For example, the newly cut field may have been predominantly planted with finger millet, while the second-year field was producing groundnuts; the third year field had been sown early with finger millet before the rains (a crop known as nkungwila) so that it would ripen before the new field, and the fourth-year field was producing beans. Variation in crop mixtures and in rotation cycles produced a large range of possibilities, especially in those situations where rotation cycles were lengthened to five years or more. Richards also indicated that the limiting factor in the length of rotation cycles was not necessarily declining soil fertility, but labor availability. If an individual cultivated a new garden every year, then they would soon have more icifwani available than they had labor to cultivate (Richards 1939:314). She also noted the importance of labor availability in determining the actual sequence of crops grown on individual icifwani. It was clear to her that this was a decision the farmer made in the context of factors such as what was being grown on other fields, the amount of labor available, household composition, and consumption preferences, in addition to a consideration of such things as soil fertility, land availability, and rainfall.

By introducing the whole dimension of management, then, Richards could demonstrate that the individual farmer could make a range of decisions on how to cultivate the household's citemene gardens, and that there was considerable flexibility within the system. Richards, like her natural science counterparts was, however, torn between demonstrating variation and flexibility on the one hand, and making large generalizations on the other. As we have already indicated in Chapter 1, one of the central themes of Richards' work on citemene was her emphasis on the relationship between Bemba identity, the Bemba chiefly system, and the male task of cutting the trees. Although there are many places in her book where this emphasis is countered by a close attention to women's labor and to the non-citemene aspects of the agricultural system, her overall message (and the one which was readily taken up by colonial administrators) was that the citemene system was centrally dependent on male labor for the cutting of new gardens, and that the development of male labor migration would thus be the death knell of the system. Citemene was, for Richards, a system in breakdown.

While the ecological scientists pointed to the physical limits of the viability of citemene, Richards pointed to the social limits. Her thesis on the effects of labor migration, combined with the already dominant idea that citemene was unsustainable, produced a picture of a system in both ecological and social crisis. Yet, just as there are alternative readings of natural scientific accounts, so there are alternative readings of the anthropological account. Richards' own careful observations, when reinterpreted outside the dominant model of "breakdown," make it clear that this was an agricultural system that could, in fact, remain viable despite the absence of large numbers of men. Because households had access simultaneously to several icifwani and several semipermanent gardens, the absence of male labor in any single cropping season was not necessarily undermining of food sufficiency. It might be ideal, but it was not necessary, to cut a new field each year. Furthermore, individual households did not operate in isolation, as Richards clearly demonstrated. Women could grow some crops that carried a high exchange value, such as beans, on the semipermanent or icifwani gardens, without male labor. Households could thus vary what they grew in order to cover for labor shortfalls, and they could also use labor exchange systems to maximize the interhousehold exchange of labor. Although Richards underplayed the role of exchange in this economy, it is clear that even in the 1930s, households had a variety of ways of gaining access to food other than growing it themselves (see Moore and Vaughan 1987; Chapter 3, this text).

If we take the citemene system as a whole, and in all its complexity, (as opposed to regarding it as simply a cutting and burning operation), then the presence or absence of male labor for the cutting of trees does not seem so crucial, even by Richards' own account. This becomes even clearer when we consider that in addition to the citemene gardens, a number of specialist subsidiary gardens, as well as the so-called village gardens were also cultivated annually. For example, Trapnell noted in the 1930s that in the large-circle system, smaller separate gardens were made for early maturing varieties of millet (mwangwe). These early types of millet were ready in March, and provided emergency rations for the hungry season that preceded the main millet harvest. Other specialist gardens included the dambo gardens for such crops as millet, maize, rice, and Livingstone potatoes, which were known in some areas (Trapnell 1953:52–54; Richards 1939:310–311). Specialist gardens of this type certainly exist in many parts of the Northern Province in the 1990s, but their role in the overall production systems is now somewhat different (discussed later; see also Table 2.6). The flexibility and sustainability of the cite-mene systems thus depended and depends on cropping patterns and sequences which are extremely sensitive to overall labor supply, but not necessarily simply to the presence or absence of male labor for cutting. In the following chapters, and especially in Chapters 7 and 8, we suggest that the citemene systems are, in fact, particularly sensitive to the timing and quantity of female rather than male labor. Furthermore, we will argue, the "gendering" of the citemene system, which ap-pears so fixed in Richards' account, has over the past fifty or sixty years undergone significant change.

The importance of the role of women's labor in these systems becomes evident when we turn to consider the question of the semipermanent village gardens that were described in detail by Richards and that are primarily women's responsibility. All citemene systems incorporate village gardens (ibala), although in the Northern Grassland and Lake Basin systems such gardens cannot really be distinguished from the main gardens, except where old village sites are actually being used as gardens themselves. This is particularly the case at the present time because of the expansion in permanent cultivation (see Chapter 8). Many different types of vil-lage gardens have been identified, with a wide variety of cropping patterns and sequences. Such gardens are mounded (mputa) and the technology for their prep-aration, which involves hoeing, piling up grass-sods, and throwing earth over them at some point during the rains or just after, has not changed much since the early colonial period (Stromgaard 1985a; Vedeld 1981:45). Once the mounds have been made, they are usually hoed over and remade every year or second year with varying degrees of thoroughness. Richards noted in the 1930s that hoeing mounds was considered unglamorous and hard work, culturally undervalued relative to the attention paid to the dramatic cutting and burning procedures for citemene (Rich-ards 1939:304). Both women and men make mounds, but women do the planting and the tending of the gardens. In the 1930s, Trapnell discussed a variety of crops grown in village gardens, including cassava, sweet potatoes, maize, millet, sor-ghum, beans, gourds, and castor oil plants. Pumpkins, gourds, maize, and tobacco were grown also on flatter beds near the village. Green maize grown in village gardens and eaten in February and March was one of the important ways of gett-ing through the hungry months (Trapnell 1953:54–55). Richards noted a similar range of crops and commented on the Bemba appreciation of the value of mixed cropping and "on the careful planting of seeds on the mounds to suit their habit of growth and time of ripening" (Richards 1939:305). The point of village gardens was to plan them and plant them to try and ensure food all year round; hence, the Be-mba organized the early and late sowing of certain plants and used quick- and slow-maturing varieties of seed (Richards 1939:306–308). The village gardens were

'Men working *imputa* (mounds)'
Photo: Audrey Richards, 1930s

thus the main way in which women made sure that they had a variety of crops to feed the family. As we show in Chapters 7 and 8, now that village gardens have suffered a relative neglect in favor of the permanent or semipermanent cultivation of maize, individual women can find themselves particularly short of relish crops and of early maturing varieties of staples designed to help them through the hungry months.

Cassava and Maize: The Changing Nature of Citemene

When colonial officials waged war against citemene cultivation, they were not waging war against a primitive and backward system, but against a complex and highly adaptive one. The almost unfathomable intricacy of the classification systems designed by colonial scientists to capture this reality was enough to make this point. Colonial scientists, although they were critical of some aspects of citemene production, certainly did not represent these production systems as in any way simple, although they did think that they were very vulnerable to change. Richards also paid great attention to the skill and ingenuity of Bemba cultivators, and to the complexity of cropping patterns, yet she also characterized this as a primitive system, that would not survive the development of wage labor and consequent changes to the social system. In some sense, the complexity and variability described by both the natural scientists and the anthropologist became buried under their own, and their readers', anxiety to generalize and systematize. The dynamic aspects of these systems were also obscured by the underlying assumption that what they were observing was the practice of a very ancient system, a vestige of the past that was by definition bound to be unsuited to modern conditions. For Richards, citemene symbolized the cultural identity of a primitive African people, now under threat from colonial capitalism and labor migration. Yet, as we argued in Chapter 1, this association between being Bemba and cutting trees was far more likely to have been borne out of the very encounter between cultivators and the agents of change—the colonial state. Citemene as practiced and witnessed by Richards and Trapnell in the 1930s, was a system that had evolved in a context of change, it was not a vestige of some distant and static past.

With the benefit of hindsight it seems that the citemene systems of the Northern Province have proved remarkably adaptable and resilient. But some of the more far-reaching changes that have taken place within the last fifty years perhaps raise the question of whether and when a point is reached at which adaptations to a system cease to be adaptations and begin to constitute a new phenomenon. These changes include the spread of cassava as a staple, the extension of semipermanent cultivation and, related to this, the development of hybrid maize production. These are all issues that we take up in later chapters, but here we provide a brief summary.

The spread of cassava production, and its increasing role in supplanting millet as a staple food, is a process that has occurred over a long period of at least one-hundred years. This was, however, a process much accelerated in the 1920s and 1930s, by colonial officials who not only encouraged but enforced the cultivation of cassava as a famine crop (see Chapter 4). When Trapnell conducted his research in the 1930s cassava cultivation was characteristic of the Western and Lake Basin systems of cultivation, but subsequently it spread over the entire province, from west to east. Cassava cultivation may imply changes not only in the methods of cultivation, but also in residence patterns because it can be grown on permanent and semipermanent fields on a three- to four-year cycle, (although it is also grown on citemene fields, without these consequences for residence). Despite nutritional disadvantages, it has been evident since the 1930s that cassava production, especially when allied with fishing and other economic strategies, is capable of supporting higher population densities than are extensive millet systems. As we show in Chapter 4, some communities in the 1940s moved from being citemene producers to producers of cassava on semipermanent fields, within a very short space of time. In the 1990s, large cassava gardens (usually intercropped with many other crops) are a widespread feature of the landscape and are often far more central to household food-sufficiency than are the citemene gardens that may be cultivated alongside them.

The spread of green-manuring (fundikila) cultivation is further evidence of a move toward permanent or semipermanent cultivation. Described by Trapnell and others as a feature of the Northern Grassland (or Mambwe) system, this has been a major cultivation practice in some areas for well over 100 years, and probably for much longer. Under this system, grass turves and vegetable matter are hoed up, thrown up into circular mounds, covered with soil and allowed to rot down as compost. In the second season the mounds are spread for the cultivation of millet (Stromgaard 1988a; Trapnell 1953:57–58). Viewed historically, the Northern Grassland system appears to have been able to support higher population densities than were the so-called "traditional" citemene systems (see Table 2.2), and it is a method for the semipermanent cultivation of staples, including millet, which does not involve tree-burning. There has been considerable debate in the literature (see above) as to whether this system is an adaptation to declining woodland availability.[13] What is clear is that forms of semipermanent cultivation using green-manuring methods are becoming more common in the Northern Province and that, over the last decades, there has been a steady spread westwards and southwards of this cultivation method from the Mambwe area in the north.

One of the difficulties, in the areas where fundikila (green-manuring) has been more recently introduced is to decide whether this is the result of an overall decline in woodland availability or whether it is a consequence of permanent settlement. The two processes are of course connected, as stated earlier, because fixed settlements combined with population concentration inevitably lead to cutting out of woodland and subsequent deforestation. When trying to determine the future of citemene and long-term trends in agricultural methods, it seems likely that semipermanent and permanent methods of cultivation will increase at the expense of citemene as the benefits (such as access to infrastructure and marketing) of fixed residence become more determining. One example of the increasing importance of both fundikila and cassava cultivation is provided by the aggregate data for three villages near Kasama collected by Holden in 1986, which shows that fundikila cultivation and cassava gardens covered 35% each of the cropped land, while citemene covered only 10%; 4% was cultivated by tractor and 16% was cultivated by other methods (Holden 1988:43). However, in traditional citemene areas it is unlikely that citemene will disappear altogether because of the role it plays in providing households and individuals with a much needed flexibility within the overall agricultural production system (see Chapters 7 and 8).

The increasing use of semipermanent cultivation methods is closely linked to the increasing production of hybrid maize, both for sale and consumption. Local varieties of maize were known in the province from at least the beginning of the colonial period. Trapnell noted in the 1930s that maize was "becoming an article of diet" among the Bisa and the Bemba of the Mpika District, as well as being grown by the Lala in the Mkushi District (Trapnell 1953:40). Maize production increased in many areas of the province throughout the 1940s and 1950s and was often grown both for consumption and for sale, especially in the areas along the Great North Road. However, while the area growing maize continued to increase in the 1960s and 1970s, it was not until the 1980s that the cash-cropping of hybrid maize really entered a boom period, as detailed in Chapter 8. In the 1990s, maize is the most important cash-crop in the province, but the amount of maize grown varies across the province, as well as between socioeconomic categories, as does the amount of maize consumed within individual farming households. For example, in the southeast and northeast of the province, maize is a very important staple, while in the west and certain parts of the central region, finger millet is still the main staple. This has to been seeen in the context of the fact that maize is the main staple among resource-rich farmers in all regions of the province.

If we attempt to analyze these recent changes within the terms set by the ecological and agricultural models discussed above, it becomes clear that variation in levels of maize and cassava cultivation are certainly linked to environmental factors, such as the presence or absence of woodland suitable for citemene. However, they are not simply determined by such factors. As we have already indicated, changes in residence and cropping patterns cannot be seen as the straightforward outcome of population pressure. As shown in later chapters, socioeconomic and political factors, such as population control schemes, agricultural policy, infrastructural provision, income generating opportunities, and labor availability all affect what people grow and where they live. The relationship between cropping patterns and residence choice is a dynamic and changing one, often mediated by cash, and it cannot simply be reduced to an argument about carrying capacity. In the following chapters, we examine the changes that have taken place in those production systems, and in so doing we link labor migration to the changing sexual division of labor in households, agricultural policy to the nutritional status of household members, and changing strategies of survival to changing experiences of self and group identity.

3

Relishing Porridge: The Gender Politics of Food

In the last chapter, we saw that citemene continues to play an important part in agricultural strategies in the Northern Province in the 1990s, but that its future will depend on a complex interplay between residence and resources. Overall, the area is still a sparsely populated one (see Table 2.5), but population concentration in fixed settlements has necessitated innovation both in cultivation methods and in cropping patterns. In particular, the increasing amount of maize produced as a cash-crop, although uneven in its spread across the province (see Chapters 7 and 8), has made people much more dependent on the market for their livelihood. In this chapter we examine the effects of colonialism and capitalism not on the production systems of the Northern Province directly, but on patterns of consumption and food security. In doing so, we begin to trace the changes that the market brought to people's livelihoods in the 1930s, 1940s, and 1950s. This is a crucial part of our study because Audrey Richards argued strongly that the citemene system was vulnerable to the forces of the market, but she saw this in terms of the exportation of male labor to the mines and the breakdown of social relations as a result of the effects of modernization. The theme of insecurity and social breakdown as a consequence of male absenteeism is also evident in mission and colonial sources, as we discussed in Chapter 1, but it became a particularly powerful theme in the discourses of priests and officials from the 1930s onward. Audrey Richards' account picks up on these discourses and works them into her analysis in anthropological voice. Local people had a different view of the changes and choices that confronted them in this period and this is explored in this and the following chapters.

The "discovery" of colonial malnutrition in the late 1920s and 1930s, and the rise of nutritional science, resulted in the development of a methodology that purported to provide a measure of the material well-being of "primitive" subsistence producers (Worboys 1988). As the result of the work of Audrey Richards, the Bemba people of Northern Rhodesia came to be a much-cited case exemplifying the problems of a primitive subsistence economy under pressure. In addition to adopting the methods of nutritional science in order to demonstrate the insufficiencies of the Bemba diet, Richards as an anthropologist was also anxious to show

46

that food was not just a matter of calories, but an integral part of the Bemba cultural system. In the same way as she saw the citemene system of production as inextricably linked to the Bemba chiefly and symbolic system, so she saw the social organization of eating as both manifesting and reproducing the kinship system. The annual cycle of production, and the daily rituals of eating were bound together, and together they were what made the society work, not just materially, but culturally.

In this chapter we re-examine nutritional data collected by Richards and her colleague, Lorna Gore-Browne, in the Northern Province in the 1930s, and compare it with similar household-level data collected in the late 1950s by Ann Tweedie. None of Tweedie's household-level data was ever published, and though Richards published some of her material in her book *Land, Labour and Diet* and in various articles, much of the data that we analyze here remained unpublished. Taken together, these sources provide us with what is, for sub-Saharan Africa, the unusual opportunity of a glimpse into the domestic economy of a rural society over time. As we shall make clear, however, the value of this data lies less in its ability to tell us about changes in levels of material well-being over time, and more in what it tells us about the social relations of consumption and exchange. In particular, it allows us to reinterpret Richards' own assertion that the cash economy would bring a breakdown in the social relations that surrounded household subsistence, for this was an area in which such social relations were to be profoundly altered, but never eliminated by, the operation of the market.

Richards' analysis of what was happening to the Bemba diet in the 1930s had two aspects, the relationship between which is not always clear in her work. First, she saw it as an example of a "primitive" diet of a "primitive" subsistence people, subject to marked season variations, vulnerable to droughts and locusts, and hedged around with mystical beliefs and taboos. This vulnerability, however, she saw as being greatly increased by economic and social changes taking place at the time, and particularly by male absenteeism, for labor migration had not only removed male tree-cutting labor, but also, in her view, contributed to a breakdown of kinship-based food-sharing systems that acted as a form of security against starvation. *Land, Labour and Diet*, then, came to be read as a documentation of the problems of a labor migrant economy. As a text it was widely cited by colonial officials, contributing not only to the discussion of the "problem of citemene," which we outlined in Chapter 2, but also to the dominant interwar discourse on the "problem of male absenteeism." Using a combination of anthropology and nutritional science, Richards was able to make her arguments directly in relation to individual Bemba households, and it is the immediacy and intimacy of these household-level observations that gives the work so much of its power.

Although she gives very little indication of this in her text, Audrey Richards arrived in the Northern Province in the middle of a what was an abnormally difficult period for the people of that area. In the late 1920s and early 1930s, seasons of over-heavy rains and smallpox outbreaks were followed by years of insufficient rain, scorching winds, and locust invasions. The problems of agricultural production were compounded by the uncertainties of employment on the mines and the vagaries of the regional economy. A general exodus of men to the mines was noted in many tour reports for 1930, but by 1932 the same administrators were recording large numbers of men present in their villages.

The Awemba Tour Reports of 1929 give us a sense of the situation in the Northern Province as seen through the lens of one agricultural season. A tour in June had attributed the poor food supplies in areas of Chiefs Nkula and Kawanda to a whole range of causes including bad rains, damage by elephants, demands on local labor made by the construction of the Chambeshi bridge, and the generalized problem of male absenteeism. In September, the northern part of Nkula's country was reported to be short of food because of poor rains, accidental and premature burning, damage by elephants and pigs, and the absence of men during the cutting season. People were apparently responding to these shortages by working for food and by bartering fish and wild fruits. In October, a number of villages in the area of Chiefs Chitimukulu and Chandamukulu were reportedly abandoned because of smallpox. In the same month, villages in the areas of Chiefs Chinkumba, Chewe, Chibesakunda, and Nkula were short of food because of poor rains and smallpox. In November, food was said to be very scarce in Chief Mwamba's and Chief Nkolemfumu's areas, and this was attributed to the outbreak of smallpox that had taken place the previous year. By December, food was scarce in Chief Chitimukulu's villages north of the Kalungu river (ZA 2/4/1: Awemba TRS 1928–29).

This summary of the food supply problems of 1929 accorded well with the general image of the Northern Province as perennially afflicted by natural disasters and social ills. But this generalized picture masked complex causation and wide variation. The long catalog of reasons given for food shortage in the northern part of Chief Nkula's area, for example, points to the fact that food shortages could occur in different places for different reasons. The comment on the absence of men in the cutting season and the reference to the construction of the Chambeshi bridge emphasize that employment was available both inside and outside the province, and that levels of absenteeism varied according to season and to levels of local labor demand. Although many individuals were employed locally on a more or less regular basis by the administration, missions, and traders, many more people were needed for specific projects, such as road and bridge building, and for labor on the only large estate in the province, Shiwa N'gandu, owned by Sir Stewart Gore-Browne. Fluctuations in local labor demand often had a specific influence on household labor since men frequently refused local work because the rates of pay were much lower than those of the mines. Thus, on occasion, women were used to substitute for certain kinds of male labor.[1] Males were not the only members of households who were affected by the demands of the colonial economy.

Despite the common understanding that male absenteeism was having an adverse effect on agricultural production, this was, in fact, a relationship that was far from easy to establish, as some colonial officers were aware. For example, Munday, a District Officer on tour in October-November in Chinsali district in 1930, noted that 66% of the taxable males were absent from villages in Chiefs Nkula, Mukwikile, Kabanda, and Chinkumba areas, but the supply of food was good. The same officer toured Shimwaule's, Nkweto's, Musanya's and Chewe's areas in February-March, where the male absentee rate was only 39%, but the food supply was very low. Munday concluded that the starvation he observed in the latter case was "largely due to inertia" (ZA 7/4/10 Awemba: TR Feb., 1930). On the one hand, there was the difficulty of explaining adequate, if not abundant, food supplies, in the absence of male labor and, on the other, there was the problem of how to account for food shortage when male labor was present. The overwhelming importance ascribed to male labor in the contemporary discourse on agriculture often led, as elsewhere in Africa, to accusations of idleness. If men were present and nothing was being produced then it must be because they were idle. Later colonial interventions into the agriculture of this area, which we discuss in Chapter 5, were designed to address this perceived problem of male lack of commitment to agriculture.

The accusation of idleness was frequently implicit and interwoven comprehensibly with ideas about labor and self-betterment and with views about men's and women's roles in society. An officer touring Mpika in March 1930 noted that 55% of the men were present in their villages, but that there as not much food. The reason given was that while the young men were present in the village during the period of the tour, they had actually been absent during kutema (tree cutting). Young men, he remarked, would probably remain in the villages consuming the rather meager supplies until the food was finished and then return to work (ZA 7/4/10 Awemba: TR March, 1930). The parish priest at Chilubula Mission in the Kasama district complained in January 1931 that young men were going to the mines and leaving their families without support, and that if they did return they took care to arrive after the tree cutting and then lived like "parasites" on others (WFD: Chilubula 30.1.1931).[2]

So prevalent was the notion of male idleness in the discourse of missionary and government officials in the 1930s that Audrey Richards found it necessary to address the issue directly in her book. She pointed out that the idea that men who had been away working on the mines would come home and spend every moment laboring in the fields was an unreasonable one (Richards 1939:404). However, much of her argument, although designed to counteract the dominant view of the Bemba and of agricultural practices in the Northern Province, actually served to bolster it. This is in essence because the argument she made is a culturalist one, and it worked to reinforce the familiar linkages between ethnic identity and mode of livelihood. These linkages between ethnicity and economics, as we have already argued, were not simply imposed on the Bemba by colonialists and others, but they were part of a mutual discourse of comprehension and miscomprehension that formed the force field within which the political, social, and economic struggles of the period were played out. Bemba men had much to gain from the idea that laboring on the mines was analogous to the warrior activities of an earlier generation of Bemba youth (Richards 1939; 29:402). This powerful assertion gripped the imaginations of all parties and served the Bemba well on the mines where they had a reputation for fierceness and bravery, a reputation that earned them the best paid jobs as deep underground workers. However, while asserting the validity of their warrior past, Bemba men were simultaneously engaged in representing citemene as central to their identity and to their very survival. The possible incongruity of an identity based both on agriculture, warriorhood, and industrial labor, was rarely directly revealed and this is, in part, because these different discourses on identity were invoked or employed in rather different contexts. Thus they worked to reinforce, rather than to deconstruct each other.

The image of the warrior and the agriculturalist are mixed engagingly in Richards's account of Bemba agriculture. She suggested that the apparent inertia of the Bemba and their lack of interest in agricultural betterment was a function both of the ideas, values, and concepts of prestige associated with their warrior past and of their experience of insecurity with regard to agriculture and food supplies. If an annual shortage of food was commonplace then there was little incentive to work hard at agriculture. Richards acknowledged that undernutrition might account for the apparent lassitude of the Bemba at certain times of the year, but her overall explanation for what she saw as their laissez-faire attitude to provisioning and trading was couched in terms of the cultural values and social organization of the people (Richards 1939: 203–216; 398–405). Richards' analysis placed so much emphasis on the relationship between the material and the cultural that it left her very little room to discuss variations in such things as ecology, cropping patterns, infrastructure, employment rates, and marketing opportunities across the province.

The reality of the situation she described was that the circumstances of agricultural communities were very variable, as were the number of men away working from any particular village (ZA 7/4/10 Awemba: TR Nov./Dec. 1930). Easy generalizations about food insecurity were thus doomed to inaccuracy. As we show in Chapter 4, during the whole colonial period, some communities of the Northern Province, both Bemba and non-Bemba, produced large quantities of foodstuffs for sale to government stations, missions, schools, hospitals, traders, and others. The demand for foodstuffs, like that for local labor, was relatively consistent, whereas the supply was not. The frustration for officials and observers of all kinds was borne of the fact that surpluses could not be guaranteed. The reasons for this were imperfectly understood and male absenteeism and Bemba idleness proved to be explanations within everyone's easy reach. But, this does not mean that they should be taken at face value. In her household-level analysis of production and consumption, Richards attempted to provide some explanation of constraints on production and an assessment of their nutritional consequences. She utilized the tools of an emerging nutritional science to attempt to measure the sufficiency or insufficiency of the Bemba diet at the level of individual households, but she also used her findings on nutrition and food-sharing to illuminate the extent of social and economic change.

Household Consumption and Diet in the 1930s

Although Richards published some of her nutritional data to provide support for her arguments in *Land, Labour and Diet*, much remained unpublished. In this section we reevaluate Richards' published conclusions through a reanalysis of her unpublished data.[3]

Three women researchers (Audrey Richards, Elsie Widdowson, and Lorna Gore-Browne) used the very new methods of nutritional science to undertake an analysis of consumption and diet in the Northern Province in the 1930s in an attempt to identify whether it could be said, in scientific terms, that the Bemba were undernourished. (Richards 1939; Richards and Widdowson 1936; Gore-Browne 1938). Richards published lists of Bemba foodstuffs and methods of preparation, an analysis of the chemical composition of the Bemba diet, work diaries for two villages, tables of daily food intake for two households over a twenty-one-day period in 1934, and a chart of seasonal changes in food supply (Richards 1939). As part of this overall assessment of the composition of the diet, she calculated that the average daily intake of millet flour was 0.3 kilos per head in the region during the period 1933–1934. (Richards 1939:40; Richards and Widdowson 1936). This average was, however, almost meaningless as Richards and Widdowson demonstrated when they calculated that in the village of Kasaka in September 1933, individual daily intake varied from 0.06 to 0.7 kilos, while in the village of Kampamba daily household consumption varied from 0.7 to 2.6 kilos (Richards and Widdowson 1936:179–180). These enormous variations within individual villages were nothing compared to the differences that all observers acknowledged existed across the province and from one season to another. But differences in food consumption at the individual and household level were (and are) crucial to questions of food security and nutrition because they point to the link between agricultural production and household developmental cycle, to intrahousehold hierarchies, and to the existence of social differentiation within villages. Richards, like other researchers, knew that some households were better off than others, and that in a community where material possessions were few, the signs of such difference were likely to be manifest in terms of access to resources, most especially food (Richards and Widdowson 1936:180; Richards 1939:208–211). She also emphasized that considerable variation existed in the number and quality of meals eaten daily, and that women and young boys often ate much less than other individuals (Richards 1939:76; 122–123).[4] These variations in food consumption patterns between individuals and groups underlined the fact that neither nutrition nor food security could be approached simply from the point of view of village, chieftaincy, or district level food supply, because such an approach easily masked important differences between households within the same village, and between individuals in the same household.

Both Richards and Gore-Browne collected data on total annual millet (male) production by household, and from this they calculated the per capital daily intake (see Tables 3:1 and 3:2), Richards collected her data in 1934 and her figures are thus based on the 1933 harvest that was believed to be a particularly poor one. Gore-Browne conducted her study in 1936 when the harvest was judged to be average (Sec 1:1039: Letter Gore-Browne to Chairman of the Diet Committee 27/1/37).[5] She calculated the average daily per capita intake of millet as 0.4 kilos per day (Gore-Browne 1938:6). This was indeed slightly higher than Richards' figure for 1933, but any conclusion is complicated by the fact that Richards gave her average daily intake as weight in millet flour, whereas Gore-Browne seemed to have been referring to weight in grain (see Tables 3.1, 3.2, and 3.3). Thus, these gross figures are of little help in determining the difference between a very poor year and an average year. Further observations on agricultural production and nutrition were made by Unwin Moffat, the Agricultural Officer at Abercorn in 1937. He estimated that annual production of grain per household in the Mambwe area in the northern part of Northern Province was approximately 1,454 kg of which 90–128 kg was probably sold, and about 363 kg was used for beer, leaving 2.7 kg per household per day for food. He also estimated that in the southern part of the province, the annual household production of millet was probably about 1,090 kg, and although little was sold, much was used for beer, leaving perhaps only about 1.8 kg per household per day for food. (SEC 1/1039, 1937). Moffat made no attempt to provide figures on household size and composition, but if one assumes an average household size of six to seven persons, his figures for putative daily intake for the Mambwe area would be somewhat higher than those given by Richards and Gore-Browne,

TABLE 3.1.   Annual Millet (male) Production Figures by Household for Three Villages

Household	Weight of Millet Harvested in Kilos					
	Kampamba		Kungu		Mubanga	
1	1,658	(410)	402	(97)	2,036	(504)
2	378	(93)	2,269	(561)	2,050	(507)
3	1,265	(313)	785	(194)	1,920	(475)
4	2,734	(676)	625	(154)	1,629	(403)
5	1,760	(435)	800	(198)	3,505	(867)
6	3,025	(748)	829	(205)	1,600	(396)
7	814	(201)	887	(219)	770	(190)
8	785	(194)	628	(154)	974	(241)
9	1,309	(270)	1,047	(259)	1,280	(316)
10	4,189	(1,036)	891	(219)	1,658	(410)
11	1,570	(388)	690	(169)	109	(27)
12	945	(234)	843	(208)	756	(187)
13	1,192	(295)	2,210	(547)	279	(68)
14	829	(205)	2,341	(579)	3,781	(936)
15	1,541	(381)	800	(198)	1,396	(345)
16	770	(190)	414	(100)	1,149	(284)
17	2,574	(637)	648	(158)		
18	683	(169)	916	(226)		
19	610	(151)	548	(133)		
20	1,396	(345)	894	(219)		
21	2,269	(561)	1,396	(345)		
22	3,098	(766)	2,530	(626)		
23	610	(151)	2,836	(702)		
24	2,967	(734)	3,083	(763)		
25	1,614	(399)	2,443	(604)		
26	2,429	(601)	1,658	(410)		
27	836	(205)	1,714	(421)		
28	1,110	(273)	610	(151)		
29	3,300	(813)				
30	3,756	(928)				
31	2,792	(691)				
32	465	(115)				
33	1,643	(406)				
34	1,018	(252)				
35	2,690	(666)				
36	2,763	(684)				
37	1,789	(442)				
Total	65,585	(16,058)	35,737	(8,819)	24,892	(6,156)

Source: AR:U106-130.) Bracketed figures give alternative weights based on Gore-Browne's formula of 1 cu. ft. of stored millet (male) = 3.6 kilos in weight (see footnote 5).

while his figures for the southern region would be lower than theirs (0.45–0.38 kg per individual per day for the Mambwe area and 0.3–0.25 kg per individual per day for the southern region).[6]

These average figures do not mean a great deal, of course, but it is important to investigate what little data exists, particularly if we wish to assess present-day

TABLE 3.2. Annual Millet (male) Production Figures by Village for 15 Villages.

Village	Adults Men	Women	Children under 15 yrs	Grain in kilos
Sampela	14	18	34	3,833
Makupula	8	14	14	2,525
Kapelembe	4	8	13	3,544
Mulenga	13	14	28	5,084
Mutubila	10	16	24	3,936
Citambi	3	8	10	1,770
Ngonge	14	21	22	6,563
Nsendamina	16	19	21	6,591
Mumamba	10	17	24	6,034
Lubemba	7	12	24	4,289
Mungalaba	12	21	24	9,829
Lukaka	17	26	42	5,825
Kapengele	14	18	49	5,453
Citula	7	10	21	1,892
Ciputa	11	17	28	4,330

Source: Gore-Browne 1938:6.

TABLE 3.3. Male/Female Ratios and Grain (male) Production.

Village	Ratio of men to women	Kilos of grain per resident male
Kampamba	0.5	698
Sampela	0.7	274
Makupula	0.6	316
Kapelembe	0.5	886
Mulenga	0.9	391
Mutubila	0.6	394
Citambi	0.4	590
Ngonge	0.7	469
Nsendamina	0.8	412
Mumamba	0.6	603
Lubemba	0.6	613
Mungalaba	0.6	819
Lukaka	0.7	343
Kapengele	0.8	389
Citula	0.7	270
Ciputa	0.6	394

Source: Based on Gore-Browne, 1938.

analyses of changes in the diet of this area (see Chapter 7). In general, it is difficult to say anything of any certainty about agricultural production or yields in the 1930s, especially in the absence of any figures on grain sales either within or from the region. It is equally impossible to make any definitive statement about average daily food intake.[7]

The data collected by Gore-Browne and Richards are extremely difficult to interpret and compare (see footnote 5). The village of Kampamba, with thirty-seven households, was large by contemporary Bemba standards. The large size of the village might be explained by its location close to the Malole Catholic mission. Richards noted that although the village stood on rich soil, it had an unusually poor supply of millet because locusts had raided the gardens for two successive years and, she added, the village had an unfavorable male:female ratio of 23:42. Of thirty-seven households only five had sufficient millet to last to the new harvest, and twenty-five households had no cultivated relish crops left of any sort, although gathered mushrooms were plentiful (Richards and Widdowson 1936:179–180).[8] When Richards' figures for Kampamba are compared with the village totals provided by Gore-Browne, the former appear disproportionately large, and even when Richards' figures are recalculated using Gore-Browne's formula, households in Kampamba do not appear to have been significantly worse off in a bad year than those studied by Gore-Browne in an average year, despite of the slight difference between their figures for average daily intake (see Table 3.3 and earlier text).

The determinants of food production levels are not easy to unravel, but Table 3.3 does emphasize once again how unwise it is to assume that the ratio of men to women was always the primary determining factor of food supply. Thus, for example, Richards' contention that food was short in Kampamba because of an unfavorable male:female ratio cannot be accepted at face value. Variations in household and village consumption were clearly related to such factors as differences in microenvironments, differences in cropping patterns, proximity to government and mission stations, household development cycles, and differential marketing opportunities, as well as to the straightforward question of male absenteeism.

The effects of some of these factors were apparent from the series of food consumption surveys conducted by Richards. In the village of Kampamba, notes were taken on six households over the period from January 1–6, 1934, and in the village of Kungu, records were kept of six households from February 28 to March 7, 1934, and these are reproduced below.[9] This was, then, a very small sample, but the results obtained are revealing.

January, February, and March, when these data were collected, were very clearly the "hungry months" when supplies of millet grow short and were sometimes exhausted. Stored relishes were also scarce, and people obtained their relish from gathered foods, especially mushrooms, pumpkin leaves, and caterpillars. In late February and early March, the first gourds and pumpkins ripened and green maize was available from the end of March. The records from the villages of Kampamba and Kungu demonstrate the composition of the diet of the hungry months vividly.[10] At first glance, they seem to show that the quality and quantity of the diet were very similar at this time of year for all households, but on closer inspection significant differences are evident. The two households in Kampamba averaged 3.05 kg and 2.90 kg of flour per day, while the two households in Kungu averaged 1.11 kg and 1.31 kg. In the case of household B (in Kampamba), the amount of grain consumed each day would have been higher than the average suggests because on three days out of six grain was given to the wife's in-laws. When the number of people eating is taken into account, it is clear that those in household B were averaging 0.43 kg of flour per day, while those in household C aver-

TABLE 3.4. Specimen Diets of the Village of Kampamba from January 1 to January 6, 1934

Household B: The household consists of a middle-aged couple (he is a traveling RC teacher) with four children. They support the wife's mother and her unmarried daughter, as well as a deserted daughter and her small son. This family is considered well-off by the other villagers.

Date	Men	Women	Children	Food	Remarks
1st	1	4	4	2.7 kg Male flour, mushrooms	Mother cooked mushrooms. Wife cooked porridge. All ate together.
2nd	1	1	3	6.3 kg Male flour, green leaf relish	Gave grain to in-laws to cook for themselves.
3rd	1	1	3	Mushrooms	Still eating yester-day's flour.
4th	1	1	3	Mushrooms	Still eating leftover flour.
5th	1	4	4	4.3 kg Male flour, mushrooms	Ate with in-laws a.m., gave them grain for p.m.
6th	1	4	4	5 kg Male flour, mushrooms	Ate with in-laws a.m., gave them grain for p.m.

Household C The household consists of a middle-aged couple with three children: a fourteen-year-old boy, an eleven-year-old girl, and a nine-year-old boy.

Date	Men	Women	Children	Food	Remarks
1st	1	2	3	2.7 kg Male flour, ground beans	Called three women to eat with her.
2nd	1	2	3	4.5 kg Male flour, beans.	Gave relatives-in-law some grain.
3rd				Leaves for a distant village to beg for grain from relatives. Children and husband fed by another household to whom she gives 0.9 kg male flour.	
4th	1	2	3	2.5 kg Male flour, 0.9 kg cassava flour, mushrooms	Brought back 13.6 kg millet. Sold 1.8 kg to another household for groundnuts.
5th	1	2	3	1.3 kg Male flour, mushrooms	Ground little flour to-day. Eat remains.
6th	1	2	3	5.5 kg Male flour, green leaf and ground bean relish	

See bibliography (AR:V37, V40–41.)
Source: (AR: V42–43.)

TABLE 3.5.   Specimen Diets of the Village of Kungu from February 28 to March 7, 1934

Household D: The household consists of a youngish man and his second wife with one child. They support the wife's unmarried sister who does not get on at home. The husband works in Kasama and the wife does "light domestic work only."

Date	Men	Women	Children	Food	Remarks
28th	1	2	1	2 kg Male flour, caterpillars	Flour from mother-in-law, whose garden has escaped locusts. Bought caterpillars with salt.
1st	1	2	1	Potatoes	No other meal. Man eats with his other wife.
2nd	She could not find any food, so she eats with her sister.				
3rd	1	2	1	2 kg Cassava flour, 0.9 kg meat	Flour bought with salt. Meat was bought for 2/s at local butcher. Refused to share except with co-wife because the meat was bought with own money.
4th	No food, she left to beg from relatives in other villages.				
5th	Husband and children are fed by wife's sister.				
6th	2	3	4	3.1 kg Male flour, pumpkin leaves, roast cassava (a.m.)	
7th	1	3	4	1.8 kg Male flour, lubanga leaves relish	Bought tobacco for 1/s some time ago and trades this to advantage with women coming in from the country villages.

Source: (AR: V51, 59–60).

aged 0.61 kg. Members of household D (in Kungu) averaged 0.31 kg and those in household E averaged 0.32 kg. These average figures for individual consumption are misleading, however, because the men ate separately from the women and children and were likely (according to the observations made elsewhere by Richards) to have consumed significantly more. In addition, men in households D and E had access to larger amounts of food because they were able to eat with their second

TABLE 3.5. (*Continued*)

Household E: The household consists of a middle-aged couple with one child and one nephew to support. The husband is engaged in trying to trade European goods for food. The wife does "light domestic work only."

Date	Men	Women	Children	Food	Remarks
28th	1	1	2	1.4 kg Male flour, potatoes, ground beans	Bought potatoes with salt from a woman passing through. Bought millet two days ago. Bought to advantage from relative in a distant village.
1st	1	1	2	1.4 kg Male flour, three small pumpkins, dried bean leaves	Still using remains of bought flour.
2nd	1	1	2	1.1 kg Male flour, dried bean leaves, mushrooms	Lucky find of mushrooms. Still using bought flour.
3rd	1	1	2	1.8 kg Male flour, beans	Millet flour exchanged for beans.
4th	1	1	2	2.7 kg Male flour. Pumpkin leaves	Finished all remaining millet.
5th				Left early with one child to go and beg or buy more millet. Husband eats with his other wife and child eats with wife's sister.	
6th	1	2	3	1 kg Male flour, three gourds, pumpkin leaves	Bought a little flour with salt.
7th	1	2	3	1.1 kg Male flour, beans, three pumpkins	Eat the millet needed for sowing for next year.

Source: (AR: V50–51, 60–63.)

wives when food was short. Gender relations, intrahousehold hierarchies, and links between households were thus crucial in determining food consumption patterns. Households were clearly not bounded units when it came to food consumption. Richards cited the case of another household in Kungu where the wife was an old and feeble woman, but insisted on sending the best part of her food to her son-in-law who was a well-paid teacher at a government school (AR:V52). As Richards clearly argued, conjugal and kinship obligations provided the content in which food exchange has to be interpreted (Richards 1939:108; 124–134).

All four households were involved in supporting kin, either at the point of consumption or through sending them raw supplies. The incentive to continue support of this kind even in times of hardship clearly had much to do with reciprocal arrangements and the importance of kin networks for gaining access to a whole range of resources. The households in the village of Kungu—close to Kasama township—experienced greater difficulty provisioning themselves than did those in Kampamba. Richards accounted for this by arguing that people close to towns were faced with the dilemma of selling their food to workers in the towns earlier in the year or saving it to eat. She also argued that Kungu provided examples not only of large numbers of old widows who had moved close to town in the hope of making money by beer brewing, but also of several isolated young couples who had cut themselves off from their kin (Richards 1939:153). In general, she suggested that people who are unable to draw on kinship networks were worse-off, particularly at this time of year (Richards and Widdowson 1936:181; AR:V50). Her conclusions, although broadly correct and producing interesting parallels with data collected in the 1950s and 1980s, can be qualified with reference to her own data.

First of all, women in Kungu in the 1930s were able to obtain food provided that they had money or goods to exchange. Cash was important to the food economy, even in the early 1930s. Richards noted this, but she could not integrate it fully into her analysis because she regarded cash as the source of breakdown in kinship relations. Cash and kinship were, for her, almost mutually exclusive. The fact that they could be simultaneously present in determining the food strategies of individual households can be demonstrated through Richards' data on household D. Richards calculated that the wife of this household had spent approximately two shillings in one week on cloth, tobacco, and salt, which she traded for food with villagers coming in from the country (Richards and Widdowson 1936:181). She used household D as an example of an isolated young couple because they refused to share their meat (see earlier text and AR:V50). However, it is also clear from the records that the wife drew readily on the assistance of her sister and her mother-in-law, and that she often went to beg from relatives in other villages. Begging from relatives could bring rewards as is evidenced by household C, but it was not always successful and, by the 1930s, neither was it always a transaction independent of money or barter exchange, as the case of Household E demonstrates. Drawing on kin networks was certainly an essential part of gaining access to resources and support, but the increased vulnerability of households in Kungu cannot be simply explained as a result of the collapse of kinship. Their overriding problem would appear to have been the degree to which they had become dependent upon, but only poorly integrated into, the emerging cash economy. Richards noted this fact, but preferred to explain its effects in terms of the breakdown of kinship links and the fact that half the male inhabitants of the village worked in Kasama and the supply of labor for making gardens was thus insufficient (Richards 1939:153). Richards did, however, mention three other relevant factors for the people of Kungu, although she did not integrate them into her analysis. First, land was in short supply around the town. Second, women were not always prepared to invest their labor in making gardens for men if they felt that the union was an insecure one. Third, the market was insufficiently developed to act as a redistributive mechanism in the absence of kinship networks (Richards 1939:153; Richards and Widdowson 1936:181, 190). Both the women householders of households D and E were recorded as being engaged in "light domestic work only" (see earlier). It is unclear what is meant by this phrase, but it does seem that these two women were not involved in cultivating either millet or village gardens. The reasons for this are unknown, but they could have been connected to land shortage or marital instability or to status concerns about the proper role of a wage-earner's wife. Whatever the reason, the vulnerability of households in Kungu was not simply a matter of the breakdown of kinship or of an insufficient supply of male labor. Dependency on low wages, the underdevelopment of the market, lack of opportunity to cultivate, and the inability to control female labor in a situation where it was essential to grow some food to supplement wages were equally, if not more, important.[11]

It is clear that by the time Audrey Richards conducted her research into Bemba agriculture, food security and household consumption were affected by partial integration into the emerging cash economy. As we saw in Chapter 1, the British South Africa Company had been promoting labor migration as a way of generating cash to pay tax from 1900, and from an early date the question of how to muster sufficient cash for tax payments had become part of the routine problem of household reproduction. Subsistence had come to encompass a great deal more than food supply, and if one takes tax payment as one component of household subsistence, then it is clear that access to cash was vital for the viability of every household. The need to generate tax revenues led to considerable confusion on the part of the colonial administration, which had to balance the need for such revenue against the spectre of social disruption and fallng agricultural yields that they took to accompany male labor migration. As we shall see in Chapter 4, the sale of agricultural foodstuffs was simultaneously promoted and discouraged because although food security was perceived as incompatible with selling stocks to raise tax monies, administrators, missions, and others were in need of food supplies in the region. The colonial administration's ambivalence went further because revenues from the sale of foodstuffs were not always of a sufficient magnitude to guarantee tax revenues, and thus it was necessary to continue to encourage labor migration while at the same time dealing with the problem of how to increase agricultural production in the face of dwindling labor supplies (Meebelo 1971:220–224).

The tour reports of the early 1930s are full of references to the difficulties faced by both Bemba and non-Bemba communities in the Northern Province in paying tax, and it is within this context that one must understand the problems of household provisioning. In May 1932, a meeting of Bisa chiefs was advised to encourage their people to make reed mats and baskets and prepare tobacco and dried fish in order to pay their taxes; and in a similar meeting with Bemba chiefs in June 1932, the advice was that they should hawk fowls, meal, and other foodstuffs on the Great North Road in order to raise money for tax (ZA 7/4/28: Awemba TR, June 1932). These seemingly sensible suggestions were actually somewhat cavalier given the inconsistencies of colonial policy and the fact that the necessary productive sources, markets, and infrastructure were only available in certain parts of the province. Migration out of the province, although often a very profitable strategy, was not open to everyone, and some found great difficulty in raising even relatively small amounts of cash. An example of this problem, and of the comparative isolation of the Northern Province, was given in a tour report of 1936. This told the story of a man who had walked to Ndola in search of work, but was turned away because he could not pass the medical examination for mine labor. While in Ndola, he had been given a pair of trousers. He returned to the province and managed to sell them on the Luapula River for 3/6d—a good deal less than their value. This money he took to Chiengi and spent on salt. By hawking the salt along the Luapula he raised 6/–. He managed to borrow 1/6d from some relations and was then finally able to pay his tax. It had taken him the best part of seven months! (SEC 2/786 Kasama: TR Nov. 1936).

It is clear then that different parts of the region experienced very different levels of integration into the cash economy, largely because of variations in infrastructure, productive potential, local labor markets, and marketing possibilities (see Chapter 4). It might be possible to argue, as Richards did, that it was only the villages close to towns that were affected by the changes that the cash economy and its accompanying social disruption brought, but this was clearly not the case because so many individuals were involved not only in supplying labor both inside and outside of the province, but also in selling agricultural foodstuffs to the administration and to missions and traders. For some, involvement in the cash economy was more profitable than for others. In 1931, for example, there were over 200 bicycles on Chilubi Island in the Bangweulu swamps. This considerable investment represented not only the technology necessary to transport fish to the Copper Belt, but the scale of profits that the local fishermen were able to command (ZA 7/4/28: Awemba TR Feb-March 1932; ZA 7/4/19: Awemba TR March 1931). As time went on, the fortunes of agriculture in the Northern Province were to become ever more closely tied to the provincial and national economy. The growing importance of cash in the struggle to secure household subsistence and reproduction was to have important consequences for diet and nutrition.

Kinship and Food From the 1940s to the 1960s

Among the Bemba it is rare for any individual to housekeep alone. Several households are grouped round one granary and even one kitchen, and are linked by close economic ties with a wider circle of relatives in the same village, with whom they constantly eat in common, and on occasion pool supplies. . . . Besides this regular unit of consumption each household is connected by bonds of sentiment and legal identification with relatives in other villages, 20, 50 or even as much as 100 miles away. In fact, these kinship obligations result in quite a considerable distribution of food.

(Richards 1939:108)

After the excited interest taken in diet and nutrition in the Northern Province in the 1930s, it is perhaps not surprising that no further work was undertaken there in the 1940s. However, a nutrition survey was undertaken in 1947 under the leadership of Betty Preston Thomson in the Serenje district just to the south of the Northern Province. Because the data were collected in different ways, it is not possible to compare Preston Thomson's empirical findings directly with those produced by Richards and Gore-Browne in the 1930s. Nevertheless, some interesting comparisons emerge. Preston Thomson conducted a survey in three Lala villages. The Lala practiced a form of citemene and their social organization, agricultural production, environment, diet, and household consumption patterns were very similar to those described for the Bemba (Preston Thomson 1954; Peters 1950). Preston Thomson noted that millet was still the staple in the Serenje district in the 1940s, but that cassava and sorghum were also consumed as staples. The relish crops recorded were exactly the same as those itemized for the Northern Province in the 1930s by Richards and Gore-Browne. Preston Thomson calculated that the amount of flour eaten per capita per year was between 100 to 147 kg (Preston Thomson 1954:Appendix VII). Once again, these figures tell us relatively little because they are averages based on man-values, but they accord surprisingly well with Richards and Gore-Browne's findings that the average intake of flour was 0.4 kg per day ($0.4 \times 365 = 146$ kg per year). But what is perhaps of more interest with regard to Preston Thomson's study is the data she collected on food transactions. She calculated that in any one year 20% to 30% of households had no millet store of their own, but instead obtained their supplies in the form of gifts from relatives or by bartering dried fish or by stacking branches, reaping, and hoeing for households that did have millet.[12] This raises the question, discussed in Chapter Two, of whether all households were in the habit of making citemene fields every year. Under the system of field rotation practiced in the Northern Province and the Serenje district, it would not have been strictly necessary to cut a new field every year. Richards noted that Bemba chiefs only cut citemene fields in alternate years, and it seems quite possible that polygynous men may have cut fields for each wife in alternate years (Richards, 1939:215). This is an important point because it clearly has implications when it comes to a discussion of the impact of male absenteeism on agricultural production. However, it also seems likely that there were no hard and fast rules on this point, and that the decision of whether or not to cut every year would have been made with due regard to labor availability, household consumption needs, and the stage of the agricultural rotation. The factors influencing such decision making have clearly changed over time as cropping patterns, labor regimes, and levels of commercialization of the agricultural economy have changed. The important point to note though is the degree of flexibility in the citemene system and its ability to adapt to changing circumstances.

In addition to interhousehold and intervillage exchanges of millet, Preston Thomson also noted that grain and pulses were sold to Boma depots throughout the district, and that sales were made to other local people at local markets, as well as to those travelling up and down the Great North Road. Of the women she questioned, 26% said they had sold maize, 22% had sold millet and beans, 18% had sold ground-beans, 13% had sold groundnuts, and a few had sold sorghum, flour, peas, and cowpeas. Most of these sales were mainly to government depots. In addition, 27% of the women sold hens, either to the Nutrition Survey or to each other, and 25% of the women sold caterpillars locally and to the Copperbelt. Other things sold in small quantities included eggs, pulse seed, root and fresh vegetables, tobacco, beer, and baskets.

The most profitable of the saleable commodities were caterpillars which provided 28% of the total money obtained; grain provided 23%, pulses provided 20%, meat provided 15%, and hens provided 5%. The sale of nonfood products only amounted to 2%, but most of the money obtained from the sale of foodstuffs was spent on nonfood products: 67% on clothes, 4% on items such as blankets, beads, soap, hoes, tobacco, baskets, and bicycle spare parts, 12% on animal products, 5% on salt, 2% on grain, 1% on pulses, and 1% on seed (Preston Thomson 1954:54–55). The amount of money spent on commodities such as clothes, blankets, hoes, and soap shows the degree to which household reproduction depend on money and commodity exchange by this time.

In addition to transactions involving cash, Preston Thomson recorded 222 cases of barter covering 37 types of exchanges.[13] She found that 33% of exchanges involved millet, and especially millet for salt. Millet was also bartered for fish, pulse seed, tobacco, caterpillars, meat, cooking-pots, and baskets. Also, 17% of the exchanges involved caterpillars bartered for salt, tobacco, millet, soap, clothes, fish, baskets, and beads. In 33% of the cases, work was exchanged for millet, meat, groundnuts, beer, hens, pulse seed, and Livingstone potatoes (Preston Thomson 1954:55). These exchanges demonstrate the necessity of interhousehold and inter-village exchange for household reproduction, and it would not be unreasonable to assume that redistributive mechanisms of this kind had always been essential for social reproduction even before the emergence of a cash economy. Certainly, they appear to be essential to the livelihoods of households in present-day Northern Province, as we shall see in Chapter 7.

It is certainly true that the Serenje district, where Preston Thomson conducted her work, was to experience far more rapid agricultural commercialization than was the Northern Province, and that ox-plows and peasant farming schemes were introduced much more successfully here than they were further north (Long 1968). However, there is no reason to suppose that the food transactions that Betty Preston Thomson observed would not have been characteristic of villages in the Northern Province. Preston Thomson, like Richards, emphasized the considerable degree of interchange of food between households, and the fact that sharing between households was most often something that took place at the point of consumption.[14] However, Preston Thomson also recorded gifts given and received outside the context of immediate consumption. She noted that 33% of the presents given came in the form of grain (mainly millet), 4% as pulses, 6% as seeds, 1% as root vegetables, 29% as animal products (mainly meat), 3% as salt, 2% miscellaneous (such as baskets, cooking-pots, beads, and tobacco), and 20% as money and cloths received from relatives on the Copperbelt. Gifts returned to relatives on the Copperbelt always took the form of food (Preston Thomas 1954:55).

Constant sharing of food was thus a feature of village life recorded both by Preston Thomson and by Richards. They both emphasized the relationship between matrilocal kinship groupings and household consumption and production (Preston Thomson 1954:29; Richards 1939:130–131). In both Lala and Bemba villages, interdependent matrilocal households were the basis for economic cooperation. Ideally, a grandmother, her married and unmarried daughters, and adolescent granddaughters would work together in the fields, collect relish, fetch firewood and water, pound, grind, and cook together. The garden work groups might not have overlapped completely with the food preparation groups, but the more important point is that the food preparation group did not correspond in any way to an individual household, and nor did it correspond neatly with the unit of consumption. The members of the food preparation or "cooking team" might eat together with their children, but they would also send food to their husbands, to the young men in the nsaka (men's shelter), to visitors, and to other female relatives and friends in the village. This system meant that men tended to eat rather better than women and children, and also ensured that daily levels of food intake were likely to vary according to the composition of eating groups (Richards 1939:129–

'Women grinding millet'
Photo: Audrey Richards, 1930s

130, 150–152; Preston Thomson 1954:47–48). Richards argued that "joint house-keeping" of this kind provided for improved food security by tying together households at different stages in the developmental cycle, and also eased the burden of domestic labor. She was particularly insistent on the difficulties of combining heavy agricultural labor with lengthy and exhausting food preparation and cooking processes (Richards 1939:102–105, 132). Richards' insight was to serve researchers well in the 1980s when they came to investigate the impact of cash cropping on nutrition and household consumption patterns (see Chapter 7).

Richards was convinced that the cooperative domestic economy of the Bemba, with its distinctive joint cooking team, would break down as a result of male labor migration, the decline in brideservice, the emergence of individual households, and the penetration of the cash economy (Richards 1939:132–134). She thought it likely that the redistributive mechanisms based on kinshp and residence would give way to redistributive mechanisms based on the market, but she was fearful of what would happen should kinship break down before the safety net of the market was in place. She was particularly anxious about the link she perceived between the collapse of the kinship network and the emergence of increasing social differentiation between households:

> This inequality as between family and family much more nearly resembles conditions in our own society, and appears to be the necessary result of the adoption of a money economy. At present there is of course the added difficulty that individual housekeeping has begun to exist without any previous organisation of trade, specialisation of urban and agricultural workers, or regular possibilities of buying food. Thus the families in these communities are at present in an intermediate position . . . Although it appears that most Africans must ultimately adopt the European system of economy, the encouragement of the single family household, before a system of trade has been established, might well lead to disaster
>
> (Richards 1939:153).

In the late 1950s, Ann Tweedie conducted a study on settlement patterns and food consumption in the Northern Province under the aegis of the Rhodes-Livingstone Institute.[15] Her marvellously detailed data on food consumption and interhousehold relations has never been published. However, her findings bear directly on the question posed by Richards two decades earlier of whether the joint domestic economy and the kinship relations that sustained it would break down under the impact of the money economy. The 1950s was a period of relative prosperity for the Northern Province. Mine wages were high, levels of employment were good, the colonial government was investing in the area for the first time, and marketing possibilities were improving (see Chapters 5 and 6).[16]

During a period of eight weeks (from February 1 until March 27, 1960), food consumption data were collected for all 18 households in Kanyanta village.[17] Six of these households, however, had few entries against them either because they were empty for much of the time or because they did no cooking at all and were regularly supplied by another household. Tweedie's specimen diets are reproduced in Table 3.6.

A simple analysis of the composition of these specimen diets from Kanyanta village demonstrates how similar they are to those collected by Audrey Richards at the same time of the year two decades earlier. Although Tweedie made no attempt to quantify daily food intake, her data do show the continuing importance of gathered foods and bush meat in the diet in the Northern Province, especially during the hungry months. Only five of the relish items recorded in the diet sheets are cultivated crops. A further interesting point concerns the ingredients of the staple porridge. All households were eating millet mixed with cassava flour, and some of the poorer households sometimes had to eat porridge made from cassava flour alone. The latter was not considered then, and is not considered now, to be desirable because porridge made only with cassava flour is thought to have an inferior texture and taste. The increasing use of cassava as a staple in the diet by 1960 reflects wider changes in cropping patterns and adaptations in the agricultural sys-

TABLE 3.6 Specimen Diets of the Village of Kanyanta February 22 to February 28, 1960

Date	Households				
	I	II	III	IV	V
M	Bwali/1	Nil	Bwali/9	Bwali/1	Nil
	Bwali/1	Bwali/4	Bwali/2	Bwali/9	Bwali/7/3
T	Bwali/2	Nil	Bwali/2	Nil	Nil
	Bwali/2	Bwali/4	Bwali/2	Bwali/1	Bwali/2
W	Bwali/2	Nil	Bwali/2	Nil	Bwali/2
	Bwali/1	Bwali/1	Bwali/1	Bwali/5	Bwali/1
T	Nil	Nil	Nil	Nil	Bwali/12
	Bwali/3	Bwali/4/5	Bwali/9	Bwali/4	Bwali/2
F	Bwali/1	Bwali/6	Nil	Nil	Nil
	Bwalil/1	Bwali/7	Bwali/6	Bwali/2	Bwali/1
S	Nil	Nil	Bwali/2	Nil	Bwali/1
	Bwali/3	Bwali/7	Bwali/2	Bwali/2	Bwali/1
S	Bwali/1	Nil	Bwali/2	Bwali/1	Nil
	Bwali/1	Bwali/8	Bwali/2	Bwali/1	Bwali/10

Days	Households				
	VI	VII	VIII	IX	X
M	Absent from village	Bwali/1/7	Nil	Nil	Bwali/6
		Bwali/12	Bwali/c/4	Bwali/c/6	Bwali/3
T	Absent	Bwali/7	Nil	Bwali/c/2	Bwali/2
		Bwali/7	Bwali/c/2	Bwali/c/2	Bwali/2
W	Absent	Bwali/7	Nil	Bwali/2	Nil
		Bwali/6	Bwali/c/1	Bwali/11	Bwali/11
T	Absent	Nil	Nil	Nil	Nil
		Bwali/12	Bwali/c/1	Bwali/1	Bwali/1
F	Absent	Bwali/12	Bwali/c/1	Bwali/2	Monthly
		Bwali/5/9	Bwali/c/11	Bwali/2/11	Period
S	Absent	Nil	Nil	Nil	Monthly
		Bwali/11	Bwali/c/4/3	Bwali/4	Period
S	Absent	Nil	Nil	Bwali/4	Monthly
		Bwali/11	Bwali/c/6	Bwali/5	Period

Days	Households	
	XI	XII
M	Nil	Bwali/c/6
	Bwali/6	Bwali/c/11
T	Nil	Nil
	Bwali/2	Bwali/c/2
W	Nil	Nil
	Bwali/5	Bwali/c/11
T	Nil	Nil
	Bwali/1	Bwali/1

TABLE 3.6   (Continued)

Days	Households	
	XI	XII
F	Nil	Nil
	Bwali/11	Bwali/c/5
S	Bwali/11	Eat with
	Bwali/11	Hut 15
S	Nil	Nil
	Bwali/2	Bwali/c/5

Bwali: mixed millet and cassava porridge
Bwali/c: cassava porridge
1: Cimpapile (dried leaves of beans)
2: Bush Meat
3: Katapa = pounded and cooked cassava leaves
4: Icilemba (beans: *Phaseolus Spp.*)
5: Umulembwe = dried leaves of bush trees, pounded to meal and then boiled:
pupwe (*fagara chalybea*), pimpa (*sesamum angustifolium*), mukonde (*sesamum angolense*)
6: Boa (mushrooms)
7: Icikanda (wild orchid)
8: Fried Groundnuts
9: Icikonko (salted groundnut cake)
10: Ntoyo (ground bean: *Voandzeia subterranea*)
11: Musalu (wild spinach)
12: Ililanda (cow-peas: Vigna unquiculata)
(*Source*: ATW Private papers.)

tem (see Chapters 2 and 4). The decline in the importance of millet as a staple was already well advanced in some areas of the province by this date, although it retained its importance as a beer crop. Thus, Tweedie's data taken together with Richards' material and the food calendar for Mpika collected in 1937 (see footnote 10), show not only the marked changes but also the continuities in the diet over the period in question.

Tweedie's data also indicate that the kind of breakdown in the village economy and redistributive system that Richards had feared would follow in the wake of the cash economy had not come about in quite the way Richards had imagined. Tables 3.7 and 3.8 show gifts of meat between households and household eating groups in the village between December 1959 and March 1960. It is clear, when this information is put together with the kinship diagram (Figure 3.1.), that households in this area cannot be considered as bounded units of consumption and that food distribution was integral to the redistributive mechanisms within the village, some of which were based on interdependent matrilocal households.

One prominent group in Kanyanta village was the "four sisters group," which included households IV, V, VI, and VII. These households lived very much in common, moving around from house to house as was convenient and cooking for each other. However, the circumstances of the four women involved were rather different. The husband of householdhead V was absent, living with another wife, although he did support her and provide gardens for her. She also had a son-in-law who worked in the mines, and provided her with occasional support. Her sister,

TABLE 3.7 Gifts of Meat Between Households In Kanyanta Village During the Period December 28, 1959 to March 27, 1960

Date	Gifts of Meat
12/28/59	IX wild pig from visitor
1/11/60	I meat from III
1/18/60	II and III meat from visitor
1/30/60	I meat from III
1/30/60	VI meat from son-in-law
1/31/60	IX, X, XI, XII meat from III
2/1/60	I, II, IX, XI meat from III
2/2/60	IX meat from III
2/13/60	VII meat from V
2/14/60	VII meat from V
2/15/60	VII meat from V
2/16/60	VII meat from V
2/18/60	IX chicken from relative
2/19/60	VII chicken from relative
2/19/60	IX chicken from relative
2/22/60	I meat from relative
2/22/60	III meat from relative
2/23/60	III meat from relative
2/23/60	IX meat from relative
2/23/60	X meat from relative
2/26/60	IV meat from V
2/26/60	IX meat from V
2/29/60	VII meat from relative
3/3/60	I meat from relative
3/3/60	III meat from relative
3/9/60	IX meat from I
3/11/60	X meat from I
3/13/60	I and II meat from III
3/14/60	I and II meat from III
3/19/60	I, II, IV, IX, XII meat from III
3/20/60	V, VIII, X, XI meat from III
3/21/60	IV, V, VIII, IX, X meat from III
3/22/60	II, meat from III
3/23/60	I and XI meat from III
3/24/60	I and IX meat from III

(*Source:* ATW Private papers.)

householdhead IV, had a permanently resident husband and no dependent children. She had two citemene gardens, both cut by her husband. Householdhead VI had been deserted by her husband, but was able to maintain the household because of the help of her two sons-in-law. Table 3.8 shows that she cooked and ate with one married daughter (XIV) and sent food regularly to the household of her other daughter (XVI). She had three citemene gardens, two provided through beer parties and the third cut for her by a son-in-law (Table 3.9). Householdhead VII

TABLE 3.8 Household Eating Groups of the Village of Kanyanta During the Period December 28, 1959 to March 27, 1960

| | Eating Groups | | |
Hut	Basic Group	Eats Regularly With	Sends Food To
I	Wife, 5 children	—	II, III, IX, XIII
II	Eats with III	III	—
III	Wife, 5 children	II	II, I, XVII
IV	Wife	V, VI, VII	II, V, X, XI, XVI
V	Wife, 2 children	IV, VI, VII, XIV	II, XVI
VI	Wife, 2 children	V, VII, XIV, IV	V, XVI
VII	Wife, 4 children	V, VI, VII, XIV, IV	XVI
VIII	Eats with XII	XII	XVII
IX	Eats with X	X	II, XV
X	Eats with IX	IX	II, IV, III, VII, XI
XI	Eats with XII	XII	IV, X
XII	Mother, granddaughter, and 3 great-grandchildren	VIII, XI	XVII

Source: ATW Private papers.)

was the youngest sister and the worst off. She had no husband, four children and had to pay someone to make a garden for her in 1959 (see Table 3.9). However, she was assisted by her sisters with whom she ate and cooked communally. She also ate with one of her nieces (XIV) and sent food to another (XVI) (ATW: private papers).

A second group in the village was the "hunter's group." The hunter (household III) used to work on the mines, but since his retirement he had made quite a comfortable living for himself with the aid of his gun. Between June 1959 and July 1960, he had shot eleven buck of various kinds. He sold some locally or in Kasama to pay for his licence, he gave some as gifts to people in the village (see Table 3.7), and he gave some as payment to people who were to build a brick house for him (emulating the example of his classificatory (MSiS) brother (Household I)). The popularity of meat in the diet is difficult to overestimate, and guns were very much desired. Richards thought that there were many men who only went to the mines solely so that they could afford to purchase a gun and a licence (Richards 1939:348). Netting, spearing, and tracking animals with dogs were all methods of hunting known in the Northern Province when Richards visited the area (Richards 1939:342–350). However, because of increasing controls and declining game populations, hunting had decreased by the 1950s, and was mostly confined to the trapping of small animals and birds, except for the fortunate few who owned guns and could kill larger animals. However, hunting still provided an important qualitative, if not quantitative, component of the diet in the 1950s. Bush meat, like other products from the bush (mpanga), was much sought after, and fishing continued to provide important supplements to the diet at certain times of the year.[18]

Richards suggested that the rules governing the distribution of meat in villages were rather different from those regulating the distribution of other foodstuffs because meat, which was perishable and only available intermittently, was distributed much more widely (Richards 1939:141). This would seem to be borne

Kanyanta Village 1959 : Kinship Diagram

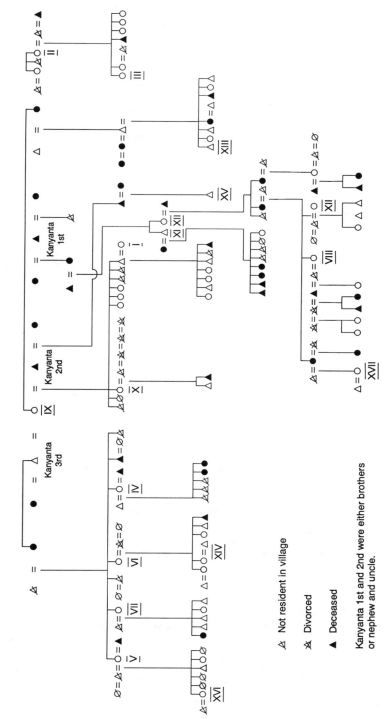

△ Not resident in village

⚰ Divorced

▲ Deceased

Kanyanta 1st and 2nd were either brothers
or nephew and uncle.

Figure 3.1

TABLE 3.9   Garden Information: Village of Kanyanta

Household	Year and Garden Type	Cut By
I	1957 cifwani (shared with V)	Husband
	1959 bukula	Husband plus labor for one week, cost £1
II	1958 cifwani	Brother
III	1958 cifwani	Husband
	1959 bukula	Husband
IV	1957 cifwani	Husband
	1959 kakumba	Husband
	1959 bukula	Husband
V	1956 cifwani	Son-in-law
	1957 cifwani (shared with I)	
	1958 bukula	Husband plus husband of household I for 1 week, cost 10/s
	1959 bukula	Brother-in-law
VI	1957 cifwani	Beer party
	1957 cifwani	Beer party
	1959 bukula	Son-in-law
VII	1957 kakumba	
	1958 kakumba	
	1959 bukula	labor cost 10/s
VIII	1958 kakumba	
	1959 bukula (shared with XVII)	Sister's son-in-law
IX	1957 cifwani	Grandson
	1958 cifwani	Beer party
X	1958 kakumba	
	1959 bukula	Son plus beer party
XI	1957 cifwani (shared with XII)	Father
	1958 cifwani	Father
XII	1957 cifwani (shared with XI)	Son-in-law

Bukula: new burnt garden; kakumba: small accessory burnt garden; cifwani: millet garden older than one year.
(*Source*: ATW Private papers.)

out by Tweedie's data, and particularly by an examination of the gifts of meat made by household III and shown in Table 3.7. Nearly every household in the village was at some point in the three month period given some meat, although we don't know how much. However, it is also clear that Households I, II, IX, and X received a larger number of gifts, and this can be plausibly explained in terms of their kinship links. Household II was the hunter's mother-in-law, Household I his classificatory brother, Household IX his classificatory mother (wife of the village headman), and Household X his classificatory mother's daughter. It is presumably

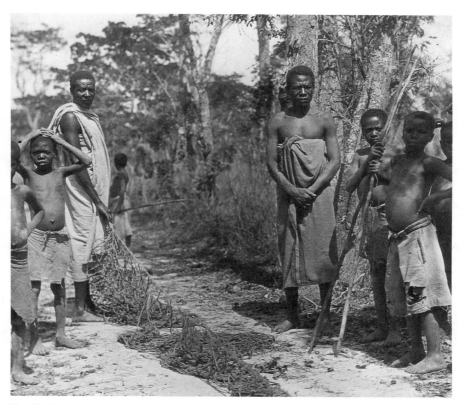

'Hunting'
Photo: Audrey Richards, 1930s

for these reasons that Household III ate communally with Household II and sent food to Household I. Households XI and XII also received a significant number of gifts of meat, and this can also be accounted for by kinship links, albeit more distant ones: the grandfather of Households XI and XII was brother to the former husband of Household III's classificatory mother. This kinship link helps to explain why Household III sent food to Household XVII (ATW: Private papers).

In addition to its own needs, Household III had to provide substantial support for the wife's mother and her sister and also help to support two of the wife's sisters. There was only one citemene garden to support all of these women in 1959, and that had been made by the wife's maternal uncle (brother of II) (see Table 3.9). The women in this group mostly earned a living by selling beer locally or by selling millet in Kasama (ATW: Private papers).

A third group in the village was labeled by Tweedie the "ex-slave" group. They inhabited a slightly separate part of the village outside the main village circle. This group seemed to be the poorest in the village and had as its center a very elderly brother and sister (Households XI and XII). The elderly brother had no wife because he refused to inherit the heir when his wife died, but he was looked after by his daughter. The daughter shared gardens with her father and one with her aunt (FSi), and her cousin (FSiDD). She earned her living by selling beer and millet. The aunt (XII) was too old to cultivate, but she was helped by her granddaughter, and they both relied on a garden that was cut for them by the granddaughter's husband before he left her for another wife. Both the old lady and her granddaughter made most of their living by selling beer. Households VIII and XVII were also part of this group. Householdhead VIII had no husband in 1959, but married a Boma messenger as his second wife in 1960. She met him when she was engaged in roadwork that he was supervising. Household head XVII was newly married and was still eating in the house of her classificatory mother (Household VIII). These two households shared a garden that was cut for them by the young husband of Household XVII.

The question of relative poverty in the village surveyed by Tweedie can be addressed through the data she provided on ukupula (food for work). Richards had argued in the 1930s that ukupula was considered a form of begging and was demeaning (Richards 1939:145). This view was still strong in the 1980s, but as we shall see, ukupula is a complex institution containing a set of meanings that link kinship, rank, and food. It is also an institution whose function, in linking relations of reciprocity with access to cash, was essential to the social and economic well-being of women in this evolving money economy. More than any other institution, then, it demonstrates the fact that cash and kinship, money, and reciprocity, rather than being mutually exclusive, were actually bound together in the rural economy.

The exchange of food for labor, outside the bounds of kinship, had historically been an expression of differential rank. It is possible, for example, to interpret the nineteenth century institution of slavery as an institution which, in part, facilitated the exchange of labor for food. Certainly this is how Richards interpreted it (Richards 1939:145), although of course slavery had many other functions and meanings. Slaves were often indistinguishable from poor relatives, except for the fact that they had no granaries or gardens of their own and were dependent on their owners for food (Richards 1939:144). In the 1950s, however, performing ukupula meant providing temporary labor in return for food during the busy seasons of the year, and this was something that most women did, regardless of rank or wealth. This can be seen through Tweedie's data in the case of Household I (the wife of the richest man in the village who was also the son of the late headman) who, accompanied by her daughter, twice performed ukupula during the month of June (see Table 3.10). In the case of households that were known to be basically self-sufficient, their occasional forays into casual labor were not likely to have been thought of as begging. They might perform this work for kin or nonkin, but were most likely to provide labor for households that were recognisably, but distantly, related to them. This may account for the fact that none of the households in Kanyanta performed ukupula for another household resident in the village, although some did work for Mwansa and Chilokoto, both of whom were former residents of the village who had broken away to set up individual settlements. Tweedie noted that ukupula could be performed both for kin and for nonkin, and that the timing and scale of remuneration varied depending on the social relations involved. Nonkin or affines tended to be paid at the time of work, whereas relatives might be given a share once the whole harvest was in. The rate of pay for ukupula during harvest in the 1950s was one basket of millet for the laborer for every two baskets filled for the employer. For carrying millet to the village, a laborer received one basket of millet

TABLE 3.10 Ukupula Worked by Women of the Village of Kanyanta Village, Harvest 1960

	Woman of Household	Employer	Remuneration
5/23/60	VIII, XII, XVII	Woman/Leshiti	One basket of groundnuts each.
5/23/60	Daughters of XII, VIII	Woman/Leshiti	3 platefuls of groundnuts each.
5/25/60	VIII, XII	Woman Leshiti	One basket of groundnuts each.
6/1/60	I + Daughter	Woman/Chasulwa	One basket of groundnuts each.
6/1/60	VIII, XII	Woman/Leshiti	One basket of ground nuts each.
6/2/60	VIII, XII	Woman/Leshiti	One basket of millet each.
6/4/60	VIII, XII	Woman/Leshiti	One basket of millet each.
6/6/60	VIII, XII	Woman/Leshiti	One basket of millet each.
6/7/60	VIII, XII	Woman/Leshiti	One basket of groundnuts each.
6/7/60	I + Daughter	Woman/Mwasha	One basket of groundnuts each.
6/10/60	VIII	Woman/Leshiti	One basket of groundnuts.
6/14/60	IX	Chilokoto	One basket of millet.
6/14/60	VIII	Woman/Leshiti	One basket of millet.
6/15/60	I	Woman/Mwasha	One basket of groundnuts.
6/15/60	IX, X	Chilokoto	One basket of millet each.
6/16/60	X	Chilokoto	One basket of millet.
6/16/60	VIII, XVII	Woman/Leshiti	One basket of groundnuts each.
6/17/60	X	Chilokoto	One basket of millet.
6/20/60	X	Mwansa	One basket of millet.
6/21/60	XII, XVII	Woman/Chasulwa	One basket of groundnuts each.
6/23/60	X	Mwansa	One basket of millet.
6/25/60	VIII	Woman/Leshiti	One basket of millet.
7/6/60	VIII, XII	Woman/Leshiti	One basket of millet each.
7/9/60	XII	Woman/Leshiti	One basket of millet.

(*Source*): ATW Private papers.

for every three carried to the employer's granary. When no garden work was being done, food could be earned by grinding millet. In return, the laborer was given a plateful of flour for grinding about 4.5 kilos of flour. In the 1950s, a plateful of millet flour would have been worth about 1/— at current prices (ATW: Private papers). All these examples demonstrate that the relative rates of pay for ukupula were very high, indicating that ukupula is probably not best understood as an emerging form of commoditized labor but as part of a system of redistribution, the need for which had increased rather than diminished with the growth of the cash economy.

Richards had argued that ukupula was on the increase in the 1930s because of the large number of women who had been deserted by husbands and needed to find ways of eking out their existence. These individuals either consumed the grain they earned with their labor, or they wandered from village to village collecting baskets of grain. When they had enough grain, they brewed beer and then invited men to a ukutumya party so that they could get the necessary granary or house built (Richards 1939: 144–147). It is true that when Richards conducted her research, women who had no need to perform ukupula did not do so, but other evidence indicates that ukupula is best understood as a form of redistribution within the domestic economy rather than simply as a sign of desperation or disintegration. Ukupula had existed long before Richards came into the region, and it was

'Women harvesting Millet'
Photo: Audrey Richards, 1930s

primarily a way of redistributing both labor and food resources. In terms of the redistribution of the former, it clearly needs to be considered alongside other strategies, such as slavery and brideservice, designed to achieve the same ends. The major constraint on agricultural production in the region has always been the supply of labor at certain seasons of the year, and ukupula was one of the mechanisms designed to ease the pressure of labor at key points in the agricultural cycle, especially during harvest time. As an institution largely confined to women, it points to the fact that, within the prevailing gender division of labor, the availability and timing of women's labor was as important, if not more important, than the availability of male labor. This was a fact obscured by the emphasis that came to be placed by researchers on labor migration and on the absence of male labor for cutting trees. Ukupula was often performed at harvest time and there is an obvious connection here between the fact that harvest is a very busy time and that women do the vast majority of labor associated with harvesting, including the very arduous and time-consuming activity of carrying baskets of grain into the granaries from distant fields.[19] However, it is also the case that ukupula seems to have been, and continues to be, part of a series of strategies of redistribution within the domestic economy carried out by women. Other examples we have also mentioned include begging from relatives in the hungry season and sharing food at the point of consumption. These strategies of redistribution were always necessary precisely because of the nature of the gender division of labor and the fact that temporary male labor absenteeism had been a feature of Bemba agriculture long before the development of a migrant labor economy.[20] Redistribution was also crucial because of the variability in production levels at the local and regional levels, as well as the unpredictability of surplus that was a part of this variability.

It seems likely that the incidence of ukupula increased with the emergence of the cash economy and the increase in labor migration, but the relationship between ukupula and the absence of male labor is somewhat more complex than as outlined by Richards. Women were not short of food simply because men were absent. Therefore, they were not necessarily forced into ukupula because of destitution. Ukupula was also a route through which women could gain access to the labor of (usually unrelated) men in order to get their citemene gardens cut. From the 1930s to the present day, women have performed ukupula in order to acquire the surplus grain required for making beer and holding ukutumya (beer parties), through which they could gain access to male labor. An institution of exchange between women thus also enabled a form of exchange between women and men. This was exactly the strategy pursued by Households VI, IX and X in Tweedie's Kanyanta survey (see Table 3.9). It is, thus, not that women were forced to ukupula through destitution, but rather that they adapted an existing strategy and made it serve changing circumstances (Moore and Vaughan 1987). The emergence of the cash economy did not obviate the need for a system of redistribution, if anything, it made it more necessary.

In addition to facilitating the direct exchange of food for labor, ukupula could also act as a route through which women could gain access to cash in an economy in which such access was extremely limited. Although it has always been difficult for women to sell their labor directly for cash, those who did not have immediate food needs might be able to sell the food they earned through ukupula to those with wages, or they might convert the food into beer and sell this. If we look at the women who performed ukupula during the harvest of 1960, we can see that Households IX and X, and Households VIII and XII were regularly involved in this activity (see table 3.10). The latter two households were part of the ex-slave group and were among the poorest people in the village. Their reasons for engaging in work of this kind are clear when we see that through this mechanism, they acquired large amounts of millet and groundnuts, that they could either store for food or use for making beer. Both of these households were involved in beer brewing and Household VIII had also been engaged in wage labor (ATW: Private papers). Households IX and X, however, were closely associated both with the headman and with Household I, the richest household in the village. These households had no need of extra food, especially at this time of year, but the women did need access to cash. Household X in particular had two children, one a schoolboy, and because she received no money from her husband or migrant relatives, she needed money for clothes and other household expenses (ATW: Private papers). Tweedie also noted that children performed ukupula on parents' farms, and that even young girls expected to be given a share of the crop they had helped to harvest. In some cases, foodstuffs secured in this way would be sold, and the children would then buy clothes and gain some measure of independence (ATW: Private papers). It is clear, therefore, that although ukupula was a necessity for poorer households, the nature and extent of this institutionalized form of labor cannot be seen as a simple outcome of destitution caused by male labor migration. Instead, it must be seen as part of a larger strategy of adaptation for women under changing economic and social circumstances.

The exact nature of these changing circumstances can be seen not only by exploring the systems of redistribution and resource flows within the village and domestic economy, but by examining the wider system of redistribution within which the village economy was situated. In Figure 3.2, we show the significant points of such a system during the late 1950s and 1960s. It is clear immediately from this diagram why households would have needed cash, because cash had become, to some degree, essential for subsistence and household reproduction by this period.

Earlier research conducted by Ann Tweedie also demonstrated this point. In 1953, she had conducted a study of five villages in Mpika District. In Kambe, a village of ten houses, she attempted to analyze income and expenditure for the various households. The data are incomplete, and because they are based on recall and are designed to provide annual totals of income and expenditure, they can be considered suggestive only. However, examples from two households are worth providing because of what they demonstrate about the importance of cash for household subsistence. The first household in question was occupied by a headman of nearly seventy years of age, who had worked in Fort Jameson and Elizabethville as a cook before 1920. He returned home and became an assessor of the Chief's court in 1931 at a salary of 32/6 per month. When surveyed in 1952–53, he had a wife and four children under fifteen years of age, and his annual income was £20.0.0, almost all of which came from his salary as an assessor. His wife was reported as earning 7/s in the same period from the sale of foodstuffs. His annual expenditure was given as shown below:

	£	s	d
Cutting Trees	1	6	0
Digging Ditch	1	5	0
Cycle Licence	—	2	6
Gun Licence	—	5	0
Trousers for Self	1	2	0
Clothes for Wife and Children	6	1	0
Blankets	2	7	0
Cassava	1	5	0
Fish	1	0	0
Salt	—	7	0
Soap	—	12	0
Parafin	—	12	0
Matches	—	3	0
Basket	—	2	0
Tobacco	—	15	0
Totals	£17	4	6

This would have given the household head a surplus of £2.15.6d, although he reported that he had spent all of the money, and could not account for the difference (ATW: Private papers). But the absolute amounts expended are not as impor-

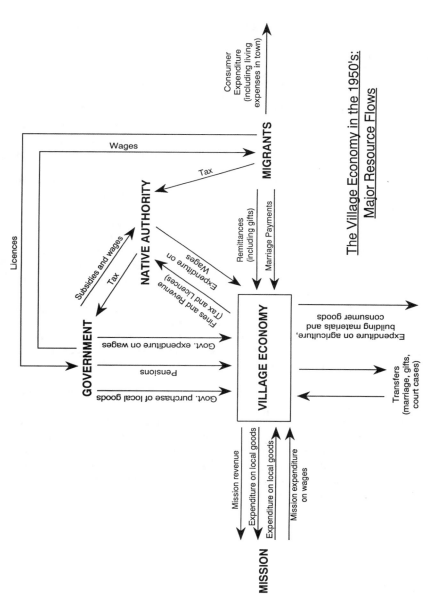

Figure 3.2

The Village Economy in the 1950's:
Major Resource Flows

GOVERNMENT

NATIVE AUTHORITY

MIGRANTS

MISSION

VILLAGE ECONOMY

Licences

Wages

Tax

Subsidies and wages

Tax

Expenditure on
Wages

Fines and Revenue
(Tax and Licences)

Remittances
(including gifts)

Marriage Payments

Consumer
Expenditure
(including living
expenses in town)

Govt. expenditure on wages

Pensions

Govt. purchase of local goods

Expenditure on agriculture,
building materials and
consumer goods

Transfers
(marriage, gifts,
court cases)

Mission revenue

Expenditure on local goods

Expenditure on local goods

Mission expenditure
on wages

tant here as the expenditure headings. It is noticeable that a third of the annual income was spent on clothes, and that approximately another third was spent on household subsistence items, including food. This particular household head was exempt from tax (10/s) because of his age, but had he been eligible for tax, he would have given over approximately 4.4% of his annual income to the Native Authority. It is also significant that he paid out 13% of his income to cover the cost of agricultural preparation.

Another household in the same village was occupied by a widow with five small children under eleven years of age. She had insufficient food supplies to last her through the year, but she still had to sell food in order to raise cash as she had no other form of income. By the middle of August, she had already sold enough of the crops harvested in 1953 to buy clothes for the whole family. Her plan was to move to the village of one of her mother's brothers for the months after harvest until the season for planting. There she would help with the digging of cassava and other tasks, and in return she would get food supplies for herself and her family (ATW: Private papers). Her case emphasizes once again the importance of ukupula in the system of redistribution, as well as the way in which cash was clearly essential for household production and reproduction in this period.

A re-examination of research conducted in the 1930s and 1950s then, gives us some idea of the problems of provisioning faced by households in the Northern Province during this period. In addition, it gives us an insight into the complex processes at work within this incompletely commoditized economy. In particular, it demonstrates that an increasing dependence on cash for household reproduction could go hand-in-hand with a continued (and even enhanced) attention to relations of kinship and reciprocity. Such relations did not crumple under the impact of the cash economy, but neither did they remain unchanged. For many of the poorer households surveyed by Richards and Tweedie, survival was a balancing act requiring the strategic operation of all of the economic and cultural resources available.

In order to appreciate the issues faced by these individual households, however, we need to direct our attention once again to a more regional picture. It was the very unevenness of the development of a market economy, and its unreliability, which accounts for the multiple strategies that households had to adopt if they were to survive. Thus, in the next chapter, we move from the level of the household to that of the regional economy, examining the postwar food economy of the Northern Province and tracing the effects of colonial government interventions and noninterventions during this period.

4

Cultivators and Colonial Officers: Food Supply and the Politics of Marketing

Richard's contention that Bemba households were food short because of male absenteeism and the breakdown in kinship relations which integration into the cash economy brought cannot be straightforwardly maintained. Even by the 1930s, and increasingly as time went on, residence and cropping patterns within the agricultural systems of the province had changed in response to a wide variety of factors. Variations in household and village consumption patterns were as much the result of differences in location, infrastructure, and marketing opportunities, as of male absenteeism and social breakdown. By the 1940s and 1950s, the introduction of cassava into the "traditional" citemene areas of the province was providing the means both for residential stability and for entrance into the market economy, as we show in this chapter. The adoption of cassava was much more noticeable in some areas than others, but in some it can be seen, at least in part, as an adaptation to the changing sexual division of labor and the absence of men for cutting trees. However, generalizations on this point have to be made cautiously because the preparation of cassava mounds is time consuming and strenuous work, and the preparation of cassava flour adds appreciably to women's domestic burden (see Chapter 7). The difficulties in assessing the changes taking place in the citemene system in the 1940s and 1950s are exacerbated by the fact that food security was so dependent on the partial and often precarious integration of communities into the cash economy. This made household and village consumption patterns vulnerable to the vagaries of government policy, as well as to other factors. On the one hand, people were encouraged to produce more food for their own consumption, and colonial sources in this period bemoaned the level of male absenteeism that they saw as responsible, just as Richards did, for social breakdown and lack of food security. On the other hand, both the government and the missions needed large food supplies, and surprisingly enough they were supplied, by and large, through the agricultural efforts of the people of the province. In this chapter then, we explore the reasons for food insecurity in the area in the 1940s and 1950s and we ask to what extent it can be attributed to male labor migration. In the process, we examine the changing nature of citemene production and the way in which cropping patterns varied across the province, as well as responding at the local level to microfactors, including infrastructure and marketing possibilities. The aim of the chapter is to link the household and village level-data on consumption presented in Chapter 3 to agricultural policy and production at the regional level. This is an extraordinarily difficult task given that we have attempted to provide a detailed reconstruction of agricultural change over an area of approx 175,000 sq km and given the uneven nature of the data available. However, we believe that it is impossible to make proper sense of household and village level data without placing them in a wider regional context, and that in order to understand contemporary changes in the agricultural systems of the province in the 1980s and 1990s, it is necessary to examine earlier transformations.

In November 1951, R. Loosemore, a cadet officer of the administrative service toured Chief Munkonge's area of Kasama district, and reported on the state of agriculture and food supplies there (SEC/792 Kasama: TR Nov. 1951). New to the area, his report reads as a puzzling aloud of the problems of production and distribution that beset this part of Northern Rhodesia throughout the colonial period and beyond. Munkonge's area, lying to the west of the town of Kasama, was by no means the most problematic part of the district in terms of food production and food security. In most years the people of Munkonge, as a whole, produced sufficient food to feed themselves. An area of traditional citemene millet production, this staple was, by the late 1940s, heavily supplemented by cassava, grown both on the citemene fields, and on the statutory mputa around the villages. Here, as elsewhere, the staple porridge, or bwali, was made from a mixture of millet and cassava flour, the proportion of cassava to millet having increased over the years since the 1930s. Over these same years, the distribution of population within Munkonge's area had altered, reflecting the increasing popularity of cassava as a crop. In 1947, for example, the touring officer had reported that many villages had moved to be close to the Lubansenshi River, where soils were particularly suitable for the cultivation of cassava (SEC2/788 Kasama: TR April/May 1947). Relish crops, as elsewhere on the northern plateau, and as reported by Audrey Richards in the 1930s, were often in short supply. In November 1951, however, most villagers had some beans and groundnuts, and all included the ubiquitous cassava leaves in their diet.

Munkonge's then, was not one of those areas of food "crisis," in which regular shortage occurred and which year after year aroused despair among officials at the district and provincial headquarters of Kasama. Yet to the new cadet, the situation there nevertheless provoked musings on large questions of historical development and economic growth. What initially puzzled the new officer was the fact that few agricultural producers in Munkonge's area produced any regular surplus or engaged in anything beyond what was termed basic, "subsistence" production. This was despite the fact that Chief Munkonge himself, a young Bemba chief, was a keen advocate of progressive farming methods. By 1951, he had one farm on which he grew a wide variety of what were termed "European" vegetables for sale to Kasama, and was opening another farm which was irrigated by furrow. There had been hopes, in the early 1940s, that Munkonge's example would inspire his subjects to imitate him, but by the early 1950s it seemed clear that this was not hap-

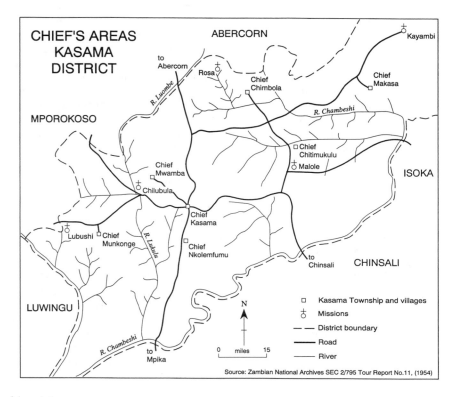

Map 4.1

pening. There were a few "independent settlements" around Munkonge. This was the term given to the newly legalized practice of individual families building their houses and farming outside the confines of a village.[1] Many of these independent settlers aspired to some form of peasant production, but few were ever thought to have earned the colonial label of "farmer."

More generally, Loosemore remarked, cultivation standards and productivity varied from village to village. It did not seem to him that this variation could be straightforwardly accounted for by the presence or absence of able-bodied males, although this was often asserted to be the case. In fact, what he considered to be the best village agriculturally was "hardly more than a settlement of old men and children," yet they were growing, in addition to the normal staple crops, sweet potatoes, onions, sugar cane, rice, pawpaws, mangoes, and bananas. In contrast, there were many other villages with most of their young men present that were neither well-tended nor well-fed. This very fact seemed to suggest to Loosemore that there could be no excuse for underproduction of foodstuffs, and that it was "within the power of most villages to produce more and more varied food with their present population." He was not unsympathetic, however, to the young men who left their villages as labor migrants. Indeed, he felt that the choice they made was a rational one, and that they should be "congratulated on having found some work that they prefer to farming." Furthermore, there seemed to Loosemore to be clear historical parallels for the situation he found in Munkonge's area. In particular, he referred to the case of the British Industrial Revolution, when many young men had seemed to prefer work as factory hands to that as agricultural laborers. Thus, the young men of Munkonge's area could also be expected to prefer industrial over agricultural labor, particularly given the very limited opportunities (as Loosemore saw it) for the development of commercial agriculture in a district that was over 400 miles from a railway line. The market opportunities for commercial agriculture in this part of the province seemed to him to be simply too limited to support the people over the longer term:

> It would be cruel to persuade villagers 40 miles or more from Kasama to grow vegetables as a cashcrop. Kasama has about 200 Europeans and an unknown number, possibly several thousands, of Africans. It is not a large market, or elastic. If the price of vegetables is reduced, the Europeans will not buy much more, and if the whole area within a radius of 40 miles tries to sell its vegetables there, prices may drop.

As far as Loosemore could see, commercial agriculture could only possibly be successful in this area if it were heavily subsidized by the government:

> I do not know where these Africans can obtain money to improve their land. British agriculture is subsidised to £23 million, raised from the rest of the population, and presumably, unless the Government of Northern Rhodesia, for reasons which cannot be economic, decides to assist the farmer at the expense of everyone else, then local agriculture will remain much as it is for many years to come.

The District Commissioner was not impressed by Loosemore's analysis of the problem, nor convinced by his analogy with the British Industrial Revolution. (SEC 2/792 Kasama: TR Nov. 1951: DCs comments). To begin with, he could not see overproduction of foodstuffs as a real problem, since the people of the area were regularly hungry and, as Richards' work had shown, there was much room for improvement in their diets. Neither did he feel that forty miles made Munkonge's area remote from the market of Kasama. The comparison with the British Industrial Revolution also seemed to him to be misplaced because Africans in the Northern Province had so much more land at their disposal, and improvement to this land could be achieved without financial assistance. The scarce resource was labor, "and although we can never hope for more than 75% resident adult male population, that percentage would be enough and would save the rural areas for future need." The urge to leave the land was not nearly so great among Africans as it was in the more crowded countries, and neither was it an urge that should be encouraged. "I have no doubts," he wrote, "that for the sake of rural development, and the avoidance of trouble if and when the "bonanza" days of copper come to an end, we should refrain from congratulating the African who leaves the land to earn his living."

In part, the District Commissioner's response can be read as a "ticking off" of a new cadet who had presumed to advance an opinion on the larger question of economic development with relatively little local knowledge behind him. But Loosemore's observations, and the District Commissioner's reply, taken together constitute a fair summary of the arguments around agriculture and economic development in this part of the province in the 1940s and 1950s. Neither are they without relevance for the late 1980s and early 1990s for, as we shall see in Chapter 8, people of the Northern Province in this period have experienced both the ending of agricultural subsidies, and the unemployment attendant on the end of the copper "bonanza."

Colonial administrative officers of the 1940s and early 1950s were still concerned, as their predecessors had been, with the question of the longer-term viability of the citemene system and its ecological consequences, but there were moments when this concern was overtaken by more immediate issues. In the first place (and partly spurred by the demands of World War II as they had been by WWI), they were anxious that the people of the Northern Province should, as a whole, be self-sufficient in foodstuffs from year to year, and that they should supply sufficient surpluses to meet the not inconsiderable demands of the administration itself (to feed its labor), the missions, the schools, and the hospitals. While Audrey Richards and her colleagues had addressed the issue of food sufficiency at the level of the household, colonial administrators in the area addressed it at the level of the province. They saw their responsibility as ensuring that no deficit occurred within the province as a whole that would necessitate the importation of food from other provinces of the territory.

The question of how this overall supply of food was distributed within the province was, as we shall see, also a matter of concern, but officials generally did not address the question of distribution of food between and within households, which had been the focus of Richards' work. In fact, the issue of staple food sufficiency had, by this time, been partly solved by the increasing cultivation of cassava, although there remained a frustrating unpredictability when it came to the supply of surpluses. The causes of this unpredictability were continually debated, and discussions revolved around assessments of supply and demand, production, and distribution. These discussions then fed into the second major issue, which was whether this part of the Northern Province could be turned from a labor reserve into an area of prosperous peasants, turning its men from labor migrants (and occasional treeloppers) to conscientious farmers. The question for the Northern Province at this time was less whether it could become a grain-bin for the industrial areas (though the export of food out of the province was a matter of heated debate, as we shall see), and more whether demand could be generated within the province sufficient to fuel a peasant revolution and to generate economic opportunities that might make labor migration less attractive. As we show in Chapter 5, by the mid-1950s this concern had taken on a more overtly political dimension, because settling the politically volatile urban Bemba men back on the land and ensuring their prosperity appeared to be the only sure way of stemming the tide of nationalism in the province, and perhaps also on the copper-belt.

Opinions on all of these issues—the sufficiency or insufficiency of production, the sufficiency or insufficiency of demand within the province, and the adequacy or inadequacy of the marketing and distribution system—differed greatly and brought forth a large amount of conflicting evidence. In this chapter we examine this evidence, and also argue that its apparently conflicting nature reveals the complexity of the agrarian economy by this period. Rather than suggest that there was one and only one correct analysis (that, for example, either Loosemore or his superior was right about the viability of a provincial market economy, or the relevance of the presence or absence of men), we show how the precarious relationship between ecological variation, labor availability, state interventions, and marketing opportunities (or constraints) created a situation in which the production of food surpluses was, indeed, difficult to predict from year to year. Officials could hardly be blamed for not coming up with a solution, but the arguments for and against labor migration that continued to occupy a great deal of administrative "memo space" at this time, generally neglected to acknowledge both the degree of variability in production systems in the area, and the degree to which these systems had adapted and responded to the absence of male labor (see Chapter 6). This is not to say that labor availability was not a crucial factor, but rather that this factor did not operate in isolation. In addition, officers paid little or no attention, in their analyses of labor migration and agricultural production, to women's work and the changing gender division of labor. This was despite the fact that Audrey Richards' study of the 1930s had clearly indicated the importance of both gender and generational divisions of labor to the functioning of the food production system.

In marked contrast to the generally oversimplified arguments on labor migration and its effects on agricultural production, are the rich and detailed accounts of the area composed by the very same officers in their tour reports. The degree of variation in agricultural strategies revealed in this material provides us with a dynamic account that can be viewed in parallel to the more synchronic typologization produced by ecological scientists which we outlined in Chapter 2. It was not simply that there were different types of agriculture practiced in the Northern Province, but rather that these types were not historically constant. Furthermore, as we have already indicated, the form of agriculture practiced was contingent, not only on ecological conditions, but on a number of factors related to the haltering development of a market economy in the province and beyond.

Our own discussion is broadly divided into two halves. The first half deals with production; the second half with marketing and distribution. We begin with a piecing together of (mostly) documentary evidence, which gives some idea of the geographical and temporal variations in the production systems of Kasama, Mpika, and Chinsali districts in the period from the early 1940s to the early 1950s.[2] This material contextualizes the data Richards presents in *Land, Labour and Diet*, and it is worth recalling here that Richards studied villages adjacent to Chief Chitimukulu's capital, Shiwa N'gandu and Kasama.[3] Thus, while she covered different areas of the province, it is noticeable that she makes little of regional level variations, preferring instead to see citemene as a definite feature of Bemba custom (see Chapter 1). In addition, however, the data presented in the following discussion show the degree to which people in the province were adopting different strategies in response to government policy and the emerging cash economy. A provincial-level analysis of this kind necessarily obscures the specific experiences of individual farmers, and unfortunately the oral histories we collected have shed little light on this matter, as they are very specific to one locale. But, what an analysis of the sources we have had at our disposal shows is the degree to which it is possible, in the absence of direct speech, to examine and interpret local people's representations through their activities. This approach, as we discussed in the Introduction, can never be a substitute for the reporting and analysis of direct testimony, yet it is absolutely essential in a situation where the dialogue between colonizer and colonized, and between chief and commoner, was most often conducted through praxis rather than through face-to-face discussion. It is also crucial to use the existing evidence for the activities and strategies of local individuals and groups to read against the grain of texts produced by officials, and missionaries, as well as by chiefs and tribal representatives. Our aim then has not been to capture or represent the experience of people living and working in the Northern Province in the 1940s and 1950s, but rather to document some of the ways in which the activities of local people at that time continued to defy the easy explanations and generalizations that others insisted on placing upon them.

Variations and Variability: Multiple Agricultural Strategies

Let us begin with some examples of production in Mpika District in the late 1930s and 1940s. This district included within it a large area of typical plateau land of largely citemene cultivation, parts of the Bangweulu swamps to the Southwest, a steep, rocky escarpment, and part of the Luangwa valley to the east.[4] The Great North Road, which passed through the district, made some areas increasingly and easily accessible, while others (such as parts of the Luangwa valley and parts of the Bangweulu swamps), were definitely well off the beaten track, then as now. To some extent it was the case that the least productive areas were the most accessible (roads being typically built along watersheds), while the more varied and productive ecological niches tended to be more remote from arteries of the colonial economy.

In February 1938, Mpika District Commissioner Thomas Fox-Pitt toured Kapoko's area, along with parts of Chikwanda, Mutupa, Saidi, Nawalya, Kambwili, and Kamwendo (SEC2/836 Mpika: TR Feb. 1938) Fox-Pitt was an unusually acute observer and this tour took him through a range of country as he descended eastward from the plateau and down into the Luangwa valley (reversing the route taken by Livingstone and other travellers in the nineteenth century as described in Chapter 1). In the Bemba-speaking plateau villages near Mpika, Fox-Pitt noticed that an increasing amount of maize was being grown, much of which was sold on the market in Mpika, or to travellers and traders on the Great North Road, along which many of the villages were sited. Maize, Fox-Pitt noted, could be grown on the edges of dambos (swampy depressions) where the ground was too sour for other crops. In addition to growing maize, however, the villagers continued to cultivate their citemene gardens of millet and sorghum (kaffir-corn), which might be up to ten miles away from their roadside village. The distance from village to field meant that most villagers built mitanda in their citemene gardens, and spent more than the usual number of months there. The earlier campaigns against the building of temporary garden dwellings (mitanda) seem, by this stage, to have been largely forgotten by officials, and Fox-Pitt mentions mitanda dwelling without any hint of disapproval. In addition to growing maize and millet, the plateau villagers also cultivated a range of sorghum varieties, which Fox-Pitt described in some detail.[5] Three types (nsambe, kanondo, and kandebele) were sown in November and harvested in April. A fourth (ntoche) ripened earlier, in February or March. Villagers here also grew a perennial sorghum (sonkwe), the existence of which was apparently unknown in the valley. Bulrush millet (bubele) was grown in small quantities, specifically for beer, and an early ripening millet (mwangwe) was also grown in small quantities. Down in the Luangwa valley, in mostly Bisa-speaking villages, the food situation was variable. In Saidi's area, on the river itself, drought had ruined the maize crop, and compounded the difficulties already caused by two years of flooding. Here the people were relying for food on two types of grass seed (lupungu), which were grown extensively near the villages. Ripening at different stages (nkwende first and zoshi later) both types of grass were harvested by swinging a basket backward and forward at the level of the seed ears. The seed then made into a fine white flour with what Fox-Pitt considered to be a "pleasant taste." Drought had apparently not affected other parts of the valley, however. At Kamb wili's and Kamwendo's, on rich valley soil deposits, the people had plenty of the maize and sorghum that they grew in the same plots year after year.

Most food shortages appeared to be localized and were usually met through local strategies, including the collection of wild foods, the cultivation of famine crops such as grass seed, and the exchange of labor for food in more fortunate areas. In 1940, however, the new District Commissioner (A.T.B. Glennie) was expressing fears of a more general shortage in the district—fears which were to be extended to the Northern Province as a whole. The causes of shortage, which undoubtedly existed in some parts of the district, were not at all clear. In some areas, the "wasteful methods of cultivation" were blamed. In other areas, over-selling to government and traders on the Great North Road was blamed. More generally, of course, the labor demands of the war were beginning to make themselves felt. The fact that in sequential reports (or even in the same tour report) the overcutting of fitemene and the insufficiency of cutting could be blamed for food shortage was not necessarily evidence for muddled thinking on the part of the administration, since both situations could occur within the same district.

In Chief Chiundaponde's area in the south of the district, the Bisa practice of cutting rather than pollarding trees was blamed in 1940 (as it had been much earlier in the century, see Chapter 2) for creating what was seen to be a very precarious ecological situation (SEC 2/838 Mpika: TR Nov. 1940). Here the District Commissioner pressed home his message that this method was wasteful of timber and unsustainable, and that in each garden a proportion of standing trees should be left. The food situation in Chiundaponde's, however, was described as "moderately satisfactory," and there was no mention of the shortages that other parts of the district were experiencing. In neighboring Chief Chimese's area, however, Mr. Glennie met a very different situation, particularly in those villages along the Lukulu River (SEC 2/838 Mpika: TR Nov. 1940). Here, the food situation was regarded as serious. There was hardly any timber left in this area, so almost no millet was grown, and only a little cassava. When he suggested to a meeting of villages that they might wish to relocate, there was some enthusiasm for this idea from the women, but none from the men. The reason for the men's lack of enthusiasm rapidly became apparent. They preferred to make a livelihood through a combination of labor migration (the average length of absence being sixty months), and fishing (see Chapter 6). The sale of fish to the Copperbelt was a profitable business. Men from the area could buy a bicycle load of fish for 5/- at the Luapula river, take it to the mines and sell it all in a single day. When fish was out of season, there was still a demand for protein in the form of caterpillars, a load of which "costs nothing but the labour of collection" and which could bring in a profit of up to £1. The fact that the wealth which apparently was available to the men of this area could not, or was not, translated into an adequate supply of foodstuffs for their households, is a question that was not asked explicitly by the District Commissioner, but is one which we will return to later. What seems clear is that the District Commissioner (and presumably the women of Chimese who had expressed an interest in moving to new land) was correct in seeing the degree of specialization of the men of this area as risky given the underdevelopment of the regional economy.

The predictions of general shortage in 1940 do not, however, appear to have been fulfilled. Some areas managed to produce a surplus for sale. The people of the small chieftaincy of Mpepo's, in the north of the district, had grown enormous gardens of kalundwe (cassava), and produced sufficient to satisfy the entire needs of the government, including rations to the hundreds of men working in road gangs (SEC 2/838 Mpika: TR May/June 1940). A total of 90,000 lbs. had been bought since November, and more was still coming in. It had been in order to ensure against shortage, both at the household, and at the district and provincial level, that the colonial administration had insisted, for some years, on the cultivation of cassava by every adult. Spurred on by the disastrous locust invasions of the early 1930s, the government had ordered the Native Authorities to pass, and to enforce, rules on cassava cultivation. In most cases, this meant that each year every able-bodied man was required to dig two hundred cassava mounds, and every adult women one hundred—a not inconsiderable investment of labor. By the early 1940s this intervention, combined with a great deal of voluntary cassava production in some areas, constituted a major adjustment to the food production systems of the district, and of the province as a whole. Cassava was grown almost everywhere, and people clearly appreciated its drought-resistant qualities, but the extent of its cultivation varied. This variation appears to have depended on a number of factors. Some soils were unsuitable for cassava cultivation. This could sometimes result in villages moving to more "cassava-friendly" areas, as we discussed earlier, but this was not always the case, especially in areas where wooded land for citemene cultivation was still readily available.

Production of cassava may also have been related to labor availability, although, as we have already discussed in Chapter 3, this is not a straightforward argument. Cassava cultivation, rather like millet cultivation, required very little labor input in production beyond the initial clearing of land and making of the mounds. Furthermore, cassava could be planted on fields that had already been used for the cultivation of millet and other crops. Thus a production strategy based on cassava appears to have been less dependent on male labor, because it required less frequent clearing of new land and this was important in a situation where only men could cut trees. The adoption of cassava as a staple (or, more commonly, as an important component of the diet), was one way in which households could adjust to the absence of men as labor migrants. Whether overall the cultivation of cassava was significantly less labor intensive than that of millet is less clear because the processing of the crop is notoriously long and arduous. Its adoption entailed a significant shift in the gender division of labor, but this was not a fact that drew any remarks from the officers who promoted its production. Their concern was to enforce the cultivation of a reliable crop that would ensure against general food shortage, and to a large extent this is what they had achieved in most parts of the Mpika, Kasama, and Chinsali districts by the 1940s.

While most parts of the Mpika district were growing significant quantities of cassava by the mid-1940s, some areas had become specialist cassava-production areas. The crop was both intercropped on citemene fields and grown on semi-permanent fields close to settlements. Though self-sufficiency and security in food-stuffs was almost certainly an important motive for people in these parts, the government's seemingly insatiable demands for surpluses to feed their labor, and their increasing willingness to open markets in areas where surpluses were fore-cast, were important factors. Cassava became a cash crop as well as a hunger crop. This was true in Chief Mpepo's area where, as we have already noted, large amounts of the crop were being sold by the early 1940s. Perhaps more dramatic was the transformation which could occur in areas that had previously caused of-ficials so much concern. Chief Chiundaponde's area, for example, which in 1940 (as we have seen) was viewed as an ecological disaster area by the District Com-missioner, was, by 1950, being viewed as "breadbasket" for the province. The tone of the touring reports shifts accordingly from disapproval of the "reckless" Bisa cultivators, to an admiration for their industry. By 1949, Chiundaponde's area was being described as a very progressive area, boasting a large number of good brick houses, a good diet, energetic cultivators, and static villages (SEC 2/840 Mpika: TR Nov. 1949).

In 1947, the District Commissioner (G.F. Tredwell) had noted that there were very large cassava gardens in Chiundaponde's area, and that this crop had almost entirely replaced millet as the staple (SEC 2/838 Mpika: TR June/July 1947). There would be a surplus, Tredwell remarked, but since access to the area was difficult, it was unlikely that much would be sold. By 1950, however, the government had opened three buying stations to encourage the sale of surpluses in Chiundaponde's area and in neighboring Chief Kopa's area (SEC 2/841 Mpika: TR Oct./Nov. 1950). Tredwell reported a "tremendous amount of cultivation" and large surpluses for sale: "The abundance of meal in these Western regions which persists year after year," he wrote, "is reassuring as it can be used to meet shortages elsewhere in the district. . . ." Further, success with cassava production seemed to have spurred a minor agricultural revolution in the area. Increasing numbers of people were grow-ing rice, and many villages boasted a profusion of mango and banana trees. The small-circle citemene system had been almost entirely superceded by hoe cultiva-tion. The cassava was grown on huge mounds, eight feet in diameter, or on long ridges (molwa), constructed on the edges of dambos. There was, it appeared, life after citemene.

In some areas, however, the production of large amounts of cassava did not herald the end of citemene. Rather, new rotations and crop mixes were introduced. Chief Kopa's area, like neighboring Chiundaponde's, produced a regular and large surplus of cassava for sale to the government in the late 1940s and early 1950s. Here, however, the citemene system continued to be practiced (SEC 2/843 TR: Jan./Feb. 1952). The citemene fields, cultivated over four or more years, produced mil-let, maize, groundnuts, beans, sweet potatoes, and squash, and cassava in the later years. But the main cassava crop was grown on mounded gardens close to the vil-lages. These gardens were said to be enormous, and much labor was expended on cultivating them each year. Elsewhere in the district the impact of cassava was much less evident. One-hundred miles to the northeast of Mpika was Chief Mu-kungule's area, another Bisa chieftainship. Remote from the road, this area was still heavily forested in the late 1940s, and was well supplied with maize and millet grown by the large-circle citemene method (SEC 2/839 Mpika: TR April/May 1948). The fields were well-fenced against animals, and traditions of cooperative labor survived. In Chief Nawalya's area in the Luangwa valley, very little cassava was grown. The villages of Nawalya's were located either along the banks of the Mun-yamadzi river or on the banks of the Luangwa itself. The very productive agricul-tural system of the area rested on the use of rich riverine soils and annual flooding, enabling double-cropping. Sorghum was grown on the higher land during the rains and maize (nyanje ya chibwesha) on the drying flats along the river banks during the dry season. In 1947, the maize gardens were said to be immense and yields were heavy (SEC 2/838 Mpika: TR Aug./Sept. 1947). The annual flood-ing maintained soil fertility, allowing the same gardens to be cultivated year after year, and allowing villages to remain on the same site for ten or more years. The people of Nawalya's also grew fair amounts of tobacco that they traded to Fort Jameson (now Chipata) to the east, along with the mats and baskets they manu-factured from the reeds which grew along the river banks. They also reared thou-sands of chickens, which they sold at "black-market prices" in Fort Jameson, and they fished the river.

This was a very different local economy than the citemene/labor migrant com-bination that was generally said to characterize the district, and the province as a whole. Successive District Commissioners looked forward to their tours of this area, viewing them rather romantically as a vestige of the "old Africa," productive in its remoteness and relative independence from the colonial economy. "Alto-gether," wrote District Commissioner Tredwell in 1947, "I think it can be said that the economic lot of these people is on the whole a singularly happy one, and the unusually joyful and contented bearing of the people, and their vociferous wel-come to the D.C. on tour, would appear to bear this out. . . . This area, far re-moved from civilisation, and devoid of educational facilities, is yet peculiarly happy, and is one of the more interesting of the Native Authority areas of Mpika District" (SEC 2/838 Mpika: TR Aug./Sept. 1947).

It was, however, the case as Tredwell pointed out, that the whole welfare of the people in this area depended on "the river and its peculiarities." The severe drought of 1949 meant that the river did not reach its normal height, and the maize crop fared badly (SEC 2/840 Mpika: TR July/Aug. 1949). The drought was followed by severe flooding in 1950, and the people were reported to be "half-starved" until the sorghum was reaped in late May and early June (SEC 2/841 Mpika: TR Aug. 1950). This was a shortage caused, not by lack of labor, lack of cultivation, or over-dependence on the new money economy, but by the same risks of agricultural pro-duction as had caused food shortage and, occasionally, famine in the precolonial period—completely different, then to the situation which officers encountered year after year in Chief Chikwanda's area on the Great North Road. Described as an "unhappy spot," shortages here were attributed to indifferent agriculture, over-selling, and the commitment of large numbers of men to wage labor of various sorts. Tredwell wrote of this area in 1948:

> Food is desperately short at the moment, but the reason is not under-
> production, but over-selling. The proportion of wage-earning, non-

> *producing adults to agricultural producers is too high, a fact due to*
> *the presence of the Boma, the PWD [Public Works Department] and*
> *the Great North Road in the area. Safe surplus is bought up by Gov-*
> *ernment or by private purchasers at Mpika in the harvest months,*
> *and the travellers on the Great North Road Transport Services also*
> *absorb a considerable quantity. By the end of November supplies begin*
> *to dry up, but the travellers are prepared to increase prices. Gradually*
> *the price of meal increases until it occasionally reaches a figure in the*
> *neighbourhood of 10/- per 4-gallon tin, or 4d per lb.*
>
> (SEC 2/839 Mpika: TR March 1948).

Therefore, according to this somewhat contradictory account, Chikwanda's area suffered regularly from food shortages, not because of insufficiency of production (despite a high rate of male wage employment), but because the accessibility of the area, and the high prices that were sometimes offered, encouraged overselling of staple foods by households. Tredwell did not make the further connections in his argument that might have made it more comprehensible and convincing. It appeared, however, that the people of Chikwanda's suffered from being both too integrated, and at the same time insufficiently integrated, into the market economy. Tour Reports from the early 1950s continued to describe Chikwanda's as suffering from the dual burden of extraction of male labor (much of it in local employment for the government) and the extraction of food (much of it sold to the government), while other areas were diagnosed as having potential that remained unexploited because of their inaccessibility. These contrasts between areas, and the apparent contradictions that they raised, fueled the heated debate on marketing that occurred in the late 1940s and early 1950s, and to which we shall turn later in this chapter.

In the Mpika district, then, there was both a large degree of geographical variation in food production systems and their productivity, and there were significant variations over time. Areas such as Chiundaponde's, as we have seen, moved from being agricultural disaster areas to being the grain bins of the district within a few years. This change was perhaps less dramatic than it appeared, because colonial officials' horror of deforestation had probably led them to overestimate the degree of vulnerability there in the late 1930s and early 1940s. Change there was certainly, and it was brought about partly by the adoption of cassava and its integration into a new food production system, and partly by the government's demand for surplus food and their willingness, in this case, to provide a basic marketing infrastructure.

Variability was also marked in the Kasama and Chinsali districts at this time (see Maps 4.1 and 4.2). Compared to Mpika, the Kasama district was less varied ecologically, and more typical, if not of the reality, of the image of the Northern Province and the Bemba system of agriculture. The 1930s in Kasama district had seen frequent local shortages of food and had inspired colonial officials to insist, in the late 1930s and early 1940s, on compliance with the Native Authority rules on cassava production. In 1940, having toured particularly problematic parts of Chief Makasa's and Paramount Chief Chitimukulu's areas, the District Commissioner (H.A. Watmore) suggested that there were only two possible ways in which the "present wasteful and inefficient agricultural methods of the Bemba" could be changed: either by famine when the timber had been completely exhausted, or by

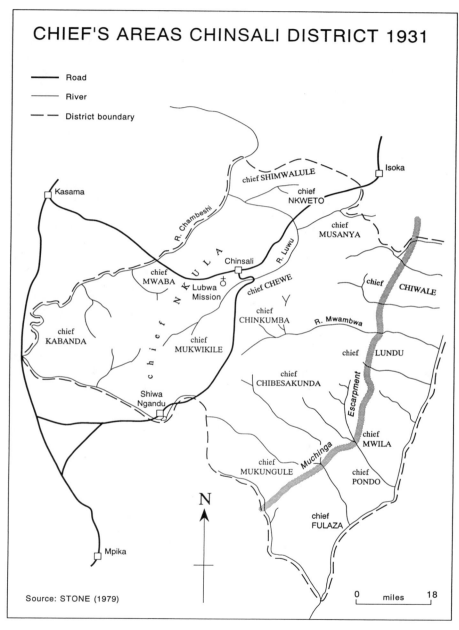

CHIEF'S AREAS CHINSALI DISTRICT 1931

Road
River
District boundary

Source: STONE (1979)

Map 4.2

the enforcement of an order to grow cassava [SEC 2/788 Kasama: TR, June/July 1940). The latter, he suggested, was preferable. The Provincial Commissioner agreed that the people should be ordered to grow cassava, because it was time that "someone made up the minds of these wasteful cultivators for them." The reform of citemene through the Chitemene Control Scheme, which regulated burning and cutting in specific areas, appeared to him to be a wasteful use of scarce technical resources, and might just "perpetuate the millet diet of past ages." Finally, he asked whether he might be told government policy on this matter: "Would it not be better to allow the natives to lay waste the forests, as the Congo Authorities did, and then get tribesmen into more confined areas and make something of them?" (SEC 2/788 Kasama: TR June/July 1940: Comments by Provincial Commissioner E. H. Cooke).

Cooke's question remained unanswered, and Makasa's and Chitimukulu's areas continued to cause concern, particularly in the context of the labor and food demands of World War II. No district official wished to find himself presiding over a local famine relief program when there were troops to be fed. After the War, in 1948, Chitimukulu's area appeared as desperate as ever to the touring cadet officer, F. Hainsworth. He had taken the opinion of the old men, who had assured him that in their youth "the food situation had been far better." Councillors and headmen confirmed this picture, arguing that the food position had grown steadily worse over the previous few years because of the absence of "the majority of young men," either on the Copperbelt on in local employment "from which they did not always come back to cut their fitemene." (SEC 2/789 Kasama: TR July 1948). Hainsworth felt that the people of Chitimukulu's area just did not appear to recognize the necessity for greater effort in order to guard against hunger: "Although they realise that cereals alone do not give enough strength to do a full day's work, they take no steps to procure any seeds of either beans or peas, and do not plant enough groundnuts." Furthermore, he wrote, the people were resistant to adoption of new methods (in particular green-manuring or fundikila), which would lessen their dependence on young male labor for the cutting of trees: ". . . if more fundikila mounds were dug, a much greater area could be dug by women and older men, so neutralising to some extent the absence of so many younger men. . . ." (SEC 2/789 Kasama: TR July 1948).

Through the tour reports of 1948 and 1949 we can follow Hainsworth as he travels to different parts of Kasama. Just two weeks after his depressing visit to Chief Chitimukulu's area, he was in the north part of Munkonge's, where he encountered a very different situation (SEC 2/789 Kasama: TR July/Aug. 1948).

Both in Bemba and in Lungu villages he found plentiful food, the product of "industrious agriculturalists," growing a wide variety of crops. The explanation he offered for this stark contrast with other parts of the district was a cultural and ethnic one:

> The reason for comparative plenty is not difficult to find. The greater part of the inhabitants are Lungu, and as their conquest by the Bemba is comparatively recent, they are still proud of their tribe and do not seem to have absorbed any of the scorn of digging the ground, which is to me, so typical of the Bemba. Their skill and industry appear to be infectious, for when I remarked to some of the Bemba villagers

that they seemed to grow far better crops than their brothers in Lubemba, they admitted that they had been glad to learn a great deal from the Lungu.

Hainsworth traveled in 1949 with the Cambridge University Bangweulu Expedition team as they made their way by boat up the Chambeshi river toward the Bangweulu swamps (SEC 2/790 Kasama: TR 1949). In this area of largely Bisa-speaking people, there was the small Bemba chieftaincy of Nkolemfumu, lying in a triangle of land between the Chambeshi and Lubansenshi rivers. The agriculture of Nkolemfumu's area confirmed and reinforced the view of the Bemba that Hainsworth had formed in his tour of Chitimukulu, and which he shared with a long line of administrators. "Until this tour," he wrote, "I was of the impression that the inhabitants of the Lubemba, immediately adjacent to Paramount Chitimukulu's village were, agriculturally, the most backward in the District. . . . Even they, however, are hardly as conservative and retrograde as the inhabitants of this small enclave of Bemba country. . . ." The contrast between the Bisa villages on the southern side of the Chambeshi (just inside Mpika district), and the Bemba villages on "our side" appeared stark. The Bemba, Hainsworth concluded, were just "not interested in growing food, and do not willingly grow any food but millet."

Variation in agricultural systems and productivity in this period could easily be explained, then, according to Hainsworth, by reference to ethnic identities and the antipathy of the Bemba people to farming. It would be easy to dismiss Hainsworth's observations and theory as evidence of misunderstanding on the part of a relatively inexperienced touring officer, or the product of a prejudiced colonial mind. But even if we do not share Hainsworth's conclusions and generalizations, his observations warrant some comment. Although legislation to control the timing of burning was still enforced in the 1940s, other forms of intervention to control the practice of citemene had largely been abandoned. Colonial officials had long turned a blind eye to the existence of temporary garden dwellings, or mitanda, and in the late 1940s, they legalized the existence of individual settlements under the new Parish System (see Chapter 5). Under the changed circumstances of the midcolonial period, different communities and different households adopted different economic strategies, giving rise to some of the variability we have been discussing in this chapter. Indeed, multiple strategies were the key to survival in this partially commoditized economy.

Some Bemba-speaking communities, it appears, did continue to resist attempts to modify their agricultural practices, insisting on the primacy of citemene production and millet, even while neighboring areas had adopted cassava production on permanent fields. The reasons for adopting this strategy of "backwardness" (in Hainsworth's terms) rather than any other may have been complex. First, we should not exaggerate the possibility of and potential for alternative agricultural and economic strategies in this part of the Northern Province. Although some communities were fortunate in being able to exploit ecological variation (river valleys, swamps, game resources), others were more firmly located on the central plateau, where citemene cultivation of millet was, and remains, the most viable food production system. Even within a citemene-based system, however, there was room for the adoption of new crops, and new rotations, as discussed above and in Chapter 2. Some communities were spurred on by their proximity to markets or to government buying stations to diversify their agriculture and to attempt to produce surpluses. In particular, as we have seen, some communities incorporated cassava as a central element of their food system. In other areas, such as parts of Chitimukulu's and Makasa's, however, there was much less agricultural adaptation in evidence. People grew the statutory amounts of cassava when the Native Authority rules were enforced, but showed little enthusiasm for the crop. Food continued to mean millet here, and much of this millet was consumed as beer in what colonial officials referred to as "the marriage and holiday season," immediately after the harvest. In these areas, the seasonality of food availability was striking, as Richards had noted in the 1930s, and hungry season shortages were common. Soil types and inaccessibility to markets partly explain why people in these areas showed little willingness to make major adaptations to their agricultural system. Colonial officials tended to be dismissive of the argument that cassava did not grow well in certain soils, but this was real factor in explaining variability in this period. The "ethnic" explanation offered by Hainsworth and others then, may well have been partly an ecological phenomenon. But it was also the case that in areas where agricultural adaptation was most difficult, and where millet was clearly the most productive crop, men in particular expended relatively little effort on agriculture, preferring to make a career out of labor migration. In such circumstances, as we have argued elsewhere (Moore and Vaughan, 1987), women may well have had an investment in refusing to adapt their agricultural practices beyond a certain point. An insistence on the importance of traditional citemene practice was also an insistence that male labor was required to keep this system running and this gave women some moral leverage over their migrant husbands. Seen in this light, the colonial officials' ethnic explanation of variation can be better understood. It was not simply the case that Hainsworth and others like him imposed ethnicity on the situations they encountered (though sometimes they did just that), but that some Bemba communities continued to actively construct ethnic identity through their insistence on the centrality of an "unreformed" variant of citemene production to their culture.

Historical sources are almost entirely silent on the question of changes in the gender division of labor in these areas, subsuming this within oversimplistic arguments on the effects of male absence. This makes perfect sense in terms of the obsessive concern with citemene and cutting trees in colonial official documents. It seems likely, however, that the variations in agricultural systems and productivity that were so noticeable in this period, rested to some large extent on changes, or resistance to changes, in the gender division of labor. Women, in particular, could be agricultural innovators, incorporating new crops and rotations, but they could also resist changes that would leave them with a greater share of labor, and let their labor migrant husbands "off the hook." We know from the interview data that we collected from a small number of women farming in the 1940s and 1950s, that at least some women were trying to retain male labor by insisting that men should cut trees. Similarly, we discovered in the 1980s that some women were against increased flexibility in the sexual division of labor because it meant that husbands could set them new tasks and demand even more work from them (see Chapter 8). In general, however, we found oral history a difficult source of information on this point, partly because many of the women farmers and migrants we interviewed were away from the area in the 1940s and 1950s and claimed to have little knowl-

edge of farming for that period, and partly because informants tended to keep em-
phasizing an ideal division of labor that stressed the importance of men as tree
cutters. When we were doing the research, we felt frustrated by this since we knew
from other evidence—much of it presented in this chapter—that considerable adaptation and innovation was going on within the citemene system and that, in the
context of men's involvement in labor migration and other activities in the province, this was likely to be in the hands of women who would be cultivating village
gardens and second-, third-, and fourth-year icifwani (ash gardens). In retrospect,
however, we realized that characterizing men as tree cutters, while unlikely to be
an accurate description of their total contribution to agriculture, was probably an
extremely succinct way of representing a key rhetorical strategy designed to retain
men's contribution to household production and reproduction under the changing
circumstances brought by labor migration and integration into the money economy. Chiefs were also keen that man should cut trees, and they had their own reasons for tying citemene to Bemba custom and identity. These reasons had to do
with questions of ruling in a situation where political power was based on the ability to command labor and food (see Chapters 1 and 3). This view of chiefly power
underwent some modification, of course, in the 1940s and 1950s, and as they were
increasingly incorporated into the colonial state through the setting up of Native
Authorities and the African Provincial Councils, chiefs drew their economic and
political power from quite different sources. In an area where the marriage system
had included brideservice performed by a young man for his prospective in-laws,
older people may also have had good reasons for insisting on the traditional obligation of younger men to cut trees and clear land (see Chapter 6). Much of this
argument must remain speculative, since it rests less on direct evidence than on a
reinterpretation and re-reading of sources which are largely silent on gender, but it
does draw attention to the fact that change at the level of the region or community
was built on change at the level of the household and the co-resident group; that
changes in the system of production rested on changes in the system of social
reproduction.

There were Bemba communities, then, that resisted some kinds of change and
whose agricultural practices continued to cause concern for colonial officials in the
late 1940s and 1950s. Although, as we have suggested, the ethnic explanation of
variation should not be dismissed too lightly, we must nevertheless point out that
more often than not it was no explanation at all. As we have seen, when
Hainsworth encountered what appeared to be industrious Bemba-speaking agriculturalists in northern Munkonge, he explained them away by suggesting that
they were either really Lungu, or were so heavily influenced by the Lungu that
they hardly qualified as real Bemba at all. Elsewhere in Kasama district in the late
1940s there were plenty of examples of successful Bemba agricultural communities
to place alongside the less encouraging ones. What is interesting about these examples is that they are all men. Women who were innovators were not mentioned
in colonial reports, since progressive farmers were, by definition, male (see Chapter 5). In Chief Mwamba's area, for example, there was little evidence of the resistance to agricultural adaptation that seemed to characterize parts of neighboring
Chitimukulu. In 1949, the touring officer found Mwamba's area well supplied with
food, both millet and cassava (SEC 2/790 Kasama: TR Oct./Nov. 1949). In 1947, during a particularly bad hungry season, people from Chitimukulu's area had crossed

the Lukupa river into Mwamba's to exchange their labor for cassava, and it seemed likely that this would be repeated in 1949, despite the heavy fines that had been inflicted there for noncompliance with the cassava rules (SEC 2/790 Kasama: TR July/Aug. 1949). Chief Mwamba himself, like his fellow Bemba chief Munkonge, was a keen agriculturalist and had recently completed the building of an irrigation furrow and a dam, with a view to growing vegetables all year round. Even in parts of Chitimukulu's area, officers came across individuals and communities whose enterprise in agriculture was impressive. In parts of Chitimukulu close to Kasama, for example, the District Commissioner in 1948 noted a number of villages that were "promising indeed." Utilizing water from the Chibile and Milima streams, villagers had built furrows and dams and were growing irrigated fruits and vegetables for sale in the town. At Kasepa, a small tributary had been dammed and farmers were growing cassava, early maize, vegetables, fruit, wheat, and coffee (SEC 2/789 Kasama: TR Oct./Nov. 1948). Farther afield, in the southeastern part of Chitimukulu's area, villagers complained to the touring District Commissioner of the difficulties they faced in growing sufficient food. But, he argued, pity was completely misplaced: "One has only to see an individual who really works or a village which is forced to bestir itself, to appreciate how much can be done" (SEC 2/789 Kasama: TR Oct./Nov. 1948). The individuals he had in mind were Pikiti and Leki Mulenga. Pikiti was an elderly man of Mpepela village, who made his living largely through fishing, but had also made himself a farm. He had a seven-room Kimberley brick house, a large Kimberley brick pig-sty and about six acres under cultivation. Mulenga was a headman and ex-regimental sargeant major with seven medals earned in two wars. He also had a large brick house, approached along an impressive tree-lined drive and surrounded by fruit and shade trees, the seeds of which he had brought from Nakuru in Kenya. He grew "acres of cassava," and to keep the pigs from eating the crop "had arranged a series of alarms which spread throughout his holding and which he could agitate with a rope whilst lying in his bed" (SEC 2/789 Kasama: TR Oct./Nov. 1948).

As these observations indicate, there was more happening to postwar agriculture in the Northern Province than could be explained by ethnicity. There was, in fact, increasing differentiation within, as well as between communities. Some of this differentiation was the result of the successful production of crops for sale, especially in areas close to a market. More often, however, economic differentiation had its roots in employment but was made manifest through an interest in farming, the planting of fruit trees, the building of brick houses, and the keeping of animals. This phenomenon is addressed in more detail in Chapters 5 and 8.

Chinsali district showed a similar degree of variation and differentiation within and between communities to that discerned for Mpika and Kasama. There were productive areas and less-productive areas; some places with poor soils, others hampered by their inaccessibility, and some that benefited from a varied ecology. Once again, the degree to which cassava had been incorporated into the cropping system varied, but overall it made a large contribution toward ensuring a minimum level of food security.

A cadet officer named Pinguet toured the northern part of Chief Nkula's area in Chinsali district in 1948 (see map 4.3). This was a Bemba area, part of which benefited from being free of tsetse. The reports of cadet officers were often extremely full, and Pinguet's report on agriculture in Nkula was no exception (SEC

2/753 Chinsali: TR June 1948). Pinguet was pleasantly surprised by the state of food production in Nkula North, his close inspectin of ifitemene, ifwani, and imputa gardens revealing "surprisingly few instances of poor agriculture." The soil was suitable for cassava production and large kalundwe gardens of the mputa type were found on the higher ground overlooking the rivers, the lower-lying land adjacent to the rivers being used by wealthier headmen for grazing their cattle. The citemene gardens, meanwhile were situated further away from the swamps and rivers.

Here, then, were agriculturalists exploiting a range of soils and sites. Pinguet inspected the gardens of each village, producing the following table as a description of what he had found. (See Table 4.1.)

Pinguet also made an estimate of the quantity and quality of food consumed in the villages he toured. People in the area bordering the Chilola River were apparently the best fed, enjoying good crops of both millet and cassava, as well as benefiting from a ready supply of fish (SEC 2/753 Chinsali: TR June 1948 Ann X). Having been impressed by the productivity of the agriculture, Pinguet was all the more surprised by what he saw as the conservatism of the African cultivator, which emerged in his discussions with local men. "The African," he wrote, "stubbornly supports his ifitemene gardens and the methods of cultivation connected with them. This is partly due to the much entertained misconception that male (millet) can only be successfully cultivated in ifitemene gardens. If the African can be persuaded to grow male—which is his main food—in furrowed gardens with compost, the practice of ifitemene cultivation may well decrease (SEC 2/753 Chinsali: TR June 1948 Ann III).

Pinguet had clearly not been around for long enough in the Northern Province to know that he was not the first to suggest that the African cultivator abandon citemene. By the late 1940s, however, the results of agricultural research were beginning to be more widely known and no one really believed that there was a more efficient way of growing millet in this area. This was gently pointed out by the Provincial Commissioner, Howe, in his comments on Pinguet's report. "It has" he wrote, "been conclusively proved at Lunzuwa [the agricultural research station], that the traditional citemene method of growing male is better suited to local conditions than other methods" [SEC 2/753: Comments by Provincial Commissioner on Annexure III].

Other parts of the Chinsali district, and even other parts of Nkula's area, were less fortunate. Chief Kabanda's area, for example, was described in 1949 as "poverty-stricken and backward" [SEC 2/754: Chinsali TR, Dec 1949: Ann 7]. The poverty of this area, it seemed to the touring officer, was due less to its "poor agricultural potential" and more to its "isolation and lack of a convenient market" for crops. It was simply too remote, neither government nor missions found it economical to open buying stations here, so there was little incentive to grow more than subsistence needs. In 1949, the people of Kabanda had cassava but little else. Very few had cultivated relish foods of any sort. In complete contrast were those parts of the district lying along the Great North Road where, as in the Mpika district, district officers found that local shortages were due to overselling.

As successive Provincial Commissioners noted, there seemed to be something wrong with a province in which some people grew surpluses that they couldn't sell, others oversold, and others grew less than the minimum required to see them

TABLE 4.1 Ifitemene and Ifwani, Chinsali District, 1948

Name of Village	No of Ifitemene	Remarks	No of Ifwani (old) Ifitemene	Remarks	General Remarks
1. Joseph Kolokondwe	19	Good amale Increase	7	G.nuts round beans, maize	Increase in crop production
2. Chipandula	16	Good amale	5	beans G.nuts	Good crops, Increase
3. Kaputula	10	V. Good amale	5	G. Nuts maize etc.	Good crops, Increase
4. Mulopa	7	V. Good amale	7	G. Nuts beans V. Good	No Increase crops excellent
5. Longe	17	V.Good amale	20	beans kaffircorn	Good, Drop
6. Kapeya	2	Fair amale	2	G. Nuts beans	Poor crops, Headman old & lazy. Village moving.
7. Chinungi	6	Fair amale	6	beans G. nuts Kaffircorn	Good irrigation for 'Imputa'
8. Mulombe	6	Fair amale	14	beans G. nuts etc.	'Kalundwe' in sandy soil far superior to' Ifitemene (amale) Cattle. Drop.
9. Malumbu	17	Good amale	17	beans, maize, round nuts	Satisfactory 7 head of cattle
10. Mwika	2	Good amale	8	Kaffircorn cow peas peas	Low number of Ifitemene 6 head of cattle. Drop.
11. Mwaba	17	Good amale	22	G.Nuts beans etc V. Good	Good. Drop.
12. Kantimbe	6	Good amale	7	beans, peas maize	Good, no increase in crops
13. Mulimatonge	2	Fair amale	3	beans etc.	Low numbers of Ifitemene.
14. Mumba	2	Poor amale	4	maize, beans, etc	Very poor Ifitemene.
15. Chibunde	—	—	7	round beans	No Ifitemene, Headman, Villagers will have no amale for rainy season.
16. Mutashya	5	Fair amale	2	G. nuts, beans	Very poor crops. Village poor & Headman absent. To be revisited by Councillors
17. Mutitima	6	Fair amale	4	G. Nuts kaffircorn	Poor agriculturalists in this village, Water supply good, Cattle. Increased crops.

18. Kampyongo	11	Good amale	12	Good	Good crops. Drop
19. Simon Ngandu	12	V. Good amale	11	Good	Promising increase
20. Katumba	15	V. Good amale	—	beans Kaffircorn G.Nuts	Satisfactory
21. Kalundi	2	Good amale	15	peas, beans etc.	Big drop in production
22. Kalanda	2	Fair amale	8	Fair	Imputa crops superior to Ifitemene. Drop.
23. Masongo	6	Good amale	5	Good	Good Increase
24. Kakotota	16	Good amale	35	plentiful crops	Very large gardens. Drop.

Source: SEC 2/753: Chinsali TR June 1948.

TABLE 4.2   Imputa Gardens, Chinsali District, 1948

Name of Village	No of Imuta	Remarks
1. Joseph Kolokondwe	17 new 13 old	Cassava. Good. Increase.
2. Chipandula	12 new 8 old	Cassava. V. Good Increase
3. Kapatula	6 new 4 old	Cassava. Potatoes. V. Good.
4. Mulopa	10 new 7 old	Cassava. V. Good.
5. Longe	13 new 5 old	Cassava. Good. Increase.
6. Kapeya	3 new 2 old	Cassava. Fair. Lazy Headman.
7. Chinungi	7 new 9 old	Cassava. Good. Good Irrigation.
8. Mulombe	22 new 20 old	Cassava. Good. Increase.
9. Malumbu	26 new 30 old	Cassava. Good. Cattle.
10. Mwika	28 new 30 old	Cassava. V Good. Drop.
11. Mwabe	17 new 11 old	Cassava. Good. Increase.
12. Kantimbe	8 new 8 old	Cassava. Good.
13. Mulimstonge	10 new 6 old	Cassava. Good. Increase.
14. Pumba	11 new 6 old	Cassava. Good. Increase.
15. Chibunde	9 new 5 old	Cassava. Good. Increase.
16. Mutashyo	2 new 0 new	Poor.
17. Mutitim	4 new 1 old	Fair. Increase.
18. Kampyongo	16 new 20 old	Good. Drop.
19. Simon Ngandu	8 new 3 old	Kalundwe excellent.
20. Katumba	9 new 10 old	Good.
21. Kalundi	14 new 15 old	Good.
22. Kalanda	14 new 14 old	Good.
23. Masongo	12 new 3 old	Good.
24. Kakokota	24 new 14 old	Good.

Source: SEC 2/753: Chinsali TR, June 1948

through the year. While few people actually starved (in part because cassava had largely solved the problem of reliable starch production and in part because the mechanisms of reciprocity that Richards described were still in place), many were poorly nourished, and some suffered severe seasonal shortages of relish crops. Yet others argued that there was a massive unmet demand for marketed foodstuffs within the province, which improved infrastructure and marketing could help to solve, others argued that there was just insufficient diversification or wealth in this area to generate any real "peasant revolution" or to keep men from migrating. Transforming this labor reserve into a peasant farming area would not happen through market forces alone, but would require heavy government subsidies. We are back to where we started this chapter—with cadet officer Loosemore's musings on the possibilities for economic development in Chief Munkonge's area.

One of the reasons we have such rich and localized data for food production from the mid to late 1940s was that this was an issue that went beyond the question of the welfare of African colonial subjects. During and after World War II, food supply was a hot political subject, both within the province and within the territory as a whole, and arguments about marketing and supply occupied a great deal of official attention. However, colonial officials were not the only ones to have views on marketing and price controls. During the 1940s and 1950s, Chiefs and repre-sentatives—such as members of African Welfare Associations—consistently argued in the Northern Province African Provincial Council that government pricing policies were acting as a disincentive to African farmers. Provincial Commissioners, conversely, tended to claim that food shortages were simultaneously caused by overselling and by backward agricultural practices. Reading the minutes of these meetings in the late 1980s, it was hard not to be impressed by the intellectual gymnastics that officials often had to perform to assert simultaneously contradictory positions. The fact of the matter was that very different situations pertained in different parts of the province, and thus single stranded solutions often appeared implausible. The African members of the council were unimpressed. At a meeting in 1944, Acting Chief Kopa of the Bisa Native Authority asserted simply "we grow the food ourselves and would like to sell it at our own prices." Chief Mwamba reported "producers say they are throwing away their food at present prices. As the price of goods in shops has increased the price for food should also be increased." The Provincial Commissioner replied to these points by defending low prices on behalf of the purchasers, seemingly unaware, as Chief Mwamba was not, that purchasers and producers were often one and the same (SEC 2/227 Vol. I: 19/5/44). At another meeting in April 1946, Chief Mwamba expressed his displeasure with the price differences set by the government between millet and groundnuts and beans. He argued that grain was the staple food, while beans and nuts were only relish crops: "people should be encouraged to grow more grain." The Provincial Commissioner managed to respond to this point by saying that the reason that the prices of beans and nuts had been raised was to encourage increases in production! (SEC 2/227 Vol I: 11–13/4/46).[6] By the 1950s, as we discuss in Chapter 5, the voices of colonial officials and European business had been joined more and more forcefully by those of African producers arguing for fair prices. In the following section we turn, therefore, from the examination of changing patterns of production, to the issues of marketing and distribution.

The Politics of Marketing: Contradictions in the Provincial Economy

The problems of food supply and distribution that were faced by the colonial administration in the Northern Province in the 1940s and early 1950s were not unlike those which, on a much smaller scale, Frank Melland had complained of in the first few years of the century. Melland, as we discussed in Chapter 1, was a young officer of the British South Africa Company, stationed at Mpika, and charged with raising taxes, recruiting labor for porterage and for public works, and maintaining law and order. As his diaries for the years 1901–1905 indicate, juggling these different demands was not an easy task. The more labor he recruited to carry loads and build roads for his administration, the more taxes he was able to raise. But at the same time, the more labor he recruited, the more food he needed to purchase in order to supply rations to these workers, and, arguably, the more labor he recruited the less food was produced locally. In part, the problem in these early days could be solved by raising some taxes in kind. But getting a balance was difficult, for Melland often required food and labor at the same time. His own need for labor, however, was variable. In some years he relied heavily on European trading companies operating in the area to provide employment, which in turn enabled him to raise taxes; in other years he found himself in competition with private enterprise for both labor and food.

In the post–World War II period the same issues, on a larger scale, arose in the Northern Province. What the dominant debate on the effects of labor migration and the inadequacies of citemene had largely obscured was the fact that the colonial government itself was both a major employer of labor within the province and a major purchaser of foodstuffs. In a sparsely populated area such as the Northern Province, the pursuance of any large-scale public works such as road and bridge-building represented a significant demand for both labor and food for rations. Although road-building was essential if the food surpluses of more remote areas were to be tapped (Chiundaponde's area, for example), in the shorter term the recruitment of labor and the provision of rations needed to carry out such public works could in themselves create food shortages. Added to this were the food demands of an increasing number of government- and mission-run schools and hospitals, the provisioning of which had always posed a problem. Large-scale private enterprise was almost totally absent in the Northern Province, but in the Chinsali district, Sir Stewart Gore-Browne's demands for food to feed his estate labor added significantly to the problem. Sir Stewart, a prominent member of the legislative council, raised the issue of provincial food supplies repeatedly in council, and in correspondence with the administration, fueling what had become, by the late 1940s, a heated political debate.

Though the issue of food supplies for the war was not explicitly raised during debates on the Northern Province, it is presumably no accident that the question of food marketing and distribution received so much attention in 1939. It was in 1939 that W. Vernon Brelsford, then District Commissioner for Chinsali, proposed a scheme for the creation of a Northern Province Marketing Board (SEC 2/264: 1939). Brelsford's scheme was based on an analysis of agricultural systems in the Northern Province that was not unlike that outlined earlier in this chapter. Recognizing the extent of diversity that existed, he argued that the real problem in the province was not underproduction so much as maldistribution; that "certain portions of the Province are rich in some commodities and that other districts are starved of these very things that they desire so much." When discussing maldistribution, however, he was referring not solely, nor even primarily, to the distribution of basic starch products such as cassava and millet, but to the additional components of the diet—salt, fish and meat—and to luxury items such as tobacco. Central to his plan was the assertion that there existed within the province, not only the desire for these products, but also the purchasing power sufficient to generate a significant intraprovince trade and, through this, provincial economic development. The results of a more "constant and regular" distribution would be several. First, he argued that nutritional standards within the province would improve; second, that more money would circulate within the province and that more employment would be created within the province. He also argued that fewer men would be likely to migrate to the Copperbelt, because "one of the reasons for the exodus is the fact that food is better on the Copperbelt and things like meat hunger can be satisfied." Finally, he argued that a more developed "trading sense" would be created in the province, giving rise to local industries (SEC 2/264:1939).

The scheme proposed by Brelsford involved the provincial administration buying commodities in bulk in one part of the province and then transporting them to areas where there was thought to exist a market. A "native employee" would, for example, be stationed in Luwingu, and then would buy quantities of dried fish there and arrange for its transportation to Kasama or Chinsali where another employee would sell it; another employee would be stationed at Isoka to buy dried meat and would send this to Mpika and Kasama, and so on. The headquarters of the Marketing Board would be at Kasama, where an official would have to work out a series of costs, including the purchase price, costs of transport to the depot, costs of transport to selling point, and the sale price of the commodity. Brelsford believed that all of the commodities he had named (salt, meat, dried fish, and tobacco) were resaleable at a considerably higher price in districts other than those in which they were produced. He also believed that if a reliable system of distribution existed, this would encourage increased production of the commodities, and hence an ultimate reduction in price. The question of whether this system would be profit-making, merely self-sustaining, or dependent on government subsidy could not be determined without "a preliminary survey of costs and transport." Brelsford was, however, generally optimistic that the scheme would pay, and where subsidies might be necessary, he felt that these would be justified by longer-term benefits:

> *Almost every station in the Province is now served by motor lorry and the Board could no doubt get a special rate for all its loads. Moreover, at the present time the native is willing to pay high prices for the few luxuries of salt, meat and tobacco which come his way. Systematically bought and sold and transported in bulk the present prices could probably be reduced. For example, Mpika salt brought by individual carrier to Chinsali is 3d per lb. Bought in bulk after being purchased at the present rate of 1d at Mpika it might be brought down to 2d and still leave 1/4d per lb margin towards the upkeep of the Marketing Board. . . . On the purely local purchase and storage of grain for local consumption a small increase in price for storage would be justifiable and help the Board pay its way when dealing with commodities in quantity and a Government subsidy might be needed for some years. But if we have at least the desire to develop the backward areas such subsidies are justified.*
>
> (SEC 2/264: 1939: W. V. Brelsford n.d.).

In Brelsford's view, the "evils of maldistribution" in the Northern Province would have to be systematically tackled by the government because "the long distances and uncertainty of sale militate against any enterprise by private individuals." Further, it was government's duty to intervene because "the constant and orderly distribution of essential commodities is one of the first steps in native development as no ill-nourished people can progress."

Brelsford's scheme generated a small amount of correspondence, and then sank almost without trace. The Provincial Commissioner, Cooke, having discussed the proposal with his District Commissioners, was skeptical. He felt that Brelsford had exaggerated the extent to which the problems of food supply in the Northern Province could be met by a more efficient distribution system, because there was relatively little specialization of production, and insufficient money in circulation to fuel such specialization. Even if people in parts of the province did want to consume more meat and fish, most could not afford to pay for these commodities. Fish, in particular, could be sold to Copperbelt traders for much higher prices than could ever be fetched within the province:

> The scheme is an excellent one and if certain parts of the province were manufacturing and others producing success would be assured without a doubt. As it is it is merely trying to foster barter of similar commodities . . . In view of the fact that almost every district of the province is a producing area I fail to see how an internal distributing organization can serve any useful purpose and cannot therefore recommend it to Government. If there were any big consuming centres here, as Broken Hill, Ndola etc, it would have my immediate support but the present scheme, analysed, will not bear close examination.
> (SEC 2/264:E.Cooke, Provincial Commissioner to chief secretary, 24.2.40).

Despite the diversity of production systems that had emerged by the 1940s in the Northern Province, and despite the readiness of many communities to respond to marketing opportunities for their produce, Cooke's assessment was that there were severe limits to the possibilities of provincial-level economic growth based on agricultural production. There was simply insufficient money in circulation in the province, despite the high levels of labor migration, and although, as Brelsford had argued, there was undoubtedly a "protein gap" in many peoples' diets, few local people could match the prices that could be obtained for such products on the Copperbelt. Labor migration, it seemed, had not fueled any local economic revolution in the Northern Province, partly because levels of remittance were low (see Chapter 6).

The issues of marketing and distribution did not go away, however, despite the rejection of Brelsford's plan. It was Sir Stewart Gore-Browne who resolutely kept them on the agenda, largely out of self-interest. Sir Stewart, as we have noted, was the only substantial employer of labor in the Northern Province other than the government itself, and he faced the perennial problem of securing enough food to feed his estate workers. He addressed this problem in part by encouraging the growth of a group of progressive farmers close to his Shiwa Ng'andu estate (a development that we will survey in Chapter 5). But, despite all efforts at self-sufficiency, Gore-Browne frequently found it necessary to buy in food, not only from other parts of Chinsali district, but also from other districts of the Northern Province. It was over this issue of the purchase of foodstuffs from other districts that Gore-Browne most often found himself in conflict with the administration, a conflict which illuminated many of the problems of agricultural production and marketing in this area.

In the Legislative Council (of which he was a member representing African interests) of September 1947, Gore-Browne asked for an undertaking that food grown in the Northern Province would stay within the province and would not be "diverted for the benefit of the Copper Mining Companies" until such time as local needs had been satisfied (SEC 1/104: Minutes of the Legislative Council meeting, 24.9.47). The secretary of Native Affairs responded that "normally the Province is more than self-sufficient in food," a statement that Gore-Browne disputed, citing

Audrey Richards' work as evidence of chronically poor nutrition, and the annual closure of schools in Chinsali district due to the nonavailability of food. Chinsali district, he argued, had a small population and very little good soil. The demand for food from hospitals, educational establishments, and his own estate was much greater than could be supplied from within the district: "In no single year is enough food obtained from local production. The mission closes year after year because they cannot get enough food. We do half the work we want to do because we cannot get enough food."

Gore-Browne wanted a commitment from the government that any surplus foodstuffs would be offered for sale within the province before they were allowed beyond its borders. This was an undertaking that the government was reluctant to give, for the simple reason that it was often difficult to move food from one part of the province to another (a fact that lent support to Brelsford's earlier argument). For instance, in 1946 there had been a surplus of 4000 bags of cassava from the Bangweulu area that had been sold to the Copperbelt, to which it could be relatively easily transported [Sec/104: Minutes of the Legislative Council, 24.9.47]. Eventually Gore-Browne obtained a carefully worded undertaking from the government. Henceforth, and in the event of food shortage in any part of the province, government undertook to "take steps to ensure that where surpluses of food in the Northern Province are so situated that their transport to other parts of the Province is feasible—having regard for practical and economic considerations—the internal food requirements of the Province will receive priority before any such surpluses are allowed to be exported from the Province." He added, however, that "Limitations of transport and expense . . . and the size of the Northern Province do not permit of its food requirements being regarded or arranged on a strictly regional or provincial basis." (SEC 1/104: Letter from E. C. Greenall (Secretary for Native Affairs) to Provincial Commissioner of Kasama, 30.9.47).

Provincial Commissioner Crawford, was irritated by Gore-Browne's demands and felt that the government should never have given such an undertaking. He was at the time in dispute with Gore-Browne over a surplus of 18,000 lbs. of cassava in Luwingu district, which Gore-Browne wanted to be taken by barge to the Chambezi crossing, from whence he would collect it. Crawford was reluctant to hold up the export of cassava from Luwingu district throughout the year on the "off-chance that the Shiwa estate might find itself short of food in the following February" (SEC 1/104: Letter from B. E. Crawford, Provincial Commissioner Northern Province, to E. C. Greenall, secretary for native affairs, 8.0.47). It was Crawford's view that the provincial boundaries were "purely artificial conceptions," and that "food sent from Bangweulu to Shiwa is as much exported as food from Bangweulu to Luanshya" (SEC 1/104: B. E. Crawford, Provincial Commissioner to F. Crawford, Economic Secretary, 5.11.47).

The failure of the rains over large parts of the province in the 1948–1949 season again brought the question of food production and distribution to the fore. Not only had the crop failed in many places but, it appeared, those African farmers who might have been able to produce a surplus sufficient to see the province through a bad year, had been discouraged by the low prices set by the Native Authorities in consultation with the Provincial Development Team. Once again, the argument that the problems of food supply in the province were caused as much by underproduction as by maldistribution was voiced. Pricing policy in Northern

Rhodesia in this period was of course largely dictated by the political imperative of susidizing the white farming sector (Vickery 1985; Palmer 1983). Remote from the white farming areas of the south, potential surplus producers in the Northern Province were nevertheless affected by this policy, which sought to discourage, on environmentalist grounds, surplus production by "unimproved" farmers. The environmentalist arguments that had been so prominent in the early colonial period, but which had been largely forgotten in the drive for food security of the 1930s and 1940s, surfaced once more with a vengeance in the postwar years as part of the policy of creating a class of African "improved farmers" (a policy that will be discussed at greater length in Chapter 5). These environmentalist arguments bore only a tangential relationship to the findings of the colonial ecologists whose work we described in Chapter 2, and from which they nevertheless drew much of their authority. Three "classes" of African producer were designated, and prices for their produce determined. The "improved" farmer was to receive "the full market price that they could obtain anywhere in Northern Rhodesia, allowing for transport costs." The "traditional farmer" would receive only part of the full market price, the remainder being paid into a farmer's improvement fund, as incentive to become an improved farmer. The third class of farmer, the "deleterious farmer," was to be discouraged from active production, "and with this class of producer often even an improved farmers' fund might offer an unwise incentive" (SEC 5/182: Extract from Administrative Conference, 1950: Control of prices for African produce).

If and how this schema was put into effect in the Northern Province is not clear. Most producers in the area would probably have merited the term "deleterious farmer," but the provincial administration could hardly afford to discourage the production of surplus by any class of farmer. In 1950, however, it did appear that low prices had compounded the effects of a serious drought to produce a crisis of food supply (SEC 5/182: Letter from Acting Provincial Commissioner of Kasama (M. A. Billing) to the secretary for Native Affairs, 16.10.50). Whether low prices had brought about a decrease in production, or had instead contributed to the development of a black market, is not altogether clear. Some officials implied that the latter was the case, and that the real crisis of food supply was actually a crisis of supply for government. Just as in the past, government officials once more found themselves with a larger population of laborers, schoolchildren, and hospital patients than they could feed from the resources available to them within the province. Villagers who had been affected by the drought mostly survived by selling their labor for food in more fortunate areas (SEC 5/182: Letter from G. F. Tredwell, District Commissioner of Mpika to the Provincial Commissioner of Kasama, 4.10.50). Hunger did not become famine for most people—the real problem for the government was not the feeding of ordinary villagers, but the supply of food to institutions and towns, a more severe version of the recurring administrative headache. By October 1950, the situation at Mpika boma was said to be "extremely serious" because there were insufficient stocks to feed government employees and schoolchildren (SEC 5/182: Letter from G. F. Tredwell, District Commissioner of Mpika to the Provincial Commissioner of Kasama, 4.10.50). A similar situation existed at Chinsali boma, where both cereal and relish crops were in short supply. Here the District Commissioner produced a table of the institutional food requirements of the district, which had to be met from local sources. He estimated total requirements for 1950 to be 536,000 lbs. of grain and 85,000 lbs. of relish, of which he had only managed to purchase 310,000 lbs. of grain and 32,000 lbs. of relish [SEC 5/182: Letter from J. E. Madocks, District Commissioner of Chinsali to the Acting Provincial Commissioner of Kasama, 9.10.50). These figures seemed to reinforce Gore-Browne's contention that, even in normal years, it was difficult to feed the district's employees and institutions from local food supplies. So anxious was the administration to rid itself of the onerous and hazardous task of the purchase and distribution of food surpluses that, in 1951, the Provincial Commissioner approved a £2000 interest-free loan to a private company, the Provincial Development Company (North) Ltd. to enable it to extend its food purchasing and transport operations on behalf of the government (SEC 5/183: Letter from G. Howe, Provincial Commissioner of Kasama to the Accountant General, 6.9.51).

In the postwar period then, the issue of food security was largely addressed by colonial officials as a provincial-level problem. In most years, and taken as a whole, the province managed to be self-sufficient in food, if by this we mean that there were no shortages that threatened to become famines. Conversely, there were many years in which one part or another of the province suffered shortage that went beyond the hungry seasons that Richards had described as part of the normal Bemba agricultural cycle, and which could not be met by the household-level strategies that she had described in her nutritional study. Usually these shortages could be met through migration, and by expert exploitation of wild foods. Nevertheless, the issue of food security continued to worry colonial officials. As we have seen, there were broadly two different theories around to account for this problem. The first, and oldest, held that overall production of food within the province was deficient, and that this was caused by a combination of poor agricultural management (seen as synonymous with citemene), the absence of many men, and the lack of commitment to agriculture by those who were present. Proponents of the second theory placed more emphasis on the problems of exchange and distribution within a sparsely populated province, arguing that if a more efficient (and government-subsidized) marketing system could be instituted, then not only would the problem of shortages disappear, but the provincial economy would be energized, allowing for greater specialization, and eventually leading to the diminution of labor migration.

Largely missing from either of these theories, but underlying the concern that produced them, was the fact that the government itself was a major employer within the province, and thus a major consumer of food for rations. While most communities could feed themselves, albeit precariously, they could not always provide the surpluses required to feed government labor. The colonial administration had disrupted, but not transformed, the local agrarian economy.[7] Some areas, as we have seen, did become regular food-surplus areas, but these were few and tended to lie on the margins of the plateau. In order to gain access to these potential food surpluses and to encourage their production, the government had to invest in road building, itself a major disruption to local economies as it temporarily removed large amounts of male and female labor. Other areas rarely produced a surplus at all, and sometimes fell short of self-sufficiency. Partial integration into the cash economy had certainly rendered some communities vulnerable because, even if they had cash from wage labor, this could not always be exchanged easily for food. More generally though, the critics of Brelsford's marketing scheme were probably right in thinking that there was simply not enough money around to generate the kind of regional exchange that he envisioned. Labor migration (discussed in Chapter 6) had not produced any large surpluses of cash invested in agriculture, for a number of reasons. Foremost among these was the fact that, despite the annual demand for food surpluses, making any significant profit out of agricultural production remained, and remains, very difficult in this area.

When investigating the degree to which agricultural production and the citemene system had changed in the 1940s, questions of ecology have to be put side-by-side with those of politics and commoditization. The rapid spread of cassava cultivation, both on village gardens and citemene fields, although uneven in its effects, was the major change to take place in this period. Variations in levels of production both of cassava and of other staples, were linked in many cases to local differences in soil type and woodland cover, but they were also influenced by infrastructure and marketing opportunities. Residents of villages who could make a living through fishing were reluctant to engage in agriculture and certainly saw no need to increase acreages under cultivation. For the women of these villages, whose access to the cash income generated by fish trading was dependent on their relations with their menfolk, the situation was not so clear-cut. Even within the same community, strategies were contested. Official discourse insisted that all should engage in agricultural production, and those who did not, whether labor migrants or traders, found themselves labeled. They were sometimes represented as lazy, but more often than not they were represented as being too interested in the cash economy and seeking to shirk their responsibilities. Underlying this oft repeated theme was the notion that social relations and social identities were breaking down: people were giving up their traditional way of life, symbolized by their lack of agricultural involvement. Many local communities resisted government attempts to control both household reproduction and social reproduction through interventions in agriculture and the manipulation of social identities. This resistance was not necessarily consciously planned or perceived in the terms in which we describe it, but what is important is that by the 1940s, local communities had to participate in the cash economy in order to survive. The colonial state, having created the conditions for this participation, then found itself with the problem of people who no longer wished to be farmers and had their own ideas about their identities and their futures.

This situation was made more complex by the fact that, in other parts of the province, certain individuals and groups were reasserting a Bemba identity based on citemene and the cultivation of millet. In some circumstances, local ecological conditions and infrastructural provision meant that there was little choice about such an assertion. The communities concerned used a particular discourse on Bemba identity to try to resist the changes that were making household and community reproduction ever more precarious as they became increasingly dependent on the cash economy. These cultivators were dubbed "backward" by colonial officials and were accused of clinging too tightly to a Bemba identity that was based on agricultural incompetence and a warrior past and, as such, they were compared unfavorably with their neighbors who were thought to have positive agricultural traditions. No one seems, at the time, to have mentioned the obvious contradiction within colonial discourse on this point, where one set of people who wished to relinquish agriculture were represented as abandoning their identity, and another set who wished to grow millet were seen as being too tightly bound to it. The evident ambiguities and contradictions in government policy and colonial discourse in this period were the result of the fact that colonial officials wanted local people to abandon citemene, but take up farming. Local people had their own strategies and most often these seem to make sense in terms of maintaining flexibility, as well as the ability to respond to changing circumstances. When colonial officials complained that communities along the Great North road were both overintegrated and insufficiently integrated into the cash economy, they came close to realizing the degree to which multiple strategies were the key to survival. Citemene production continued throughout this period, although modified in various ways in different times and places, most notably by the introduction of cassava. Citemene remained a viable option because, as a system of production, it contains multiple strategies that allow cropping patterns and field rotation systems to change with labor availability.

In the next chapter we discuss the efforts of the late colonial government to encourage certain individuals to abandon citemene in favor of a "modern" agriculture. Their aim was to create a prosperous peasantry through the development of "settlement schemes." This policy was a highly gendered one, being aimed exclusively at men, and it brings us back once again to the enduring theme of the politics of residence, a politics which became as explosive in the late 1950s as it had been in the days of the British South Africa Company.

5

Developing Men: The Creation of the Progressive Farmer

In the period after World War II, colonial policy in the Northern Province, as elsewhere, underwent a number of changes. These changes were influenced by a number of factors, among them the rise of the nationalist movement that led some officials to suggest development as a means for preventing possible political unrest in the rural areas. A great deal of concern was expressed about the lines of communication between urban and rural areas, and local officials within the province worried about the pernicious effects of urban agitators who, it was claimed, came back to the rural areas to stir up trouble. A focus on development in the Northern Province inevitably meant a focus on agriculture and on citemene. The changes that had taken place in the citemene systems in the 1940s and 1950s, and which we described in the last chapter, were connected to the increasing integration of the province into the cash economy. Labor migration drew male labor away from agriculture, while the market demand for foodstuffs encouraged production for sale, thus depressing foodstocks in some households and changing cropping and labor use patterns in others. The scale of marketed production was never enough, however, to generate adequate growth and expansion in agriculture. The result was that the province remained economically underdeveloped. In this context, citemene continued to be an integral part of agricultural strategies in the province. We have no data for the 1940s and 1950s on citemene acreages compared to size of village gardens, thus we cannot say whether the average size of citemene fields was decreasing or not. But, we do know that the "traditional" citemene areas continued to be millet dependent in this period, and combined the production of staple with that of relish crops on the main citemene fields, as well as growing maize, beans, pumpkins, and other relish crops on village gardens. It was in the more geographically peripheral areas of the province that villagers began to grow huge cassava fields. These fields were, however, with the exception of some of the swamp areas, integrated into citemene systems, and their purpose frequently was to produce cassava meal for the market. The dominant discourse on tree-cutting, subscribed to by both locals and outsiders, obscured the fact that the citemene systems had always involved permanent cultivation, and the introduction of cassava fields, although being acknowledged as a change, was not in any way conceived of as a permanent alternative to citemene. Cassava was part of an adaptation within the existing citemene systems, and it was perceived as such. The semipermanent character of the cassava fields did not prevent farmers from engaging in citemene, nor did it appreciably increase settlement stability or change settlement patterns. The colonial anxiety about the shifting nature of local populations associated with citemene tended to neglect the fact that villages were only really forced to move every ten to fifteen years for environmental reasons, although they might move more often for political reasons.[1] The idea that citemene necessarily involved constant movement was one that the people of the province had visited on officials, experts, and missionaries. This discursive strategy was perhaps an obvious one because, by the turn of the century, it was already part of an ongoing dialogue between chiefs and commoners that was really about rights over people and food (see Chapter 1).

The debate that took place between chiefs and colonial officials in the 1940s and 1950s was similar because it was about who should control land, as well as people. From the 1940s onward (see later), the government permitted the setting up of individual settlements, subject to regulations about housing and farming practices, within what were called parishes. This changed the previous system in which people had had to reside within a registered village (Tweedie 1966). The goal of this policy was to encourage the emergence of modern farmers who would occupy a certain area of land permanently and build a permanent, modern house on that land. This policy, which would amalgamate housing, sanitary, residential, and agricultural goals in one package, was to be combined with a system of peasant farming blocks, in which individual farmers would use improved methods, but would work together in cooperative units. The two schemes revealed, once again, the contradictions inherent in government policy. On one hand, individual farmers were to be encouraged, but on the other, local people would not be allowed to scatter across the landscape, signaling a breakdown in settlement patterns and thus in social relations. These points will be discussed further. Most of the chiefs in the province were against both schemes, despite the fact that some of them had invested heavily in modern farming methods (see Chapter 4 and following text). Individual settlements were not favored by chiefs because they gave control of the land to the farmer and thus removed the individual from the powers and dominion of the chief.

At a meeting of the Northern Province Provincial Council in May 1944, the Provincial Commissioner asked if the Native Authorities would be prepared to allocate sufficient land to those who wished to build permanent houses and to farm. He mentioned, in particular, soldiers (askari) who would be returning from the war. In reply, Chief Nkula simply asked "How could they rule themselves?," and Chief Nkolemfumu said ". . . we are in favour of permanent villages for everyone, not only for ex-askari. Many of them will not put up good houses, and we don't agree to setting aside land for the sole use of any one village" (SEC 2/227 Vol. I: 19/5/44). The emergence of a class of men who would not be under chiefly control was not favored by the chiefs. The relationship between their dominion over land

and their chiefly powers, which included rights over people and food, was clear to them, and they did not want it encroached upon. They were against peasant schemes for much the same reason because they did not want to cede control over land permanently. This gave rise in the course of the 1950s to a series of debates about the proper use of land on peasant farming blocks and about inheritance (see later). However, much of the anxiety on the part of chiefs and officials about residence, land, and political control in this period was directed toward individual improved farmers living on individual farms. Both parties saw these men as members of an incipient middle-class who would be unwilling to submit themselves to chiefly control and who might develop into a group who could challenge the colonial government's legitimacy. In this, they were proved partly correct because, as it turned out, improved farming often had little to do with farming per se, but was part of a wider political strategy.

In 1950, the Commissioner for Native Development in Lusaka received a letter from Mr. Kenneth Kaunda, an ex-schoolteacher who wrote in his capacity as secretary and treasurer of the Chinsali Youngmen's Farming Association (SEC 1/428: Letter from Kenneth Kaunda to Commissioner for Native Development, 28.3.50). Two months later he was to receive an almost identical letter from Mr. J. R. Mboo, manager and secretary of the Kamanga Farmers' Association in Chief Chibesakunda's area (SEC 1/428). Both letters began with this opening sentence: "Dear Sir, I have read and re-read the article or Talk that you broadcast on 15th March 1949 on development, and I have at last decided to write to you for wider advice, Sir." They then went on to describe the composition and activities of their respective organizations, enclosed copies of their constitutions, and asked for further advice: "On reading our constitution, sir, do you think there is anything which comes under the Native Development that would help further our work in this District? If there are any useful bulletins or the like, I would like to buy some for the Body, Sir. I believe according to your talk our aims in our constitution that we will work along smoothly Sir. Your advice will be highly appreciated. Wishing you best of luck in your work. . . ."

Kaunda's letter received a cautious reply from the Commissioner for Native Development who wrote: "If you are making a real effort to improve the agriculture of your area and to farm your land properly, I shall make every effort to assist you" (SEC 1/428: Letter from Commissioner for Native Development to Kenneth Kaunda, 22.4.50).

Meanwhile, the Commissioner for Native Development had written to the District Commissioner for Chinsali, R. R. Stokes, who had provided him with more information on the Chinsali Youngmen's Farming Association. This had been formed, wrote Stokes, by Kenneth Kaunda and his brother Morrison, both sons of the late David Kaunda, an African missionary of the Church of Scotland mission at Lubwa. They had been joined by a group of ex-Lubwa employees and by one or two Lubwa clerks, the latter rendering financial help. They had applied to, and received from Chief Nkula a piece of land in the fertile chipya soil along the Luvu River, about one mile from Lubwa mission. They had also received the advice and approval of the Agriculture Department and had began cultivating about ten acres of beans and groundnuts, not entirely successfully. The District Commissioner pointed out that Kaunda, the "leading light of the venture," had been well-known at Mufulira as an urban councilor and was "very active in political life, and has formed a branch of the African Congress at Lubwa." The political anxieties of co-lonial officials are evident in this correspondence.

There could be nothing to object to, however, in the draft constitutions of ei-ther the Chinsali Youngmen's Farming Association, nor in that of the Kamanga Farmers' Association. Both aimed to "set a practical example to our fellow Afri-cans . . . by applying modern methods of farming and thereby helping to uplift the area in farming," and to "work in collaboration with Development Officers, fol-lowing their methods of doing things and doing all in our power to enlighten their work in our district by advising them as to how Africans should be approached." Both ruled that "All Government regulations should be followed" and that "should there by any misunderstanding following the Government regulations, a Committee will meet to try to settle the trouble or misunderstanding with any 'Body' concerned" (SEC 1/428: Chinsali Youngmen's Farming Association: Draft Constitution, 13.6.49).

These were men who, in the 1950s, engaged in, and contributed to, an emerg-ing discourse of development and progress. At times, they appeared to be doing the work of colonial officials for them—not because they were "stooges" of any sort, but because they genuinely identified with, and saw their interests lying within, the colonial government's postwar developmental strategy. But, the way in which they saw that strategy working was quite different from the way it was per-ceived by colonial officials and chiefs. However, they organized meetings with the Native Authorities and with villagers to persuade them of the advantages of living in larger villages and of abandoning citemene cultivation. In March 1950, for ex-ample, John Mboo, the instigator of the Kamango Farmers' Association, had chaired a meeting at Chief Chibesakunda's headquarters for all headmen, Native Authority staff, and schoolteachers. The meeting opened with the chief asking his people "if they really wanted to develop, and they all replied in a chorus that they want to." Then the agenda was followed, beginning with (1) "Living in Big Com-munities," through (2) "Build Big Houses and Clean Villages," to (3) "Learn New Ways of Agriculture." What is interesting about the audience at this meeting is that it consisted entirely of men who were either educated or in positions of authority. There were no women involved. A further point is that the interests of educated men like Kenneth Kaunda and John Mboo were far from identical with those of local chiefs and headmen, although the latter were sometimes educated men and vice versa.

Local headmen most often encouraged participation in government develop-ment policies because they saw advantages in terms of access to government re-sources, which would help them to gain prestige and build up a large following within their villages. Most headmen feared the undermining of their positions through the absence of young men and the break up of villages. The educated men tended to be more politically active, and their interest was in challenging the gov-ernment on its own terms, and in using whatever resources were available to im-prove local conditions. These men had an investment in modernity and in the benefits of modern education and modern agriculture. However, ten years later, many of the most enthusiastic advocates of "peasant farming" were involved in political campaigns that actively rejected those interventions of the late colonial state that had been made in the names of "Development" and "Progress." Farming schemes were subverted, machinery destroyed, hygiene rules were broken, and mitanda took on a new meaning as individuals previously labeled "progressive" took to living in shelters in the bush or in the swamps of the Chambesi, leaving their neatly laid-out peasant farms and Kimberley brick houses in the care of their wives. For the colonial authorities, citemene gardens and mitanda once again came to represent the essential ungovernability of this area.

In this chapter we tell the story of those ten years. This includes an examination of the contradictions inherent in colonial attempts to create a politically docile peasantry in this area and the failure of this project. Central to this story are several long-standing themes in the history of this region—the relationship between agricultural production and Bemba ethnic identity, the question of residence patterns and political control, and the economics of production. There are many continuities, we shall argue, between what happened in this decade and the situation in the Northern Province in the 1980s and 1990s. What is most interesting, however, is the way in which citemene remains a key issue throughout this period, not only because of its continuing role in a repertoire of economic strategies, but also as a metaphor for an ongoing dialogue between colonial officials and local people, chiefs and commoners, and women and men.

Colonial Development and the Aspirant Farmer

As we indicated in Chapter 4, the problems that colonial officials encountered in this area in the postwar period were essentially similar to those they had encountered in early times. In particular, they continued to worry about the question of food security in the Northern Province and they continued to relate this directly to the phenomenon of male labor migration. Following from this was the problem of the regional economy, or rather the lack of it. How could one develop agriculture within the province to a sufficient extent that it would generate the kind of wealth that local people either earned, or believed it possible to earn, on the Copperbelt? Because economic development was seen to depend entirely on male labor and male commitment, the question of how to create a buoyant regional economy was easily reduced to the question of how to keep men in the area.

Although none of these questions were new, the manner in which they were addressed in the postwar period was. The postwar colonial state here, as elsewhere in British-ruled Africa, was more interventionist, more technocratic, and more "bullish" than it was in the interwar years.[2] The key to this new mood was money. The funds granted in the 1950s under the Colonial Development and Welfare Act, although not vast, allowed for some badly-needed expenditure on infrastructure in the Northern Province and for the possibility of realizing the plans of some of the more visionary colonial officials of the time.[3] In the late 1950s, funding for what had become known as "Development" in the area was significantly increased by a grant of £2 million from the mining industry, which was specifically earmarked for projects in the Northern Province. With money came a new optimism among colonial officials that this underdeveloped region could be transformed into a thriving peasant economy, its men transformed from labor migrants to heads of rural families, from disaffected miners to prosperous and politically conservative farmers. This new colonial project of social and economic transformation sat sometimes uneasily alongside the policies of Native Authority administration, which the government continued to insist on as the basis of local

'The family of an employed man' Kasama, early 1930s
Photo: Audrey Richards, 1930s

government and participation. Many of the long-standing contradictions of Indirect Rule policy surfaced clearly during this period. The agents of the new colonial "development state" had actively fostered social and economic aspirations, and had inadvertently introduced the notion of accountability in the process. When development failed to produce results, the colonial state was left more vulnerable than it had been in earlier periods, when simple impoverishment had limited its scope of action.

In the 1930s and early 1940s, District Officers in the Chinsali, Mpika, and Kasama districts had noted the existence of the occasional "progressive farmer" (see Chapter 4). These included chiefs such as Munkonge, and pensioners such as Mr. Pikiti. In the case of members of the chiefly elite, colonial officials were keen to support these activities and often provided help with marketing and infrastructure, effectively subsidizing what were usually market-gardening enterprises. This was all part of a strategy of bolstering the status of Native Authorities, and any sign of interest in progressive farming on the part of chiefs was seized upon eagerly by colonial officials. The colonial notion of a progressive farmer was deeply ideological. In this area in particular, it implied a rejection of "backward" citemene methods of cultivation and a commitment to full-time, settled agriculture. It carried with it assumptions about lifestyle that went beyond the question of agricultural methods. A progressive farmer would be a man who wished to separate himself from the network of kin relations that dominated most peoples' lives; he would build himself a "decent" brick-built house, and he would be modern without being too urbanized; he would educate his children, and his wife would be keen to learn the rudiments of "domestic science." Progressive farmers themselves, however, often had other agendas.

For the most part, progressive farming in the Northern Province was a form of conspicuous consumption, rather than a source of accumulation. Multiple strategies of household reproduction remained as essential in the 1950s as they had been in earlier decades, and very few people could ever afford to rely on agricultural production alone. Individuals who had been particularly successful in their migrant careers, and especially those skilled workers who could find some occasional employment in their local areas, demonstrated their success by building impressive brick houses, planting fruit trees, keeping some livestock, and growing a wider variety of foodstuffs than was normally the case in village gardens. Becoming a farmer was an indication of wealth and a sign of status, it was rarely in itself the source of such wealth. Nonagricultural sources of income remained essential to the maintenance of a progressive farmer lifestyle.

A number of factors contributed to a growth in peasant farming after the war. There had been a slow realization on the part of the colonial authorities here, as elsewhere in Africa, that reforming the agricultural practices of the masses (which in this case meant persuading them to abandon citemene) was a near-hopeless task. It now seemed that the best hope for the Northern Province was to concentrate attention and resources on progressive (male) individuals, whose successful farming enterprises might eventually create local employment and fuel the development of a regional economy. This new shift in policy was not without its contradictions, as we shall see, for the profound colonial suspicion of individualism among Africans remained, leaving officials with the problem of how to encourage progressivism at the same time as insisting on "collectivism."

Closely related to the shift in thinking described above, the postwar period saw the introduction of the Parish System, which legalized, under certain conditions, the existence of individual settlements outside village groupings and which eliminated the "ten taxpayer rule" under which it had been illegal to reside in a settlement of less than ten taxpayers. The introduction of the Parish System was, to some extent, a recognition of the failure to eliminate the building of mitanda outside of formal village settlements, and a recognition of the enduring fragmentation of settlement patterns and the unwillingness of young men to submit to the authority of headmen.[4]

The postwar period also saw the return of demobilized soldiers to the area, with pensions and raised expectations. As we have already indicated, however, by far the most important factor at work was the availability of colonial funds that provided for the first time for the possibility of colonial intervention on a scale that might arguably make a difference to the development of the rural economy.

In 1943, a series of discussions took place at the district and provincial level directed towards the formulation of a development plan for the Northern Province.[5] Central to the visions of the officers who contributed to this plan was the idea of the creation, or re-creation, of rural society in the area through the introduction of Peasant Farming schemes.[6] These schemes would enable resources to be directed toward Africans who had demonstrated a commitment to sustainable and progressive farming methods. However, these resources would only be directed to progressive peasant farmers as long as they formed and continued to belong to cooperative groups (cooperation being more or less loosely defined). In this way, the more productive members of society could be encouraged to produce even more, while the supposed dangers of atomization and individualism would be avoided. Although increasing economic and social differentiation would take place, it would do so within the wider framework of the tribal collectivity and without bypassing the structures of Indirect Rule. Whereas labor migration was thought to have fragmented Bemba society, this form of economic development was envisaged as cementing it.

Partly as a result of their own propaganda, colonial officials were soon approached by aspiring progressive farmers for financial and technical help, even before the Peasant Farming schemes had been formally established. In July 1948, for example, Matthew Chipondo from the village of Kaniki in the Mporokoso District wrote to the District Commissioner that he had heard that the government was introducing "schemes of Development, Peasant Farming, and Rural Development." He came straight to the point. "I am writing to you, and through you to the government for help . . . I would be grateful if government would lend me some cattle, make me a water-furrow, and buy me some orange trees" (SEC 1/428: Copy of letter from Matthew Chipondo to the District Commissioner of Mporokoso, 5.7.48).

Chipondo, who described himself as a "retired steward of the Seventh Day Adventist Church," and "Mabemba by tribe" had, in 1947, opened a village store and had made three gardens (two citemene gardens and one cassava and vegetable garden). His store had then been robbed of some goods and all of his cash, which amounted to £32.11.8 ½d. He was appealing to the District Commissioner to help him start over again. His goal was "to make a big farm, covering about 2 miles, for which I intend to plant vegetables, orchard, and African foodstuffs, from which I shall be able to supply both Mporokoso and Kasama populations, as this place is on the main Mporokoso to Kasama road, thus the transport will be convenient. I would also be keeping pigs, fowls, even cattle. All these could be easily done if I had a water-furrow to start with. . . ." Provincial Commissioner G. Howe gave Chipondo's appeal short shrift: "Government have no intention," he wrote to the District Commissioner for Mporokoso, "of assisting in the creation of a large landholding class of Africans." Individuals establishing farms which would cover "two miles" was not at all what the government had in mind. Under the proposed peasant farming schemes the size of a holding would be "restricted to 40 acres per family" (in fact it was eventually restricted to half this size) and it would be essential for "eight to ten families to be established on adjacent holdings at each block so that farm implements can be used communally." (SEC 1/428: Letter from Provincial Commissioner G. Howe to the District Commissioner of Mporokoso, 15.9.48).

Thus, if aspiring farmers were to receive government help, they would have to form themselves into groups. This is precisely what the Kaunda brothers and John Mboo did by creating the Farming Associations referred to at the beginning of this chapter. As Mboo quickly discovered, however, the success of such an enterprise depended as much on political skill as it did on farming skills. In 1950, there were twenty-five members of Mboo's Kamanga Farmers' Association, two of whom were women. Mboo, as Manager and Secretary of the Association was the only member actually resident on the farm, all the others being in employment elsewhere. Mboo assured the Commissioner for Native Development that "every member has a mind that on resigning his/her post will come and erect up a splendid building here," and Kaunda provided the same assurance that the sixteen members of his association would eventually settle down as full-time farmers.

It was clear that colonial officials required the assurance of full-time commitment to farming on the part of these groups if they were to qualify for government help, but it was also clear that a nonagricultural source of income remained essential for most of these aspiring farmers. Both Mboo and Kaunda attempted to follow the technical advice and instructions of the Agriculture Department, but this had not always been easy. The Agriculture Department had approved Kaunda's group farm, offered advice," and most of their advices if not all are followed." The District Commissioner had helped by lending a plough and oxen, "but unfortunately we had no experienced ploughman so very unfortunately they were taken back and we had to employ hand labor." All in all they had only managed to cultivate ten acres, and much of that poorly. Mboo's group had cleared sixteen acres in 1949, but had been late in planting because they had had to wait for the Agricultural Officer to visit, approve the site, and log the plots. Only five acres had eventually been planted. But the success or failure of Mboo's group farm rested not only on agricultural performance, but also on his ability to tread a delicate path through local political relations.

Colonial officials hoped that there would be no difficulty in persuading Native Authorities of the advantages that could be had through the presence of group peasant farms in their areas. But things were not always this straightforward, and it was left to individuals like Mboo to negotiate or avert problems. The land occupied by groups like the Chinsali Youngmen's Farming Association and the Kamanga Farmers' Association was customary land, for which permission to cultivate had to be obtained from the Native Authority. Colonial officials, as we shall see, were reluctant to create separate tenancy arrangements for "progressive farmers," because such arrangements might undermine Native Authority power. Nevertheless, it seems that chiefs did not always feel entirely happy about the occupation of the most fertile swathes of land in their areas by groups of "modern" educated men like Kaunda, who then appeared to command valuable resources from the District Commissioner, such as ploughs and livestock. In 1950, John Mboo was clearly having to negotiate a few difficulties with Chief Chibesakunda who had himself been an aspiring "peasant farmer" who had sought help with setting up a seed farm in 1948 (SEC 1/428 Chinsali: TR No. 4, 1948). Chief Chibesakunda, argued Mboo, should have his own Native Authority farm, the proceeds of which would help to enrich the Native Authority treasury. Labor for this farm would be provided by the Chief's subjects on a quasitraditional basis, or would be paid for through "subscription" by these subjects: "I should tell the chief to persued (sic) his people to subscribe or each village to come and work on the N.A. farm for a day or two. Commenting on the latter, I find that not every person will come and work for some are away seeking employment and some are business men who will not like to leave their stores etc. and go to work on the farm. But if the ruling was everyone to subscribe, I find that everyone at home and abroad, an ordinary villager and a trader, will all subscribe" (SEC 1/428: Letter from J. Mboo to Commissioner for Native Development, Lusaka, 14.5.50).

Around the same time as Mboo and Kaunda were forming their farming groups and seeking government support, W. Vernon Brelsford was writing a report on a rather more established group of African farmers close to Gore-Browne's Shiwa-Ng'andu estate. His report raised issues of land tenancy arrangements, of relations with the Native Authority and of the economics of peasant farming (NR 17/131: W. V. Brelsford, Report on African Farms Close to Shiwa Ng'andu, n.d. [1948?]). The Shiwa African Smallholders' Association had apparently been the brainchild of Capt. Crawford, Gore-Browne's estate manager. It had been formed with a view to solving the recurrent food supply crisis that the estate, hospital, and school, and other institutions in Chinsali faced in the 1940s (and which we discussed in some detail in Chapter 4).

There were two men who had been farming close to the Shiwa estate since the early 1940s. They were Anderson Mulenga and R. S. M. Chishina Petro. Mulenga had been well-known in Lusaka and on the "line-of-rail" as a skilled upholsterer. He had begun farming in 1940 on a piece of land close to the estate, to which he sold all of his produce. By 1947, the farm was failing because of the poverty of the soil, and he moved to an area of richer soil close to the Great North Road. Mulenga was more highly capitalized than most peasant farmers (presumably from his earnings as a skilled worker), and he owned a significant amount of equipment, including a plough, a planter, a ridger, and a mill. He also owned six trained oxen, three cows, and five heifers. At the time of Brelsford's survey (probably 1948) he had ploughed twenty-five acres, and was employing four laborers throughout the year, who lived and ate with him. Brelsford calculated that Mulenga had invested in £150 worth of equipment, and that labor cost him about £50 per year. Chishina Petro had begun farming in 1945 on a plot seventeen miles from the Shiwa estate. He had spent twelve years in the army and had retired at the rank of Sergeant Major, using his gratuity of £65 to invest in the farm. In 1948 he had £15 of this gratuity left. Like Mulenga, he sold all of his produce to the Shiwa estate. Unlike Mulenga his farm was entirely hand-worked because it was situated in a tsetse belt. In addition to these two men there were, in 1947, pensioners of the Shiwa estate who were farming and selling their produce to the estate. Some members of this group, like Mulenga, and Chishino Petro, resided on the customary land of Chief Mukwikile, others resided on land held by the estate.

The Smallholders' Association, formed in 1947, included Mulenga and Chishina Petro as well as the Shiwa pensioners, and at the time of Brelsford's visit had thirteen members. Chishina Petro was the chairman of the Association, and the secretary was Jeffrey Mulenga, the son of Chief Mukwikile who ran a farm on behalf of the Chief. Brelsford provided details of ten of the thirteen members, and his short biographies are indicative of the kinds of men for whom farming appeared attractive. Apart from Anderson Mulenga and Chishina Petro (who operated on a more ambitious scale than the others) the group included two men still working for the Shiwa estate, one pensioner of the Shiwa estate, and two others who had worked for long periods on the estate but had no pensions. In addition, one man had been employed as a postal runner, one as a lorry driver, and one as an ordinary laborer for the Department of Public Works. Solomon Sinkamba, the only 'non-Bemba' in the group was the one individual still employed by the Shiwa estate as a capitao, but was about to retire after twenty-one years of service with a pension of £1 per month. He had begun his farm in 1947 and he was planting an average of 25 acres. He had spent £6 on buying his own plough, 30/- on the purchase of twenty sheep, and 70/- on a Kimberley brick house. He had also spent £15 on clearing and stumping the land, and he employed his younger brother to run the farm for a wage of 25/- per month. Sondashi Mwela had been a house servant on the Shiwa estate for twelve years, but he had no pension. He had ploughed about eight acres of land by borrowing oxen from the Shiwa estate, and he had purchased three heifers for 45/-. His younger brother and son both worked for him for no pay, and what Brelsford called an "odd collection" of relatives resided on the farm, creating a small village.

Most of the farmers lay somewhere on a spectrum between Anderson Mulenga and Sondashi Mwela. Some owned their own ploughs, others borrowed from the estate; some employed significant amounts of non-family labor, others relied heavily on a range of family members who usually resided on the farm creating, as in the case of Sondashi Mwela, a settlement that might be easily as large as the average Northern Province village. All had some access to nonagricultural income, and all depended to some degree on the Shiwa estate, which lent out trained oxen free of charge, and sold cheap ploughs by installments to any members of the Association. In return, the farmers agreed to give the estate first call on their produce at the government-controlled price.

Clearly some of these farmers were more likely than others to be successful. Some were much more heavily capitalized than others, and there was also the question of marked variability in soil fertility. Some had chosen to cultivate pieces of land that were barely wooded, presumably because this made the task of clearing in preparation for the plough much easier. But, as was often remarked, the most heavily wooded land was usually the most fertile, although it was also the most costly to clear and stump. Then there was the vexed question of labor, the significance of which went far beyond the issue of cost and availability. Only Anderson Mulenga and Jeffrey Mulenga employed nonfamily labor. All of the others relied on either paid or unpaid family labor, and consequently some farms came to resemble small villages. Brelsford voiced his concern over this spawning of new villages, and over the related problem of inheritance. These were issues, Brelsford wrote, on which the government had not yet provided rulings for the official peasant farming scheme in Serenje district. The Shiwa farmers, being outside immediate government control, had "taken matters into their own hands." The questions of land tenure and inheritance were related, in Brelsford's view, to this provision of family labor, as well as to the issue of the relationship between these farmers and the Native Authority. Chief Mukwikile, himself a beneficiary, through his son, of the Shiwa Smallholders' Association, told Brelsford that he was quite willing for every farmer to be given a lease stating that he agreed to the farm and that the tenure was secure so long as the man used the land in accordance with government agricultural rules. Brelsford felt that the Shiwa farmers should be given such leases, not least because the chief himself was sometimes inclined to use his powers to move individuals on if their presence threatened to undermine his status in the community. Anderson Mulenga, for example, had first set up a farm in the chief's village, where he had built a Kimberley brick house. The chief had immediately removed him "because he said only the chief could have a Kimberley brick house."

On the issue of inheritance, Brelsford thought some clarification was required. Many of the farmers, he argued, were being helped by their sons or brothers. Meanwhile the "real heir" according to Bemba custom was "probably away at the railway line." Brelsford felt it would be important to avoid a situation in which this matrilineal heir (which he characterized as an elder nephew or brother), having contributed nothing to the farm, would come and claim it at the death of the farmer, while the younger men who had helped develop the enterprise would receive nothing. The solution proposed was that farmers should make wills, one copy of which would be deposited with the Native Authority, the other with the District Commissioner. Nowhere in Brelsford's report (nor indeed in any of the discussion of peasant farming schemes at this time) was there any mention of the role of women's labor in these enterprises, nor of women's rights to land or property. Peasant farms were, by definition, male enterprises.

Brelsford considered the Shiwa Smallholders' Association to be something of a test-case for peasant farming in the Northern Province, from which lessons might be learned for the formulation of government schemes. Brelsford emphasized that in comparison with the peasant farmers of Serenje District who had received massive government subsidy, the Shiwa farmers were very much standing "on their own two feet" and had sunk their own capital into the enterprise: "The Serenje people are having almost everything done for them: here the men have started doing things for themselves. Anderson (Mulenga), for example, must have sunk about £150 of his own capital into his farm." Of course, this characterization of selfsustainability was not entirely accurate. The Shiwa farmers had, in effect, been subsidized by Gore-Browne's estate, which had supplied advice, implements and transport, and, perhaps more importantly, this group had a major advantage in knowing that a market for their produce was assured by the estate. Brelsford's calculations demonstrated that the potential market was in fact extremely limited, notwithstanding the yearly complaints of Gore-Browne over the problems of provisioning institutions in Chinsali district. Taking the Shiwa estate, the hospital on the estate, and the Timba School as the immediate market, Brelsford calculated that an annual 300,000 lbs. (or 1500 bags) of grain were required:

> *Taking an average yield of three bags to the acre over all crops—grains, beans, nuts, etc.—it seems that 500 acres of cultivation would satisfy the wants of Shiwa in any one year. If we add ½ that acreage again to allow for ground under fallow or non-edible catch crops, then 750 acres are needed. Thus 19 or 20 farmers each with 40 acres should be able to feed Shiwa. The farmer must eat some of his produce—so at present a maximum of 25 farms of 40 acres are needed.*
> (NR/17/131:W. V. Brelsford: Report on African Farms Close to Shiwa Ng'andu n.d. [1948?]).

Brelsford's calculation was a cautious one because other institutions existed within Chinsali district that would have helped to constitute an expanded market. Nevertheless, the basic point remained that there was only room for a very small number of (successful) peasant farmers in the district, because exploiting more distant markets would involve heavy transport costs, and would probably render the entire exercise uneconomic.

The Economics of Farming Schemes and the Politics of Settlement

Many of the issues raised in Brelsford's report surfaced shortly afterward with the creation of peasant farming blocks in the Northern Province.[7] The question of the economic viability of peasant farming and the levels of subsidy required to make these schemes work, constantly recurred. So too did the question of what constituted a peasant farm, how to prevent the proliferation of individual settlements "masquerading" as peasant farms, and the attendant problem of dispersal of population. It was partly to combat this tendency that the farming blocks were created. In return for government subsidy, farmers on these blocks would have to abide by the regulations of the Department of Agriculture and sign agreements with the Native Authority.

By 1953, there were seven peasant farming blocks in the Northern Province as a whole. One of these was in the Kasama district (Ngululu) and one in Mpika (Lwitikila) (NR 17/152: Provincial Commissioner Kasama to Commissioner for Native Development, 11.9.53). Chinsali had no formal peasant farming block until 1955, when the Chasosa block was formed incorporating some, but not all, of the members of the Shiwa Smallholders' Association. Expenditure on peasant farming blocks had gradually increased over the years since 1949, although as the following table shows, the Northern Province had received a relatively small share of total expenditure compared with the more favored (and fertile) areas of the Eastern Province (Petauke and Fort Jameson) and the Serenje district (NR 17/152: Minutes of the Second Meeting of the Peasant Farming Revolving Fund Committee, 11.10.54) (Table 5.1).

More schemes were opened in the late 1950s, and government expenditure on peasant blocks continued to increase. By 1959 there were 141 peasant farms in the Northern Province, although, as John Hellen's table makes clear, this was a very small number in comparison with the Eastern Province at the same period (Table 5.2):

During the period 1959–1960, expenditure on peasant farming blocks in the Northern Province totaled £9695.00, of which £7885.00 was considered to be "recoverable." Recoverable expenditure included the cost of clearing land and provision of equipment, for which the farmer took out a loan from the government [NR/17/106: Letter, Provincial Agricultural Officer to Chair, Provincial Development Team, 14.4.59]. How much of the recoverable expenditure was actually recovered, however, remains in doubt. In 1955, it was reported that the recovery of loans from peasant farmers in the Northern Province "left something to be desired" [SEC5/393: Minutes of Third Meeting of the Peasant Farming Revolving Fund Committee, 22.2.55].

George Kay's detailed study of the peasant farming blocks of the Luitikila basin in Mpika district made this point, among many others. He had studied the development of the Luitikila scheme from 1956 (when the Chalwe block was opened) to 1961 [Kay, 1962; 1967]. Kay reported that farmers on the scheme had "no conception of even elementary farm economics" and that "Government loans are cheerfully accepted and willingly increased, though the average loan is more than twice the present average gross annual income" [Kay, 1962:43].[8] The result was that, in spite of very generous government assistance and subsidy, loan repayments remained a "heavy burden." In the three blocks of the Luitikila scheme surveyed by Kay, average loan repayments (if fully met) were £18.8 per farm on the Malashe block, £19.7 at Chishinga, and £12.5 at Chalwe. Taking these loan repayments into account, Kay calculated that every farmer on the scheme was nevertheless better off than most villagers, but added that this measure of success "was obtained at enormous cost to the government, not only in initial capital investment

TABLE 5.1 Schedule of Warrants of Expenditure (in £) for Peasant Farming since 1949

	Abercorn	Kasama	Serenje	Serenje (cattle)	Fort Rosebery	Fort Jameson	Petauke
1949	200	200	4000	—	—	1600	—
1950	200	—	2800	800	—	1575	—
1950 May	—	560	—	—	—	—	—
1950 Aug	—	—	—	350	—	—	—
1951 May	—	—	1000	750	—	2150	—
1951 Sept	309	318	—	—	—	—	—
1952 June	—	—	2000 (880 R)	1500	—	2630 (R)	1800 (R)
1953 April	960	—	2188	1500	150	4700	8100
1953 Oct	351 (234R)	407 (390 R)	188 (R)	—	—	—	—
1954	1460	415 (R)	650 (R)	—	—	2500	—
TOTAL	3480	1900	12826	4900	150	15,155	9,900

	Mumbwa	Mkushi	Broken Hill	Mpika	Kawambwa	Chinsali
1952 June	1250	—	—	—	—	—
1953 April	1250	1620	688	—	—	—
1953 Oct	—	—	688	351 (234 R)	120 (R)	—
1954	—	500 (R)	—	351	—	370 (R)
TOTAL	2500	2120	1376	351	120	370

R z Recoverable

Source: NR 17/152 Minutes of Second Meeting of the Peasant Farming Revolving Fund Committee 11.10.1954

TABLE 5.2   The Peasant Farming Scheme in 1959:
Northern and Eastern Provinces

Province	Farmers	Average loan per farmer, £
Abercorn	41	83
Chinsali	17	70
Isoka	20	125
Kasama	23	118
Luwingu	20	156
Mpika	11	88
Mporokoso	9	88
Eastern Province		
Chadiza	162	60
Ft. Jameson	585	95
Katete	315	67
Lundazi	266	67
Petauke	576	110

(Adapted from Hellen 1968: 140.)

but also in recurring costs of supervision, administration and subsidies in one form or another," adding that "Perhaps it is inevitable that the difficult northern areas will remain relatively undeveloped and progress be concentrated in more favourable areas." [Kay, 1962:48].[9]

The "difficult northern areas" were to become more, rather than less, difficult in the years which followed, but Kay was referring to the technical rather than the political difficulties of these schemes, and these were considerable. Rightly or not, the agricultural experts had thought it necessary to totally transform the landscape of the Luitikila area to produce what they hoped would be a viable farming scheme. Such a transformation turned out to be extremely capital-intensive. An ecological survey was carried out and land use planned down to the last detail. As Kay remarked, the farming system and the layout of the fields were both planned "in close relation to the physical resources and their limitations," but "adjustment to the human resources received relatively little attention" [Kay, 1962:40]. There was no question of building on local knowledge or on existing agricultural techniques since these were regarded as part of the problem to be addressed. Blocks were laid out in strict accordance to "the principle of soil and water conservation." Each block was then divided into plots of twenty acres each. Each plot was enclosed by an uncleared belt of woodland, designed to act as a windbreak and to help prevent erosion. Erosion within each plot was also controlled by contour banks and furrows, the spacing of which varied according to the gradient of the plot. Since, in the interests of conservation and sustainability, blocks were located on "crest lines" away from river banks and dambo areas, this meant that farmers had to be supplied with some other source of domestic water. Furrows were built (at great expense) to serve the Chalwe and Chintu blocks, wells supplied the Chisanga block, and until mid-1961 the domestic water of inhabitants of the Malashe block was actually from tanks hauled into the area by tractor and trailer [Kay 1962:40].

The plots having been laid out according to scientific principles, a farming system was then devised with the aim of preserving the soils and increasing their fertility, as well as providing intending farmers with subsistence and a cash income. Farmers were allocated a twenty acre plot, half of which was cleared and stumped for them, and the remainder of which they were expected to stump within four years. They were required to build their own houses and cattle kraals, and were provided with oxen and seeds on government loan. A strict eight-course rotation was to be followed, which allowed half the plot at any one time to be under grass or fallow. In addition to the formal plot a "house-plot" was provided which fell outside the rotation and on which the farmer was expected to produce any subsistence requirements not satisfied by the main crop rotation [Kay 1962:41]. As Kay pointed out, the ultimate practice of the eight-course rotation depended on the clearing of an area equal to that tilled within four years of commencing farming. But farmers had shown no sign of clearing this additional land and "gave the impression that they expect any further land clearance to be done by the government" [Kay 1962:42]. Kay felt that the government, having paid close attention to the physical resources of the area and their limitations, had paid very little attention to the "human resources." He identified two related characteristics of the Luitikila farmers which he felt were salient. First, very few of them were the young men which the government had hoped to attract away from labor migration and into farming. Most were relatively old and some very old, having spent a lifetime on the Copperbelt. This urban experience appeared to have affected their attitude to farming "which they likened to wage labour rather than village life," and gave rise to the expectation "that office hours and a five-and-a-half day week should be sufficient to ensure a satisfactory income" [Kay 1962:42]. At the same time, most of the farmers on the scheme were local men who maintained strong ties with their villages and none of whom appeared "fully divorced from village life." Two men were, in fact, village headmen [Kay 1962:41].

It is here that Kay puts his finger on one of the fundamental contradictions of the government's peasant farming policy in the postwar period. On the one hand, participants in the peasant farming schemes were expected to be progressive men, whose wish to escape the restrictions of traditional village life could be usefully channeled into farming. As farmers rather than subsistence producers, they were supposed to act according to the principles of basic agricultural economics, calculating their costs, planning for long-term benefits, and reinvesting in the enterprise. On the other hand, they were expected to recognize that what they were participating in by joining a scheme was a way of life that required more commitment than wage labor, and whose costs and benefits could not be calculated simply in material terms. The attitude of the industrial worker toward time and money was thought to be entirely inappropriate in the farmer. Colonial development policy assumed that engagement in farming involved an ideological and political commitment, and not merely an economic calculation. There was also a further contradiction. As Kay indicated, farmers on the scheme faced considerable difficulties in acquiring labor, especially at peak times, and some spent significant sums per acre on labor. Others relied more heavily on whatever family labor they could muster, although Kay alleged that such labor was often unreliable. Dependence on family labor beyond the immediate household implied, of course, that the farmers would remain "village men" because it was only through the maintenance of those social ties with kin that the labor could be commanded at all. However, employing kin could give rise to the situation described by Brelsford for the Shiwa farmers in which numbers of relatives intended not only to work, but also to reside, on the plot. On the official government schemes this was heavily discouraged, but the problem of how to become a farmer without also being a village man with wide-ranging social obligations, remained.[10] The approach of individual farmers to this situation varied in Luitikila, just as it did on the Shiwa farms. Some of the more highly capitalized farmers attempted to do without more than immediate family labor and employed additional labor on a strict contract basis.

The issue of labor and its employment was significant, not only from the point of view of its cost and the viability of the farms, but also apparently from the point of view of social status. The farmer who paid for labor acquired a new social position as an employer. In contrast, farmers on the peasant schemes who worked their plots only with the help of relatives came to be regarded as dupes or "slaves of the government" because, in most peoples' eyes, there was nothing to distinguish them from an ordinary villager except that they had willingly submitted to the rules and regulations of the Department of Agriculture (Kay, 1962:42; Long, 1968:17). This suspicion of the government schemes was not new when Kay and Long carried out their research in the early 1960s. In 1958, for example, it was reported that the permanent posting of an Agricultural Supervisor to the Luitikila scheme was regarded with suspicion in the area, and this suspicion had been fostered by members of the African National Congress (SEC 2/849 Mpika: TR Nov./Dec. 1958, Lindsay-Stewart in Chief Chikwanda's area). The problem, as perceived by many district officers, was that the schemes offered very little in the way of social amenities to attract the younger men back from town, and the economic returns to participation in the schemes in Northern Province were insufficient to compensate for the inconvenience of abiding by the regulations imposed. In making this point, officials were frequently drawing a contrast (implicit or explicit) with schemes in Serenje or the Eastern Province, where the economic benefits of participation in government schemes were greater.

Throughout the 1950s, many colonial officials continued to believe that controlled peasant farming schemes of one sort or another represented the way forward in the Northern Province. In theory, they would allow for the economic advancement of more progressive individuals without the evils of individualism. One official talked of the promotion of "platonic communism," and many extolled the virtues of a version of the collective farming practiced in the Soviet Union. Certainly there were reasons for thinking that some kind of cooperation would be needed if the form of agricultural development advocated by colonial experts was to be pursued, because this involved heavy expenditure on the clearing and stumping of land, and the introduction of ploughs and animal traction. But most aspiring peasant farmers in Northern Province appeared to think otherwise, preferring to go it alone as "individual settlers." This was the designation given to individuals who, under the Parish System, were residing outside of the confines of an established village and attempting to grow crops for the market using methods other than (or more often in addition to) citemene. Although many officials remained profoundly uneasy about this development, by the mid-1950s it had achieved large enough proportions for most to feel that it could not be ignored. In a meeting of the Peasant Farming Revolving Fund Committee in 1955, for example, the Development Commissioner reported that there were 593 individual farmers in the Northern Province. The meeting expressed concern that this development was uncontrolled and that many of these farmers were producing cash crops by unimproved methods. The Committee agreed that henceforth suitable areas (in terms of soils, supervision, and marketing) should be set aside for individual farmers who would be entitled to financial assistance from the government, subject to their signing an agreement to practice "good farming" (SEC 5/393: Minutes of 4th Meeting of Peasant Farming Revolving Fund Committee, 25.5.55). "Individual Farm Rules" were passed in 1956 by the Bemba Superior Native Authority, which required individual farmers to sign an agreement that allowed the Native Authority to demarcate the boundaries of the farm, committed the farmer to obeying the instructions of the Department of Agriculture (through the Native Authority) on agricultural practices, and contained penalty clauses (SEC 2/797: TR enclosure, Jan. 1956: Rule No. 1/56 made by Bemba Superior Native Authority). There was general skepticism, however, concerning the possibility of enforcing this rule, and a general disappointment with the performance of "individual farmers." Touring officers reported that most individual settlers were "old men of limited intelligence" (SEC 2/796 Kasama: TR March/April, 1955), and that the Native Authority rules were unlikely to have much effect on their practices: "having seen the age and decrepitude of many farmers I begin to despair of their ever being able to understand the necessity and significance of them." The main problem, as articulated again and again by touring officers, remained the absence of so many younger men as labor migrants. Older men were not regarded as good material for an agricultural revolution: "It is uphill work teaching such very old dogs such new tricks."

Colonial officials continued to hope that improved farming on individual farms would stimulate economic development in the province, they never seemed to understand that the setting up of individual settlements was about prestige and the demonstration of wealth. In this sense, improved farming was an expression of success and not its cause. This is why it continued to attract older men and did not bring young men back from the mines. There were at least two things that older men were likely to have access to that younger men did not, and which the official analysis ignored. The first was sufficient capital to invest in the enterprise; the second was a command over labor. Although government did provide some assistance to aspiring individual farmers in the 1950s, it did so only when they had provided sufficient evidence of their commitment to improved farming methods and to permanent settlement on a piece of land. Demonstrating such a commitment involved a prior investment in a brick house, some fruit trees and, possibly, some livestock and equipment. Older men, who had spent most of their lives as migrants, were much more likely to have such funds at their disposal than were younger men. Similarly, older men were more likely to be able to command the labor of their kin than were younger, newly married men. This is not to say that commanding labor was necessarily easy for these older men (we have already indicated that it was not), but they had a better chance of being able to command it than did their younger counterparts. They also had an investment in maintaining a political following and developing their own prestige and status within the community. Very often these individual settlements controlled by older men became, in fact, new villages, following earlier and well-known patterns of fission within Bemba settlements. Colonial officials were probably right to suspect that "individual settlements" were often no more than attempts to found new villages. Ann Tweedie reported in 1959 that villages in this area were becoming ever smaller and that "there was a rather nebulous distinction between being the headman of a small village and the leader of a large farm" (Tweedie (1966:75); see also Long (1968: Ch. 5)).

For all of the government's efforts to promote peasant farming in the province, there seemed little likelihood that agriculture would tempt back the young men from the mines unless it could be made more profitable. Estimates of the incomes of individual farmers in the 1950s were not encouraging. In 1956, for example, the District Commissioner for Kasama carried out a survey of the incomes of individual farmers in the Ituna section of Chief Mwamba's area and compared them with estimates of what could be obtained by ordinary "subsistence cultivators" (SEC 2/797 Kasama: TR Jan. 1956). The mean acreage cultivated by ten "individual farmers" was 10.7 acres, on which they grew maize, groundnuts, beans, potatoes, vegetables, and cassava. Half of the gross income from sales of crops was estimated to be spent on household needs, leaving an average of £7.3.7d net profit per season. This fell far short of the cost of a bicycle (£12), which was one of the standards set by local people to assess the profitability of any enterprise. It also, according to the officer, fell short of the incomes produced by some "subsistence producers." District officers could always come up with one or two examples of farmers who were "doing well" by growing and selling vegetables—usually those in well-watered areas who were fortunate enough to be close to a market for their produce. But, as one sympathetic officer in the Kasama district pointed out in 1959, there were real limits to the replicability of these examples. Referring to the accusations of "apathy" and "indolence" frequently levelled against the Bemba, and their tradition of "cultivating with the spear," he pointed to the lack of any real economic incentive to engage in farming within the Kasama district: "the southernmost parts and the area of the Chambeshi, probably the most fertile, are rendered economically inaccessible by distance from the main road and from Kasama . . . Five farmers already exploit the restricted opportunities afforded by a few nodal points in the area, but far from underlining the apathy of the populace which does not produce more farmers, this fact surely underlined the absence of economic incentive" (SEC 2/800 Kasama: TR Sept./Oct. 1959). A viable cash-cropping economy would depend, he argued, not on an external market but on the development of an internal one through a "dietary revolution." But this self-sustaining revolution was not to happen in the 1950s, any more than it had in the 1940s when Brelsford had first discussed the possibility of a Provincial Marketing Board [see Chapter 4].

Discourses of Development

The colonial government's most ambitious peasant farming scheme (and its most expensive failure) came in 1958 with the development of the "rural township" of Mungwi, just outside of Kasama. Any understanding of the failure of Mungwi and its political consequences, however, depends on some analysis of the rise of a more general discourse on development in the postwar period. Although we have been describing the history of peasant schemes in the Northern Province very much in terms of the formulation and implementation of colonial ideas about economic and social development, this is only part of the story. The letters from Kaunda and Mboo to the Development Commissioner, which we quoted at the beginning of this chapter, are examples of the way in which the discourse of development had become a shared discourse between colonial officials and certain groups of African men by this time. This participation in the construction of the idea of development had its origins in the 1930s but became evident in 1943 during consultations with members of African Welfare Associations and others in the formulation of a Native Development Plan for the Northern Province. What was most evident in these discussions was the frustration felt by many educated Africans with the "laissez-faire" policies of the interwar period. These were men with class interests to promote; frustrated by the poverty and backwardness of their home areas, by the lack of opportunities for advancement, and the lack of any substantial government investment to promote it. These African spokesmen argued strongly for more, rather than less, government intervention, for a social as well as an economic transformation, and crucially, for more money to be spent on themselves, the "improved" Africans.[11]

The Abercorn spokesmen appear to have been only too eager to offer their views on how the government might spend its money in the Northern Province. Among other things, they proposed the establishment of an Agricultural School, an improvement in midwifery and infant welfare services, the training of sanitary inspectors, the improvement of roads and of education, and the creation of local industries. But in offering their suggestions, they also offered their criticisms of existing government policy and what they saw as the deliberate retardation of development inherent in Indirect Rule:

> *We are under the indirect rule, the sort of it; where we get the worst of it, all because the so called (we wonder if it is real so) direct rules were empowered on a savage people. For this reason the work is just that of the indirect rule, but then we get the worst out of it. Nothing else because there was no education for the people to whom the power of ruling his people was supposed to be given. We take it in this way the Government delegated the power of indirect rule in the chiefs because the Government knew these people were not educated and the Government for a long time did not want to educate them. . .*
> (SEC 2/281 Vol. 1: Minutes of the African Welfare Association, Abercorn, April 4, 1943).

What these men saw as the colonial government's deliberate underdevelopment of education for Africans, and the lack of recognition accorded to those who had achieved some education, was contrasted unfavorably with what they had heard of the French colonial policy of "assimilationD:

> *African[s] learn to be French, they learn and become French all over with the exception of the colour of their skins. The English people can do more better than these French attempts. To arrive at this demand, in this subject, European must show us what kind of a people do we think they are . . . Something must be done for the progressing African, and that it is them who shall have to look to the masses of their people.*
> (SEC 2/281 Vol. 1: Minutes of the African Welfare Association, Abercorn, April 4, 1943).

As members of the Abercorn Welfare Association put it: "Government must call upon progressing Africans, allow them the loans free of interest, supervise their movements until such (cooperative) societies were able and proved to stand for themselves" (SEC 2/281 Vol. 1: Minutes of the African Welfare Association, Abercorn, April 4, 1943).

These were clearly men whose views had been deeply influenced by the events of World War II: alive to British hypocrisy on the subject of freedom, they spoke passionately on the injustices of what they called "colour-bar prejudice," a phenomenon they attributed less to the British and more to the whites from Southern Rhodesia and South Africa. For them, the question of how the government should plan development in the Northern Province was inseparable from issues of freedom, justice, and the advancement of an African middle class. It was later to become inseparable from the whole question of the imposition of the Central African Federation.

Meanwhile, some members of the colonial administration saw development as a way of buying the loyalty of large numbers of rural dwellers who might otherwise be tempted by the language of "freedom and justice" employed by the educated elite. The Commissioner of Native Development expressed this view very clearly in 1950 when he wrote that "building up the Rural African economy on the basis of peasant agriculture provides the best guarantee for the European section of the Community in Northern Rhodesia, that European Government within the Territory will persist for as long as we can reasonably foresee." His reasoning was based on the "well-known" fact that "a stable peasant community in any country always consists of the most conservative elements of the Community." The "rural masses" were self-interested, they would respond to "political agitators" only if these agitators could appeal to "legitimate complaints." By investing in rural areas and increasing rural incomes, the government would keep the people on their side: "A progressive policy in these Areas will keep the masses loyal to constituted Government for no other reason than that the peasant is shrewd enough to appreciate that his lot under the Government is as pleasant as could be reasonably expected" (NR 17/131: Commissioner for Native Development to Secretary of the Development Authority, 1.6.50).

Despite these very different political views on the purposes of development, there was a short period between the end of the war and the mid to late 1950s when the availability of government money, the ambitions of aspiring farmers, and the language of social and economic betterment seemed to act as a unifying force between colonial administrators and certain groups of Africans.[12] This was an extremely fragile unity however, because the more the government promised material benefits from development, the more it laid itself open to direct criticism when these benefits failed to materialize. Of course, the rise of nationalist politics in the Northern Province was influenced by many factors other than the failure of government development policies. By the mid-1950s, opposition to the Central African Federation was crucially important in creating this nationalism, and the politics of labor on the Copperbelt (in which Bemba-speaking men were heavily involved) constituted a powerful motor of nationalist feeling.[13] However, what is important about the failure of development schemes in the Northern Province is the way in which they acted as a vehicle for the expression of nationalist sentiment in far-flung rural areas. Here were direct government interventions of a sort that had not existed in the interwar period; far from deterring the "political agitators," they became sitting targets for them.[14]

District and Provincial Commissioners in the North had long voiced the opinion that the success of any agricultural revolution in that area would depend not only on the provision of agricultural inputs, but also on the provision of social facilities that might attract the town-dweller back to the land. The "half-educated," urbanized African man would have to be enticed back to his rural home with the promise of good housing and amenities, as well as the promise of a reasonable agricultural income. Eventually this process would be self-generating because, with a rise in rural incomes, the local market for other goods would improve, and specialization would increase. Small-scale service industries, for example, would thrive in this new economic climate. The example given by one administrator was a prospective rise in the shoemaking and shoe-repair industry: eventually shoemakers would "abandon even subsistence agriculture and will sell shoes and buy local food with the money" (SEC 2/281 Vol. 11: Submission for Chinsali district (n.d) 1943). Another officer expanded this vision in his submission of plans for development in Mpika district. It was crucial, he argued, that rural incomes should "have some relation to the higher standard of living which it is the principle object to inculcate." Many officials felt that they were dealing with two groups of Africans. One group was made up of the urbanized Africans whose expectations could hardly be fulfilled in rural areas; the other included the villagers who had no expectations whatsoever and who could not be roused to improve their lot. The officer from Mpika felt that it was the duty of the government to arouse the latter group of Africans "to a sense of dissatisfaction with his present precarious existence," while the expectations of the former group would have to be satisfied through the creation of a local labor market (based, initially on public works) and a minimum wage. The familiar "chicken and egg" argument was produced concerning the relationship between agricultural practices and the establishment of better standards of housing and village life. Development of the type envisioned, with its increasingly complex division of labor, could only take place when there was some "stabilization" of village life: "The improvement of agricultural methods is closely connected with the improvement of village planning and housing and will help to stabilise village life and thus will ultimately affect problems connected with administration and education . . ." (SEC 2/281 Vol. 11: Memo on Five-Year Plan for Northern Province, Kasama District (n.d) 1943).

We have seen that the peasant farming schemes that were established in the Northern Province from the late 1940s did not produce the kind of internally generated economic and social development sought in these submissions, nor did they attract any significant number of younger men back to the land. They could not, therefore, perform the political role expected of them. Some officers felt that this failure lay as much in the absence of social amenities as in the difficulty of making agriculture profitable. Regarding the Luitikila scheme, for example, it was felt in 1958 that "wider social amenities such as rural townships will obviously be needed to go hand-in-hand with agricultural development if the semi-urbanized African is to be persuaded to return home" (SEC 2/849 Mpika: TR Nov./Dec. 1958). The opportunity to create just such a "rural township" came, in fact, in 1956, when the Rhodesian Selection Trust group of mining companies lent £2 million to the Northern Rhodesia government specifically for the development of the Northern

Province.[15] The chairman of the group, Sir Ronald Prain, expressed the view that the mining companies held a duty and a responsibility to redress the "state of imbalance which the activities of the copper companies in Northern Rhodesia have helped to create in the Federation." Clearly he felt that this imbalance between the prosperity of the Copperbelt and the impoverished rural areas was unsustainable and potentially politically explosive. Furthermore, he argued that such an imbalance could not be rectified overnight, but would require a sustained injection of significant funds:

> It is obvious that a healthy and stable African community cannot be built as long as such extreme conditions exist within the borders of the Federation. This imbalance is of a nature which can hardly be adjusted, except over a long period of time, by the normal rate of economic development. A preliminary to any redress of the situation must involve the expenditure of very large sums of money on capital works which are unlikely to produce any return of revenue, except on a long-term basis (Prain quoted in Johnson).
>
> (Johnson 1964:46).

In 1957, the Intensive Rural Development Scheme for the Northern Province was created and a Development Commissioner (Magnus Halcrow) was appointed to implement "as speedily as possible a co-ordinated plan of development on whose foundation the future economic, social and political life of the Northern Province can be safely built" (Government Circular No. 1 of 1957 quoted in Johnson 1964:47). In 1958, Chief Munkonge, from Kasama district, was sent on a tour of the urban areas of Kitwe and Nkana in an attempt to promote the idea of development and to encourage Bemba men to return to their homes. He wrote in his report:

> I told them wonderful opportunities existed for those who wanted to make money at home by farming and for this purpose funds to loan are available those who have saved some money could come back and begin trading in goods or fish. I pointed out to them that the two million pounds which the mines lent to the government is to be spent in the Northern Province where most of the rock-breakers come [from]. The time in which it is to be spent is 4 years but already one year is gone, hence only 3 years left. This was the time to make hay while the sun was shining or people will regret their indecision
>
> (SEC 2/799 Kasama: TR April 1958).

The officer accompanying Munkonge on his urban tour was not optimistic that this message would bring results unless some large investment was made in delivering urban services to rural areas. In his view, the Bemba were still essentially a warrior people, easily bored by the humdrum existence of the agriculturalist. They would be more likely to fall prey to "political agitators" in rural areas, he reasoned, than on the Copperbelt with all its entertainments: "Peace is a poor substitute for the excitements of war . . . we have taken from the Bemba their primary occupation and left them to grow food. It is interesting to note that by far and away the most popular film shows are the Westerns, or the blood and thunder type show" (SEC 2/799 Kasama: TR April, 1958). Ethnic stereotypes of the Bemba were clearly alive and well in the late 1950s, and fuelled by the involvement of north-erners in the nationalist movement.

The Mungwi Scheme

The Mungwi scheme was devised to satisfy the urban expectations of Bemba men, and at the same time to entice them back to an essentially rural livelihood. Mungwi was conceived explicitly as a Bemba township, a place that would become the focus of ethnic pride and identity. But it would also be a "modern" place with "modern" amenities, the existence of which would render migration to the Copperbelt less attractive, both economically and socially. The Mungwi ridge, fifteen miles east of Kasama, was chosen as the site of this Bemba township. This was a relatively fertile stretch of land (considered to be typical of Bemba country in general), and it lay within the area of Chief Chitimukulu (and therefore could be felt to be unambig-uously Bemba). On this 20,000 acres, an intensive agricultural project would be combined with the creation of a township, complete with an electricity supply and piped water, a secondary school, a farm institute, and a dairy.

A farming system was devised specifically for the agronomic conditions of the Mungwi ridge. This system was not unlike those applied to earlier peasant farming schemes such as Luitikila, but was if anything more fertilizer-dependent. Farming units would consist of twenty-one acres, of which at any time eighteen acres would be available for cropping. Of these eighteen acres, nine would be down in a grass ley and the remaining nine would be subject to a rotation system of finger millet, followed by maize, then groundnuts, then maize again and finally crop grass. The finger millet would be dressed with 100 lbs. of sulphate of ammonia, and the maize would be dressed with 120 lbs. or more. A small plot of land would be taken from the first maize of the sequence to be used for growing a cash crop of Turkish to-bacco. The main cash crop, however, would be maize, which would be used to sup-ply a roller mill in Kasama (Johnson, 1964:56). Prospective settlers would be allocated a seven acre plot and undergo training for one year, before being allo-cated a full twenty-one acre holding. Four of these twenty-one acres would be felled, and help would be given with stumping and contouring. The farmer would then be expected to clear the remainder of the land, complying with technical ad-vice given by the Department of Agriculture. He would also be responsible for building himself a house and kraal, and for updating the house to a Kimberley brick structure within two years. Seed and fertilizer would be supplied on credit, to be paid back after the harvest (NR 17/106: Letter from Agricultural Officer Frank Tait to the principal of Mungwi Farm Institute).

As the editorial in the Northern News put it, Mungwi was, in 1959, "the most comprehensive and up-to-date planning scheme in Africa" and was expected within two to three years to have a population of some 3,000, many of whom would be housed in the model houses designed by the Northern Rhodesia African Hous-ing Board. (Excerpt from Northern News, 4.12.59, enclosure in KDH1/1 Vol. 4). The developments in Mungwi would also have a knock-on effect on the surrounding area. Villagers would be able to participate in training schemes at the Institute, which would include farm training for men and a comprehensive "domestic sci-ence" training for their wives. Eventually, large numbers of villagers would be persuaded to abandon citemene cultivation, and villages would be stabilized (KDH 1/1, Vol. 4: 1961–66, Mungwi Development Plan).

The Bemba response to the Mungwi scheme, however, was far from overwhelming. Despite numerous "come back" appeals in newspapers, and through publicity tours on the mines, Mr. E. M. L. Mwamba, Minister of African Agriculture was, in 1959, expressing his disappointment to a group of thirty-three "leading residents" of Kasama at the poor "Bemba response to the urgent call for trainee farmers at Mungwi." Eleven training blocks were still vacant. At Mungwi itself he attempted to reassure existing trainees, telling them that the rumor that Europeans would take over the land when it had been prepared and improved was false. Rumors of this nature, he said, were "usually disseminated by loafers who have failed to succeed on the Copperbelt as well as in their original homes" (Nshila, October 13, 1959: enclosure in KDH 1/1, Vol. 4). Of the first batch of nine trainees on the scheme (1958–59), none elected to stay on for another year. Whether this refusal was due to the political campaigns by UNIP or dissatisfaction with the scheme, is not clear. But in 1959–60 a further twenty-one trainees were taken on, of whom nine applied to stay on (and three more came back in the following year). In 1960–61 a further twenty-one trainees were inducted and the nine survivors from the previous year moved out onto their plots and proceeded to cultivate an average of fifteen acres (more than the recommended ten acres). As Johnson pointed out, this meant that farmers were severely over-stretching themselves in terms of credit:

> *The fertiliser and seed advance on all of this was tremendous as 60 lbs. of Rhodes grass seed and 800 lbs. of sulphate of ammonia was required for the recommended ley. The rate on maize was 100 lbs. of supers and 200 lbs. of sulphate of ammonia per acre. The rate on groundnuts was 200 lbs. of "D" as before, and millet was given 200 lbs. of sulphate of ammonia. All this, plus herdboys which had to be provided and extra labour was charged to the farmers and secured as a seasonal loan. The average loan came to £125. As this was far beyond the farmers' ability to pay, and also because of their complaints, the Rhodes grass programme was written off . . .*
>
> (Johnson 1964:62).

Despite these difficulties, farmers on the scheme produced yields that roughly accorded with expectations, and by 1962–63 were achieving a gross return per farm of £100. Johnson estimated in 1963 that once the very heavy overhead costs were taken into consideration the ratio of all revenues and benefits to all costs of the scheme was 0.64 (Johnson 1964:91) and knock-on effects in the region had been negligible because there had been a failure to use local suppliers of goods and fuels, and local transporters.

Certainly in the construction phase of Mungwi there appear to have been some benefits to local villagers in the form of increased local employment, and the expansion of a local market for foodstuffs. This was the "Mungwi effect" that had been hoped for, but it was not to last, for two reasons. First because Mungwi was not an economically viable project: its capital-intensive agricultural system could not be replicated or sustained; its dairy failed entirely, as did the Turkish tobacco scheme. With the economic engine missing, the township could not develop or grow. But the second factor was perhaps more important, and this was a political one. Mungwi became a target for political agitation by the newly constituted United Independence Party (UNIP), whose leadership included the former advocate of peasant farming, Kenneth Kaunda. In 1960, headmen in the vicinity of Mungwi had refused to take part in an official visit to view its development projects (SEC 2/801 Kasama: TR Jan. 1960). Chief Munkonge was meanwhile sent on another urban tour and spoke on the Federal Broadcasting Company on the subject of development, and specifically on Mungwi, in an attempt to counter criticism of the fact that "Mungwi was planned and built by Europeans" (SEC 2/801 Kasama: TR Jan. 1960). But by 1960 development had become completely politicized. In Mpika district, for example, the District Commissioner went on tour in Chief Chiundaponde's area, close to the Bangweulu swamps, in an attempt to track down UNIP supporters. He stopped in key villages where he knew UNIP to be active and pressed home his message. First, he painted a general picture of how the area might develop with the help of a new road that the government proposed to build. He said there would be the possibility of a small market town developing, but that this "depended on their attitude." He said that the government planned to create more schools and dispensaries in the area, and that the "bad political leaders" active in the area would only "deter development." He insisted that the people must show respect for the Native Authority, and must obey the Native Authority rules. Finally, he impressed on them the importance of "not confusing politics with development."

The District Commissioner's conclusion after visiting the area was, however, indicative of the way in which politics and development had become inextricably bound together. The area should be watched, he argued, and the most sensitive barometer of the political climate would be "peoples' reactions to development" (SEC 2/851 Mpika: TR Aug. 1960).

The Politics of Development

Some officers continued to believe that "development" would be the solution to their political problems: "Development," wrote the Provincial Commissioner for the Northern Province, "is the probable solution to any subversive political action" (SEC 2/851 Mpika: TR Aug. 1960: Letter from Provincial Commissioner L. Bean to the District Commissioner of Mpika, 22.9.60). By 1961 the word "development" had, in fact, become a euphemism for the exercise of control over peoples' movements and residence, reminiscent of the days of the British South Africa Company. Inaccessible areas such as the Bangweulu swamps and the Chambeshi marshes were hiding places for political activists, and therefore prime targets for development. Touring the Chambeshi swamp area in 1960, the District Commissioner for Kasama argued that the area would have to be "opened up" by the building of roads and bridges, and by the introduction of development projects: "There is little doubt," he wrote, "that such action could go a very long way towards settling the area" (SEC 2/801 Kasama: TR July 1960). "Individual settlers" once again came under suspicion, with "farms" being seen as the site of "disturbances and subversive activities" (SEC 2/801 Mungwi: TR August 1960). The long-standing hostility toward multiple residency and mitanda also resurfaced. It was indeed hard to keep track of political activists when they were not resident in their villages, but moving around in the bush, and living in mitanda. During the disturbances of July to October 1961, it was estimated that some 500 houses and huts were burned, most of them in Chinsali district, the center of UNIP activity in the Northern Province. This action was often undertaken in the name of Native Authority hygiene regulations and the illegality of mitanda (NR government, 1961: An Account of the Disturbances in Northern Rhodesia, July to Oct., 1961, Lusaka: 71). As the UNIP campaign was stepped up, participation in government development schemes increasingly came to be seen as a political statement. Commenting in 1961 on the popularity of food-buying depots in Chinsali district, the District Officer explained that "The great attraction is that the villager . . . is not bound by any rules or regulations, he does not have to sign any forms which might imply, absurd though it may seem, that he is agreeing to Federation . . ." (SEC 2/766 Chinsali: TR June 1961).

It was perhaps inevitable that in this context Mungwi, the largest and most symbolically significant of all government interventions in the Province, should have become the target for direct attack. On August 4, 1961, an "armed gang of some 300 persons" gathered near the Mungwi development center "with the object of beating up the settlement." Another 200 people were reported "lurking" near the road between Mungwi and Kasama, with a view to cutting off the settlement. On the same evening, people began gathering in the periurban villages around Kasama and Mungwi. On the following day a platoon of the Mobile Unit was despatched to the area and arrested several people. A patrol in Kanyanta village, near Mungwi, was attacked by a group of 100 men wielding spears and other weapons. The police then fired a number of shots and a direct attack on Mungwi was averted (NR government 1961:9).

In their political campaign for Independence, then, UNIP activists in the Northern Province reactivated a long history of evasion and resistance to external control over settlement patterns and mobility. In the construction of nationalist feeling, the experience of labor migration and the urban color-bar were of course central, but UNIP was also successful in mobilizing the support of these migrants once they were back in their rural homes, and this they did through a reconstruction of the notion of Bemba identity. As Michael Bratton has remarked, the Northern Province produced most of the militant "freedom fighters" of the nationalist period, and the "role in which Kasama peasants cast themselves accorded closely with the militarist tradition of the Bemba" (Bratton 1980). The Bemba-speaking people continued to participate in the construction of an ethnicity built around a warrior history and an affinity with the bush, but this time these myths served to advance a nationalist cause.

This was not the end of attempts to resettle, stabilize, and reform agricultural producers in this area, however. During the first decade of Independence (1964–74) village regrouping, as it was called, was a central tenet of government rural development strategy and ideology (Bratton 1980:125; see also Berry, 1993, forthcoming; Chipungu 1988; Gertzel 1980; Makings 1966). The technical and social rationales for this policy were almost identical to those that had been offered by the colonial government, but this time with the added gloss of Kenneth Kaunda's philosophy of "humanism," which extolled the virtues of cooperation and village life. Once again, the notion that the Northern Province was suffering from the disintegration of village life, evidenced by scattered settlement, was a central assumption underlying rural policy. The possibility that village life in the Northern Province *was* scattered settlement, was no more acceptable to the post-Independence political leadership than it had been to the colonial government.

Of course, there were good reasons (as there had always been) for wishing to regroup people in this area. The new government was committed to redressing the rural/urban imbalance in Zambia and to providing improved services to the rural population, including social, health and educational amenities. Advising the government in 1963 was a joint UN/ECA/FAO mission, headed by Dudley Seers, which advocated rapid modernization of Zambian agriculture as an aid to the diversification of the economy away from copper (Makings 1966:238). Mechanization was seen as an essential component of this proposal, and the concomitant of mechanization was resettlement. The Seers Report proposed the establishment of over fifty resettlement units, averaging 200 farms per unit. These recommendations were incorporated in the First National Development plan, which ran from 1966 to 1970. As several writers have pointed out, this period saw a confusing proliferation of rural institutions, including the new organs of local government (Rural Councils), Provincial Development Committees, Cooperative Societies, Ward Development Committees, and Village Productivity Committees (Bratton 1980; Dodge 1977; Berry 1993, forthcoming).

The strategy of village regrouping was reinforced in 1971 through the Village Regrouping Act, and this was supplemented under the Second National Development Plan (1971) by the establishment of Intensive Development Zones, of which there would be one in each province (see Bratton 1980: 125; Gertzel 1980: 241, 248). In 1975, both of these strategies were declared unsuccessful and were replaced by a new Rural Reconstruction Programme run by the National Service and consisting of agricultural camps in which urban youths were to acquire the skills of farming. By 1975, when the collapse of the copper price threw the Zambian economy into recession, it had become clear that none of these initiatives had significantly raised rural incomes, least of all in the Northern Province.

As Sara Berry recently argued, the chief effect of the setting up of a succession of institutions was to provide "new nodes of network formation." Rural people participated in government rural initiatives if they gave them access to significant sources of new wealth, but ignored them when they were ineffective. Although some writers have suggested that the UNIP government's development strategy in the Northern Province was a direct attempt to buy political loyalty,[16] the relationship between development and political allegiance was probably more complex than this. Another set of related factors was more institutional and political. As several writers have indicated, the proliferation of institutions set up in the post-independence period with a view to providing services to rural people, more often acted to obstruct than to enable. They became sources of income in their own right and centers of local political patronage networks. Many rural people were disillusioned, their expectations unfulfilled. By the late 1960s, the local politics of development had become inseparable from developments in national politics, just as they had been during the Federation.[17]

Michael Bratton's detailed case-study of the local politics of rural development in the Kasama district in the 1970s indicates that peasant expectations of rapid and dramatic rural development had, indeed, been engendered in the early 1960s during the election campaigns (Bratton 1980:207) but, within ten years, the majority of peasants in the Kasama district had been disillusioned enough to withdraw their support from UNIP. By the 1970s, government interventions in the area, which had previously been justified as encouraging "grass roots" political participation as well as raising rural incomes, looked more like strategies for the containment and control of peasant political activity.[18]

In Bratton's description of the fate of village regrouping schemes in Kasama district in the 1970s, the continuities with colonial attempts to promote rural development through resettlement are again clear. Although peasants in Kasama apparently recognized that village regrouping might hold certain advantages in terms of access to social services and employment, the basic problem remained that the economic transformation required to raise rural incomes and provide employment in these new settlements never materialized. The UNIP government, like the colonial government before it, was never able to live up to its promises. Poorer peasants in Kasama simply could not afford to risk resettling unless the economic viability of this strategy was assured. Even in the more successful of the two schemes studied by Bratton, most residents could not afford to abandon citemene cultivation, but the concentration of population meant that their citemene fields became more and more distant from their homes. In a passage strikingly familiar from colonial reports, Bratton wrote that "At moments in the agricultural cycle, Nseluka became deserted as men, women and children migrated en masse to the distant fields" (Bratton, 1980:178). Citemene remained part of the repertoire of agricultural activities even after Independence because the flexibility it offered continued to be essential, as it had been in all the decades since Richards's study, to household reproduction. More importantly, citemene continued as a metaphor for that very strategy of flexibility itself and for a continued and continuing dialogue with local (both chiefs and Party representatives) and government officials about the politics of residence, control, and food.

Neither the late colonial, nor the post-Independence government was able to successfully engineer the creation of a class of politically loyal and settled farmers in this area. This failure was the result of a number of interconnected social, economic, and technical factors, which displayed some striking historical continuities. Although colonial ecological science had clearly demonstrated the adaptability and the productivity (under certain conditions) of local agricultural practices, including citemene, generations of agricultural "experts" have premised their advice on the assumption that no truly profitable agricultural system could be built in this area without mechanization and its concomitant resettlement. Of course there was much in this analysis that made a great deal of sense. If rural incomes were to rise significantly in the Northern Province, then a profitable cash-crop would have to be introduced. In the 1950s, both maize and Turkish tobacco were regarded as potentially profitable crops, but on the poor soils of the Northern Province their cultivation was very dependent on the supply of fertilizer and other inputs. There is no reason to suppose that the people of the area were averse to the cultivation of crops on permanent fields (for there were many instances of their doing just this within local agricultural systems), but they preferred to continue to cultivate their citemene gardens in addition to their permanent fields.

There were several reasons why citemene might have remained for many an important component of the system of production. Settled agriculture was a risky business, the success of which depended on a number of factors, including the (usually subsidized) supply of fertilizer and the guarantee of a market. It also demanded a labor regime that was much less flexible and more demanding than that applied in the local agricultural systems (an issue we address in Chapter 8). Under these conditions, the continued cultivation of citemene gardens provided some insurance against household food deficits and, importantly, supplied the central ingredient for beer-brewing which, in turn, continued to constitute an important mechanism for gaining access to extrahousehold labor. In most cases, settled agriculture was simply never sufficiently profitable to allow households to forego other sources of food and income. Thus they continued to engage in a set of multiple strategies including the cultivation of citemene gardens and wage labor. This was true of the individual farmers of the 1940s and 1950s, and of those engaged on peasant farming blocks and the later resettlement schemes. Whenever scrutiny was rigorously enforced and it became impossible to cultivate citemene gardens, the farmers simply left whatever scheme they were on. More often, they carried on cultivating citemene fields, but in secret. Citemene cultivation thus remained compatible with several different residence patterns, but in order for this to be possible there had to be changes in the sexual division of labor and in cropping patterns. We do not have enough accurate field data for the 1950s, 1960s, and 1970s to describe these changes in detail, and oral history has proved no help here because people's recall is insufficiently precise. We do, however, take up these issues again in chapters seven and eight, where we describe recent changes in the agricultural systems of the province. There are, we argue, a number of continuities between the 1950s and the 1980s. Farmers in the Northern Province have grown accustomed over many decades to being told that they would receive benefits and support in return for giving up citemene—we found this story to be common enough even in the 1980s—but they have also always known that they must diversify their risks and their opportunities if they are to make a success of agriculture. Farmers were reluctant in the past and continue to be reluctant now to gamble on the promise of the benefits and the support coming through, and they have been proved right. Citemene continues at the present time, as it did in the past, to be part of farmers' processes of diversification, and, more importantly, it remains a metaphor for, and a symbol of, the potential of that very diversification; acting as an orientation point on a changing horizon.

In the next chapter, we reconsider changes in food production and residence from the point of view of labor migration. In so doing, we return once again to the theme of breakdown, which dominated official discourses from the turn of the century onwards. This theme was given much of its impetus by the development of labor migration. Labor migration itself was a central component of the historical experience of the peoples of the Northern Province, and we reinterpret the evidence for its social and economic consequences.

6

Migration and Marriage

The biologist once told me of a theory that when a species of animal in an area had been nearly exterminated, even if it were then protected it might never re-establish itself in its former numbers, and might even die out. I think a similar theory could be applied to village life. When an area loses, say, 75% of its tax payers, it more or less dies.

. . . When men return from the mines, they do not find a thriving rural life into which they can enter. Most of them do not bother to make gardens, but stay until their savings are spent and no one is inclined to feed them, and then go back to the line of rail.

(Loosmore, SEC 2/793 Kasama: TR 1952).

Since the 1930s, colonial reports of the Northern Province speak of social breakdown, exodus, apathy, and despair. If the language of these reports is sometimes biblical, this is only an indication of the anxiety and apprehension that this issue invoked. This anxiety focused on the question of social disintegration, seen as the inevitable result of the impact of a modern economy on a traditional way of life. The spectre of ruined villages, full of old women and dogs, was paralleled by the fear of the deracinated male "native" who, incompletely urbanized, was a menace in the town and a liability in his rural home. The evolutionary overtones of this discourse are unmistakable. The idea of there being a "natural state" of a society and the fear of the consequences of disturbing that state raised issues that occupied the minds of both colonial administrators and anthropologists from an early date. Anthropological writing from the 1930s through the 1960s was much concerned with the possible breakdown of traditional tribal society brought about by westernization, the impact of a modern economy, and the experience of urban living.[1] The anthropologists were particularly anxious about what they termed "detribalization." This possibility was of equal concern to colonial administrators attempting to operate a system of indirect rule. A "detribalized native" would, in this context, literally be an "ungovernable native."

There is no doubt that labor migration and the rise of the industrial mining centers brought about enormous changes in the lives of people living in the Northern Province. Capitalism and commoditization were forces to be reckoned with. They transformed the very fabric of life, including food, dress, transport, work, and time itself. The experience of life in the mines was brought near even for those who never visited the Copperbelt, and its effects were felt not only in agricultural production and the household economy, but in more amorphous areas such as concepts of status and wealth, and in people's most intimate worlds, including sexual relations. There is no doubt then that there were changes, but the question we wish to address is how these changes are to be assessed and understood. In this chapter we argue that the dominant discourse of "breakdown" obscures a great deal more than it illuminates, making it more difficult for us to specify the complex and variable effects of labor migration in this area. We show that even within one geographical area and at one historical moment, there were several different labor migrant strategies in operation. The social consequences of these strategies, which we address largely through an analysis of marriage, could be equally variable. In general, we suggest that it is necessary to move away from the dominant discourse of "breakdown", particularly as it is applied to kinship and marriage systems in this area. Rather than posit a model that stresses breakdown over continuity or vice versa, we would prefer instead to examine the changes that have taken place in the Northern Province in terms of the sets of solutions people have proposed to specific problems. Both solutions and problems are recursive, in that they emerge and re-emerge in changing forms over long periods of time. It is this fact that makes an historical anthropology so appropriate for their analysis. As we discussed in the Introduction, when we speak of people's solutions we do not imply that such solutions are necessarily consciously formulated or spoken aloud, most often they take more concrete, practical forms, being revealed in the working out of lives lived.[2] We base our conclusions on a reading of the colonial archives, on a reanalysis of work by Richards and others, on participant observation, and on the results of sixty life histories collected from men and women migrants.

Cycles and Circulation: The Real Nature of Labor Migration

The conventional view of the development of male labor migration in Northern Rhodesia is that it followed a series of stages. It was said to begin with short-term circulatory migration, where workers returned home frequently and were involved in agricultural production, and then to have moved on to a second phase referred to as one of partial stabilization or temporary urbanization, where workers stayed for years in town, accompanied by their wives and children, only returning to the country on retirement. This phase was followed by a third, characterized as permanent urbanization, where people born and bred in town had cut their ties with rural life and become proletarians. There is considerable disagreement as to the historical periodization of these three phases, with a number of writers decreeing that the third phase did not begin until after Independence and the abolition of restrictions on urban settlement.[3] James Ferguson (1990a; 1990b) has recently published an excellent critique of this phase model of labor migration, and we do not wish to recapitulate the details of his argument here. However, Ferguson's main point is that the actual patterns of migration and urban settlement do not support the conventional phase model of the development of labor migration. In reality, he argues, individual men and women pursued a variety of different strategies in an effort to make use of the opportunities provided by the rural-urban divide. Some of these strategies involved either long or short stays in town; some workers married on the Copperbelt and others did not; yet others retired to the rural areas after a life of moving from job to job; and some, the amuchona, were lost forever. These different strategies are not to be seen as typical of particular historical phases, as some have suggested, but rather they should be understood as coexisting as a diverse set of strategic alternatives (Ferguson 1990a:411–412).

The questions of how long men stayed away at work, either on the Copperbelt or elsewhere, of whether they took their wives with them, and of whether they retained links with the rural areas or not, are crucial, to the question of the effects of male labor migration on agricultural production. If circulatory migration in the early period really had been replaced with permanent urban residence, then this would have been a very significant change with long-term consequences for food security and social reproduction. If married women were routinely abandoned, as some seemed to suggest, then this too might have been disastrous for rural communities. If, however, remittances were high, then the whole meaning of abandonment might have to be rethought. These questions and many others were in our minds when we began our work on labor migration. However, the first thing that became apparent was that the phase model of the development of labor migration would not hold. Our research findings, like Ferguson's, indicate that different individuals pursued diverse strategies from an early period.

Our material also suggests that many of the difficulties that arise in trying to comprehend and analyse these strategies derive from an inadequate understanding of what people were actually doing. For example, colonial officials often castigated individuals for failing to engage in agricultural production during visits home from the mines, and the missionaries did the same. The confusion here is clearly about the different purposes and meanings attached to the idea of a visit home. Colonial administrators and others held a clear view that rural Africans were rural Africans and that the purpose of a visit home was to engage in agricultural production. In other words, they subscribed to the notion of circulatory migration, and well they might given their anxieties about detribalization and declining rural productivity. However, this view of the situation was not necessarily shared by the visiting worker. He might labor in the fields in recompense for his maintenance, or because he wanted to retain good relations with his in-laws or parents, but this was not always the case. Besides, an individual who thought of himself as working and living in town would not automatically assume that the purpose of a visit home to see family and friends was to engage in subsistence agriculture.

A further point that our material demonstrates is the powerful symbolism of the Copperbelt. It is as if the image of industrial mining and urban living as the very antithesis of rural life provided such a potent set of metaphors for social change that there were few other ways to conceive of it. Many men went to work on the Copperbelt, especially in the 1950s and 1960s, but many did not. Of those who did, a considerable number did not work in the mines. The Copperbelt was approximately three weeks' walk from the Kasama district, whereas the mines and sisal plantations of Tanganyika were about one week's walk.[4] From an early period, men from some areas of Northern Province went north to work not south. Many of them did not travel far at all, and spent all their time working in the province itself. The overwhelming association between migrant labor and the Copperbelt is very misleading for certain parts of the province and this unduly narrow focus distorts our understanding of the different strategies that individuals pursued under the general label of labor migration.

Departures and Destinations

Audrey Richards made her position on the nature and effects of labor migration from the Northern Province in the 1930s quite clear, and it is on the basis of her account that much of the conventional wisdom has grown up. However, a close reading of archival sources and an examination of oral histories tells us a very different story from the one provided by Richards and summarized in the quote below:

> . . *wage earning possibilities are scanty for the natives of this area, and to pay their Government tax, now 7s.6d. a year, and to purchase the European goods to which they have become accustomed, large numbers of the adult men of the tribe—from 40 to 60 per cent—are obliged to leave the territory annually to look for work, mostly in the copper mines of Northern Rhodesia, but also in the Katanga mines, Southern Rhodesia, and even in South Africa. It is rare to find a man who has never left his country to work abroad, and the majority migrate to and fro between the mines and the villages and only finally settle in their home districts in their old age. Thus in this respect many of the food problems of the Bemba are typical of those that exist in all such manless areas in Africa . .*
>
> (Richards 1939:23).

In this passage, Richards states that the characteristic form that labor migration took in the Northern Province was one of circulatory migration between the mines and the rural areas. She made this statement boldly even though (in the later chapters of her book) she had to admit that this was often not the case, partly because of the enormous distances between the mines and the home areas (Richards 1939:398). Richards (1939:133) was also well aware that men often went away for a very long time and that they often took their wives with them, but she seems to have chosen not to integrate this information into the picture of the area she paints in the 1930s. The only contemporary writer to acknowledge that circulatory migration was not an accurate way of characterizing the migrant labor process was Godfrey Wilson. Writing in 1940, he argued that the pattern of spending short periods in town and country, and thus circulating between the two, followed by retirement to the country in middle-age, might have been typical twelve or fifteen years earlier, but that by 1940 it was not typical at all of the workers he surveyed. In his view, the rising demand for labor and the possibility of bringing wives to town had changed the nature of migrant labor and transformed it into temporary urbanization. The temporarily urbanized (69.9% of the working population of Broken Hill) were those born in rural areas who had spent two-thirds of their time in town since first leaving their home areas, with only occasional visits home (Wilson 1941:46–47). Wilson demonstrated rather convincingly that although African laborers may

'The importance of clothes'
Photo: Audrey Richards, 1930s

not have stayed long in any one job, they did not tend to go home when the job came to an end, but rather took another one, sometimes in a different town. This produced a situation of interurban rather than rural-urban circulation; a fact that was disguised by mining company figures showing high turnover rates (Wilson 1941:56–57).[5] The same point was made equally forcefully by a number of missionaries who lived and worked on the copper-belt in the 1930s (Parpart 1986:143; Moore 1943:59–60). Taken together, the available evidence suggests, as Ferguson argues (1990a), that circulatory migration was not characteristic of the prewar period as so many have argued (see footnote 2). In fact, it may not have been straightforwardly characteristic of any period. In making this latter statement, we do not wish to imply that circulatory migration from the Northern Province did not exist, it most certainly did. However, we do wish to stress that even from a very early date, circulatory migration to the mines was only one of a number of coexistent strategies employed by a variety of individuals under the label of migrant labor. In addition, circulatory migration tended to occur at very specific stages in an individual's life cycle, as Wilson (1941:46–48) and others have since pointed out (Mitchell 1987:82–83; Epstein 1981). The story we would like to tell about labor migration can find no better beginning than in the following abbreviated and compressed versions of two of the stories told to us:[6]

> Mr. Mulenga was born in 1919 in Luwingu District. He first migrated when he was fourteen because he didn't have enough support from his parents. His parents were "too old" and they "didn't work enough to feed him." He worked at the Ipusukilo mission in the Luwingu district first in 1933 as a bricklayer. He stayed for one month (October) and he went home at weekends to his parents. In 1934, he went to Kasama and worked for a white man. He was employed as a bricklayer on the building of the Ituna Primary School for three months. He then went back home and gave his parents some money. In 1935, he started working as a cook for the mis-

sionaries at Luena. He stayed four months. He didn't go home, but his parents came to visit him. In 1936 (August), he went to Tanzania. He and his companions spent a month traveling there. He went to work for a white man in a diamond mine. He stayed seven months. He went back home and brought a blanket for his father and a cloth for his mother.

In November 1937, Mr. Mulenga went to Mufulira on the Copperbelt. Before he went he had become engaged. He paid his future in-laws the equivalent of K2 as mpango. He worked in Mufulira for three months for a white man, but the work was too tough and he "ran from it." He went home and got married in 1938. In 1940 he went back to Mufulira and worked as a miner. He worked there for two years and then went back home. In 1943, he went with his wife to work as a miner after his father had died. He was "chased" from work without doing anything wrong, so he went to work as an electrician. He then worked as a miner in Chingola and Mindola. In 1948, he changed his job again and started working as a foreman with the Public Works Department building roads in the Luwingu District in the Northern Province. He continued until 1959, but in 1960 he was "chased" from the job for being a member of UNIP. In 1964, he went back to work for the Public Works Department and retired to take up farming in 1969.

Mrs. Chanda was born in 1901 near the Chilubula mission. She was married in 1926. Her husband paid her family 2/-nsalamu and 1/- mpango. She first migrated to Lubumbashi in the Congo 1934. She had to follow her husband who had come to fetch her. She went with her two children and her husband. They went on foot and the journey took about two weeks. They carried bunga (flour) and money to buy food on the way. When she got to Lubumbashi, she did not do any work apart from "household chores." She could not remember whether the husband paid rent for their four-roomed house. The husband did not receive any rations, but he earned enough to afford them a decent living. She stayed in Lubumbashi until 1956. They came back because the contractor her husband used to work for left Lubumbashi. While in Lubumbashi, they referred to themselves as amuchona (the lost ones). She said she used to miss home, but she had no way of maintaining home ties. She never made any visits home while in the Congo and never made any remittances. They returned home in 1956 to start farming. They settled in her home village, where her brother is the headman, because her husband's relatives had all died by the time they came back and he had no one to welcome him. Her husband died in 1970 and since then she has been dependent on her children.

On an initial reading, these two stories seem very different from each other, but perhaps the first thing to note is how very different they both are from the picture painted by Richards in the passage given earlier. Mr. Mulenga's story seems closest to the conventional picture of circulatory migration. Between 1933 and 1938, he had five jobs and none of them lasted longer than seven months. However, three of those jobs were within the district in which he was born, and his primary responsibility during those years was to his parents. It is also worth noting how young he was; when he finally went to work as a miner on the Copperbelt in 1940, he was still only twenty-one years old. When he first married in 1938, he stayed at home with his wife, then he left her there for two years while he migrated again. But after his father died he took his wife with him and stayed away for a further four years. He returned to his home area in 1948, but he continued in full-time employment, with one break, until he retired at age fifty. Mr. Mulenga, then, had a fairly long history of employment within the Northern Province and in Tanganyika before he finally set off for the Copperbelt. In his case, a pattern of "circulatory migration" seems to have been stabilized on marriage.

Mrs. Chanda's experience appears very different from that of Mr. Mulenga, but there are a number of similar features. Mrs. Chanda's husband had no history of local employment, but he did migrate before he married. He left his wife at home in the early stages of marriage, and only later took her and the children with him. Once Mrs. Chanda left with her husband, she stayed away from her home area for more than twenty years. During this time, she and her husband made no visits home, and they did not make any significant remittances, although they did send a few small gifts. What is interesting about this account is the fact that Mrs. Chanda accompanied the husband after an initial period at home in the early stages of the marriage. This pattern is repeated in many of the life histories we collected. Twenty-seven out of thirty women migrating in the period 1939–60 had accompanied their husbands. Men like Mrs. Chanda's husband were not involved in classic circulatory migration, and they made no contribution to the rural economy either in labor or in remittances after becoming employed. The fact that many women accompanied their husbands to urban areas was a fact noted by Richards (Richards 1939:133), and by a number of other contemporary commentators. A District Commissioner on tour in the Luwingu district in 1929 noted that of 403 men away working, 79 (19.6%) were accompanied by their wives (Wickins, ZA 2/4/1 Awemba: TR 1929). The District Commissioner of Mpika, touring in 1938, noted that about half the migrant laborers from Kambwili's area took their wives with them (Fox-Pitt, SEC 2/836 Mpika: TR 1938). Wilson (1941:21) calculated that by 1940, the working population in Broken Hill was made up of 41.2% single men and youths, 13% married men with wives at home, and 45.8% married men accompanied by their wives.[7] The fact that nearly half the employed men had their wives with them is significant, but it becomes even more so when we consider that a survey such as this, conducted at one point in time, gives a very particular slant to the data. For example, the material we now have available strongly suggests that a considerable number of the men recorded as single or youths would later form part of the accompanied workforce. This biographical pattern, like that of rural residence for women in the early years of marriage, should not surprise us, particularly given the extreme youth of many of the parties involved. Our data suggests that many men started employment at the age of fourteen or fifteen years. Thus, Mr. Mulenga is in no way atypical.[8] In addition, our material indicates that age at first marriage for women was often as young as fourteen or fifteen years of age, and it is therefore not surprising that men should have left their very young wives at home for a year or two, perhaps longer. In some cases, men left their wives at home for long periods of time, and only brought them to town when one or two children had been born and/or when they found a secure job that they liked and that paid enough to support a family in town. This latter point was emphasized by the informant who narrated the following employment history.

Mr. Luchembe was born in 1926. He became a general worker at the Chilubula mission in 1943. In 1947, he left the mission and went to work in Kasama making bricks. He didn't like the job, so he came home to help his parents with the fitemene. In 1948, he started work on a local government project, and he later joined the Public Works Department and worked on the Kasama-Luwingu road. During this time, he sent some cash to his parents and he gave them some clothes. He married in 1949. In 1954, he went to Kitwe to work for a contractor. He left there after a short time and went to Ndola to work as a busboy. In 1956, he returned to Kitwe and began work on the mines. His wife joined him there because he had found a good job. He worked there until 1982. His wife stayed with him the entire time. They have eleven children. During this period, he sent remittances home of £8–£30 per year. The savings he made during his working life and his pension helped him to start farming when he retired.

Thus Mr. Luchembe left his wife at home for the first seven years of their marriage. During this time, his wife lived with his parents, and he maintained contact by sending small amounts of cash and some clothes to his parents. In contemporary tour reports, Mrs. Luchembe would have been recorded as having been left behind by her husband. Once again, the collection of data at a single point in time, whether by anthropologists or colonial officials, gives a particular bias to the outcome. Mrs. Luchembe was not strictly speaking a "stay-at-home wife" because she actually lived on the Copperbelt for 26 years of her life. One conclusion that can be drawn from this discussion, however, is that the processes and strategies that together make up the package of labor migration were often obscured by the way in which data was collected, as well as by the assumptions underlying analysis of the phenomenon.

The question of distortion or misunderstanding as a result of methods of study is raised by Ferguson in his critique of theories of labor migration in Zambia. He argues that Richards's (1939) and Watson's (1958) studies were inevitably biased towards workers involved in some sort of cyclical labor migration because these were the men who were available for interview by the anthropologist in the villages. Men who had gone away for longer periods of time were necessarily absent. Ferguson sees this as "a methodological consequence of the village-level unit of study" (Ferguson 1991a:411). He rightly argues that there were men who behaved in the way these studies suggest, but that we should be wary of treating these individuals as typical of the majority. Our research findings support Ferguson's arguments, but at the same time we would want to stress that Richards (1939:115, 133) and Watson (1958:45–46) were well aware of alternative labor strategies, but they chose not to integrate this evidence into their explanatory models. They were as convinced as any colonial administrator that cyclical labor migration was the order of the day, despite the occasional difficulties they had in reconciling their observations with their theories.

The evidence from archival sources on this point is clear: cyclical migration was only one of a number of strategies employed. A District Commissioner on tour in the Mpika District in 1939 conducted a survey of twenty-five Bemba villages. He found that of 414 taxable males, 192 were away working.[9] Of these 192, 30 had been away less than six months, 10 had been away more than six months, 56 had been away more than one year, 34 had been away more than three years, 38 had been away more than five years, and 25 had been gone more than ten years (Sec 2/837 Mpika: TR Feb./March 1939). These figures indicate that nearly one-third (32.8%) of

TABLE 6.1 Migrant Labor Rates

	Percent of absent males	Prov	CB	TT	SRC
			Percent Working in		
1937					
Mwamba (North)	60.4	12.5	35.4	12.1	0.2
Misengo	42.2	0.9	34.0	6.3	0.9
1938					
Makasa	69.0	9.7	5.8	53.1	0.2
Fwangila	75.0	10.7	4.7	59.5	0
Changola	72.0	14.2	0	58.2	0

Prov: Northern Province; CB: copper-belt; TT: Tanganyika; SRC: Southern Rhodesia and Congo.
(*Source*: Based on figures in SEC 2/786 Kasama: TR 1937–1938.)

the men away had been absent for five or more years. Another colonial official touring in the Mpika District in 1940 noted that 50% of the taxable adult males were away, but he remarked that "Many of these men have been away so long that the headmen do not expect them to return." He recorded meeting some mine employees who were home for short visits, but commented that "Those men who come home on leave at fairly frequent intervals still represent only a small proportion of the mineworkers of the area" (SEC 2/838 Mpika: TR May 1940). It might be imagined that the number and proportion of men staying away for long periods would increase over time, and there is some evidence to suggest that this was the case, although a 1952 tour of Chief Chikwanda's area in the Mpika District showed that only 38.2% of the absent men had been away for 5 years or more, a figure very close to that for 1939 (SEC 2/843 Mpika: TR Nov./Dec. 1952).[10] The evidence is not in any way conclusive, but it does appear that a wide variety of strategies continued to be employed under the label of labor migration throughout the whole period under consideration in this study, some involving long stays and some short, some involving local employment and some not.

The critical assumptions in both Richards' and Watson's work about the nature of labor migration gave rise to a situation in which that nature was thought to be known but was in some ways very little investigated. For example, it is clear from archival sources that rates of labor migration varied greatly from one part of the province to another, and from one village to another. This variation was apparent even by the time Audrey Richards conducted her research, but it has never been analyzed. A young colonial officer touring the Chinsali District in 1930 noted that ". . . the number of men away varied considerably from village to village . . . In one case, there were as many as twelve away out of a total of nineteen. In other cases, only one or two." (ZA 7/4/10 Awemba TR: Nov. 1930). The reasons for this variation were several. Critical factors included the destinations of migrant laborers, the availability or nonavaliability of local employment, and the opportunities for earning cash without entering waged employment at all. Table 6.1. is illustrative of the degree of variation that could exist within a single district (Kasama). In the areas of Chief Mwamba and Chief Misengo, the overall rates of absenteeism varied dramatically despite the fact that approximately the same proportion of men absent from both areas were at the Copperbelt. In the areas of Chiefs Makasa, Fwangila, and Changola, approximately 10% of the men were employed locally, but over 50% of absent males were in Tanganyika rather than on the Copperbelt. Marked preferences for particular destinations were often apparent in figures of this kind, and particular areas, and even individual villages, tended to build up an association with a specific place. For example, an officer touring Chief Kopa's area in 1952 noted that this part of the Mpika District had a strong connection both with the Broken Hill and the Roan Antelope mines (SEC 2/843 Mpika: TR Jan./Feb. 1952). Factors such as distance, working conditions, pay, the nature of recruitment, and the employer's reputation all played an important part in creating and cementing these associations, but so did links of kinship and friendship. Many of the male migrants we interviewed reported following an elder sibling or getting help from a family member to find a job.

Building Bridges and Constructing Roads: The Importance of Local Employment

As we have indicated, it is difficult to make generalizations about the nature and extent of labor migration across the province because of the variation that existed at both the regional and the village levels. However, one point that has been little addressed is the importance of work available in the province. It is certainly true that rates of pay in the province were nowhere near as high as those at the mines. It is also true that because people migrated for reasons other than the procurement of cash (the attractions of a "modern" urban life were very real), they did not always want to work within the province even when work was available locally. Most of the waged labor in the province was associated with government bomas, road and bridge building, missions, food-buying depots, and one or two ranches and estates. A District Officer noted in 1930 that almost every man recorded as working in the Chinsali District was working either in Chinsali town, at the Lubwa mission, or at Shiwa N'gandu (the estate owned by Sir Stewart Gore-Brown) (ZA 7/4/10 Awemba: TR Oct./Nov. 1930). A number of individuals worked for storekeepers and private households, and a few were in business. The numbers employed locally grew over time, most particularly in the 1950s when the regional economy was at its most bouyant. The expansion of schools and hospitals, the extension of government through Native Authorities, and the increase in colonial government functions (agricultural research stations, locust control centers, etc.) all involved a demand for labor. A corresponding demand developed in the subsidiary areas associated with these developments. Cooks, guards, and other service personnel had to be taken on, and businesses grew up to provide them with clothing, food, and essential items. The numbers employed in the province might never have been enormous when compared with the effects of capitalism and commodification on local labor markets in other parts of Africa, but they were still significant. The archival sources suggest that, on average, anywhere from 10–35% of the employed taxable males were working in the province during the period 1930–1960.[11] In Makasa's area in 1939, 14.1% of employed males were employed locally. In 1947, the figure was 32.1% (SEC 2/786 Kasama: TR July 1938; SEC 2/788, TR Aug./Sept. 1947). Once again, we have to remember that enormous variation existed across the province, and that areas close to urban centers, missions, and estates tended to have larger numbers at work locally than the more distant areas.[12]

Much of the work provided locally in the province was of a temporary nature. Special projects, such as the building of the Chambeshi pontoon or the construction of a school tended to come and go, and most of the people employed were casual labor. This means that figures for the proportion of men employed locally could vary dramatically from year to year, and this makes time series comparisons very difficult. In addition, much of the labor required locally was seasonal. This applied particularly to road and bridge building, general building, and maintenance work of various kinds, much of which was better done during the dry, cool season. This results in figures for employment that vary greatly depending on the season in which they were collected. Thus, making comparison over time very hazardous.[13]

The concentration of certain kinds of work in the cool, dry season meant that workers were often required when they might have been cutting their citemene (June–October). Thus, it is difficult to make any straightforward assertion about whether certain kinds of local employment could be combined with agricultural production or not. There is, however, some evidence to suggest, as in the account given above by Mr. Mulenga, that at least some of the workers employed in the province could return home on weekends or make visits on a fairly regular basis. There are a number of reports of workers doing this from the 1930s to the 1960s. For example, an officer touring Chief Mpepo's area in the Mpika district in 1951 noted that 29.9% of the men away working were employed in the province:

> "The bulk of those engaged at work within the Province are at work, in fact, in the road gangs stationed within the area, on the Kasama road. Some of the road camps . . . are almost entirely staffed by the males of particular villages, and it might almost be said that the villages themselves have transferred to the road camps. The advantage lies in the fact that such labourers are within one days walking distance of their true village, can get home at week-ends, and generally keep social life going there"
>
> (SEC 2/842 Mpika: TR March 1951).

Keeping social life going is not necessarily the same thing as cutting citemene trees, but these workers may well have been able to engage in agriculture later in the year or pay someone to cut for them, or perhaps they did not cut citemene in that particular year. It is worth noting in this regard that the touring officer recorded elsewhere in his report that there was "every hope of an abundant harvest," and this was despite the fact that nearly 74% of the taxable males were absent.

Another officer touring Chief Mpepo's area in 1952 noted the large number of people working on the roads, but offered the opinion that men employed locally at the boma or on the roads considered a working period of three months to be the maximum if they were to fulfill their obligations in the village. He claimed that a system existed whereby a man worked on the roads for three months and then his place was taken by a brother or another relative, allowing the first man to return to the village to cultivate his garden and do other necessary jobs. In this way, a family

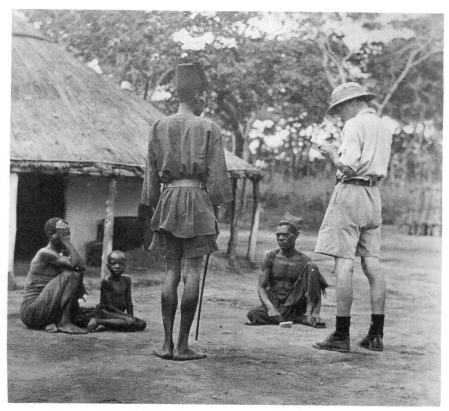

'Government census, Bemba village 1933'
Photo: Audrey Richards, 1930s

was assured of its income both in terms of money and of food. The officer also observed that nearly all taxpaying males had to leave their home for work at some time or another during the year. Apparently, a few of them even worked for their better-off neighbors, digging gardens or preparing fitemene (SEC 2/843 Mpika: TR May 1952).

The population and labor figures collected by colonial officers on tour are notoriously unreliable and, as a result, it is fruitless to submit them to sustained or complex analysis (see footnote 10). However, an examination of the figures collected for a sample of twenty of Chief Mpepo's villages in 1951 proves suggestive when it comes to a consideration of the relationship between agricultural production and labor migration. The 1950s was a time when the mining economy was bouyant and large numbers of men were away at work. In this particular area, as noted above, 74% of men were away working. When the numbers are averaged for the twenty villages in question, the figure is still 68.4%.

Table 6.2. shows the enormous variability from village to village in the proportion of men employed in the province. This means that area averages (see above

TABLE 6.2 Chief Mpepo's Area 1952: Percentage of Men Away Working and Percentage of Employed Males Working in the Province in Twenty Villages

Percentage of all taxable men absent	Percentage of employed males working in the Province
56.6	29.4
66.6	40.0
83.3	40.0
96.4	33.3
61.1	77.7
46.1	0
50.0	25.0
77.7	21.4
65.0	7.6
62.5	100.0
95.4	4.7
80.0	31.2
69.2	55.5
60.0	40.0
28.5	75.0
65.0	23.0
81.2	30.7
44.0	50.0
79.1	31.5
42.8	0

(*Source:* Based on figures in SEC 2/842 Mpika: TR March 1951.)

and footnote 9) disguise a great deal of local variation. Some of this variation is linked to demography and to the life-cycle of villages. The average population of the twenty villages sampled was 70.2 persons, with an average of 22.7 adult males, 19 taxable males and 18.3 adult females; although many of these adults were not resident, of course, in the village. This seems to have been a fairly average size for commoner villages from the 1930s through the 1960s. Thereafter, village regroupment resulted in larger villages in some areas. The average size of the villages tells us nothing about their overall composition, but we do know that age and marital status were key determinants in labor migration strategies, a fact remarked on in many tour reports, especially in the 1950s. We must assume, therefore, that some of the variation in the proportions of men away, as well as the proportions working in province, are related to the demographic composition of villages, and to their life cycles of fission and fusion.

The question of the relationship between agricultural production and labor migration is thus further complicated by the issue of the individual demographic composition of each village. We have little insight into the details of village composition because tour reports simply give figures for adult males, adult females, and children.[14] They give no information on relative ages, kinship links, or polygyny. They do record the number of exempt males (nontaxpayers), and this gives some clue as to the number of elderly men in the population, although age was not the only reason for exemption. If we do some calculations based on the figures given for the sample of twenty villages in Mpepo's area in 1952, some interesting points concerning sex ratios emerge. The ratio of registered (present, absent, and exempt) adult males to registered adult females on average for the twenty villages was 1 man to 1.2 women. The total number of adult females for each village compared to the number of resident taxable males (those not exempt through age or infirmity) gives a ratio of 1 man to 3 women. We might assume, therefore, that one of the effects of labor migration is that it increases the sex ratio from 1:1.2 to 1:3. Even if this were the case, it is by no means certain that these proportions would have disastrous short-term effects on agricultural production given the adaptability of the citemene system (see Chapter 2). But, if we do further calculations and include the exempt males in the figures for resident males, the ratio drops to 1 man to 1.8 women. In addition, we know that calculating these ratios on the number of registered adult females is going to distort the results because many of these women were not actually resident in the villages—they had gone away with their husbands—but the colonial records do not tell us how many of them were absent. If we assume that 25% of the registered adult females were absent (a conservative estimate) and then compare the reduced number of females with the number of taxable males resident in the village then the sex ratio works out at 1 man to 2.2 women.[15] If we assume still further and include the exempt males in the figures for resident males, the ratio drops to 1 man to 1.4 women. We can continue even farther and assume that all men recorded as working in the province were able to contribute to production at some point in the agricultural cycle. If we include these individuals in the number of resident males and compare it with the number of registered females less the 25% assumed absent, the ratio is 1 man to 1.3 women. Once again, the addition of the exempt males to the total number of resident males (including those working in the province) would bring the ratio down to the rather amazing figure of 1 man to 1 woman.

Given the size of the sample, the unreliability of the data in general, and the unverifiable nature of our assumptions, we cannot draw definitive conclusions from these figures. However, the figures do draw our attention to the fact that, because of the way in which the data were recorded (with taxation, rather than social analysis in mind) we actually know very little about the effects of labor migration on sex ratios and on village level agricultural production in the Northern Province. Certainly, the tour reports and other colonial sources are full of dire descriptions of villages filled with old men, abandoned women, and sickly children. These images of social disintegration may have been symbolic of other fears for many colonial administrators, but there is no doubt that people occasionally went hungry, villages broke up, dissatisfaction and demoralization were palpable, and that sometimes food aid had to be organized. These were the material realities of the Northern Province in the colonial period; they were not invented. However, the question we have to ask ourselves is how many of these misfortunes were attributable simply to the absence of men. Colonial officials throughout the whole period tended to be most dismissive when told by local informants that poor crops were the result of wild animals, insufficient rain, village fission, and other forms of social and ecological disruption. Although there were many occasions when local people complained of an absence of labor, they often did not.

Fishes and Furs: How to Make a Living Without Leaving Home

It is not just the neglect of the role of local employment that has made the impact of labor migration on rural production so difficult to assess, but also the general neglect of any consideration of the complex dynamics of the local economy. It is certainly true that in comparison with many areas of Africa, and with other parts of Zambia, trade and markets over most of the Northern Province have always been poorly developed. However, it is also the case that many individuals did not need to leave their homes to earn money in formal waged employment because they had other ways of raising the cash to pay taxes, gain access to commodities, and ensure household reproduction. The sums of money involved were not necessarily large, but they were indispensable. From an early date, there were communities that, by virtue of geographical location, had ready access to certain resources and were able to exploit them for cash instead of leaving for work. Many of these communities were situated on the margins of the northern plateau, or off the plateau altogether. Many, but not all, were areas of Bisa-speaking people who had been pushed to the edges of the plateau during the nineteenth centeury expansion of the Bemba state, a situation that came to have distinct advantages in the context of the colonial economy.

In 1929, an officer touring the Luwingu District noted that only 82 (10%) of 805 taxable males were away working. "There is little need for these people to go away to work," he wrote, "as their food, fish and meat (from lechwe trapping) supplies are, together with otter skin trading, a source of wealth" (ZA 2/4/1 Awemba: TR Sept. 1929). A similar observation was made by a District Officer on tour in Mpika the same year. ". . . they make a sufficiently good livelihood out of the sale of fish and otter skins to obviate the necessity of going away to work" (ZA 2/4/1/ Awemba: TR June/July, 1929). Villages close to the swamps and rivers in the southern parts of the province were engaged in exporting food (mostly cassava) and dried fish to the Congo and to the Copperbelt (Brelsford 1945; 1946:Preface). The men involved in this trade made a very good living and had no need or desire to involve themselves in waged employment. One District Commissioner estimated in 1956 that a canoe load of dried fish could be sold for £60 (SEC 2/847 Mpika: TR June 1956). A more modest claim made in 1959 suggested that the annual income of a fisherman in Chief Kabinga's area was between £7 and £50 per annum. Fish was not only sold on the line of rail, but it was traded across the plateau, carried by bicycle to Luchembe, Chikwanda, and Mpika. Some was also sold at the Chambeshi pontoon, as well as being marketed in Kasama, either directly or through the offices of middlemen (Sec 2/850 Mpika: TR April/May 1959).

Fish and skins were not the only items traded. A number of villages grew tobacco for sale; others made baskets, raised chickens, and produced salt. Even in 1938, villages along the Great North Road were making a living by selling food to travelers, and those close to towns were doing the same by supplying the small urban populations. As we have seen in Chapter 4, this situation, in some areas, resulted in a large increase in the amount of maize grown especially for this market (SEC 2/836 Mpika: TR Feb. 1938). As the economy expanded, the volume of this local trade grew, apparently reaching a peak in the 1950s.

As we have indicated in Chapter 4, the increase in the sale of agricultural produce from the 1940s onward was not inconsiderable. The Provincial Commissioner exclaimed in 1948 that "there is no doubt that very much more food is grown in the Northern Province than was the case five years ago" (SEC 2/839 Mpika: TR Feb./March 1948). The exact magnitude of this increase is impossible to assess, although it is clear that it had a very uneven distribution across the province partly because of ecological reasons and partly because it was stimulated by demand and thus remained crucially dependent on infrastructure and access to markets.[16] However, the fact that there was an increase in agricultural production at all in the province in the 1940s and 1950s is instructive because this was a boom period on the mines, and some areas had as much as 70–80% of their adult male population away working.

The demand for food was stimulated not only by the growing urban centers, but also by the government, by the missions, and by a few private enterprises, which had to feed large numbers of laborers and supply schools and hospitals. An officer on a 1949 tour of Chief Kopa's area noted that the increasing surplus available year after year in the area, and in others areas like Kabinga and Chiundaponde, was certain proof that people were steadily extending their acreages. He also commented that most of this surplus had been "absorbed by increased consumption at centres like Mpika Boma" (SEC 2/840 Mpika: TR April/May 1949). In 1956, an officer touring Chief Mpepo's area commented that the district produced more food than it required and that it sold much of its surplus to the boma. This amazing feat was achieved while 74.2% of the adult males were recorded as away at work, although 16.1% of their number were employed in the province (SEC 2/847 Mpika: TR April 1956). How are we to explain this, given that labor rather than land has always been the scarce resource in this agricultural system?

On the surface, it makes no more sense to us now than it did to colonial officers at the time. Their puzzlement still streams off the pages of archival documents nearly forty years old. One point again relates to the collection of data and its almost exclusive emphasis on the male population. Very often when it was said that large numbers of adult men were away from their homes, this obscured the fact that large numbers of the total population, both male and female, were actually absent. The focus on male migration and the overemphasis on cycles and circulation means that there is little record of all the people who left the province who were not adult males. Two exceedingly rare examples of migration figures for all sections of the population suggest that in Mpepo's area in 1953, 36.8% (64.3% men, 27.6% women, and 27.7% children) of the total population of the area was absent, and in 1958 in Chikwanda's area, 44.2% was absent (65.5% men, 33.7% women, and 36.4% children).[17] This means that when we think about increases or decreases in food production, we have to consider the effects of general migration rather than simply male absenteeism. Local communities did not necessarily suffer severe imbalances in sex ratios, leading to high consumer/producer ratios. On the basis of some of the information available, it seems that they were more likely to be suffering from population depletion—up to 40% in some areas. However, this cannot have produced a disabling labor shortage, at least in the short term, if the very same areas were apparently producing large amounts of foodstuffs for sale.

As we argued in Chapter 4, some of the increase in agricultural production was due to communities putting large acreages under cassava. There are many different types of cassava with varying maturation times, but one common variety grown in the province was harvested only after three years. Such a crop is clearly less labor intensive than is the finger millet grown on citemene fields. However, it must be remembered that cassava was grown on mounds, and that the labor required for mounding was very substantial, particularly if acreages were large and continuously being extended. It is not enough, therefore, to claim that cassava solved the problem of labor shortages, although it is true that its production did not necessarily require male labor.[18] What we can say is that changes in cropping patterns and labor regimes were one of the many ways in which communities in the province were responding to commoditization and labor migration, but the exact nature of that response varied from area to area.

A further point that requires emphasis is that some of the areas selling foodstuffs were genuinely disposing of surplus, while others were engaged in what might be termed "overselling" (see Chapter 4). Those who were selling more than they could afford either suffered shortages later in the season or, more commonly by the 1950s, bought food back in again later in the year. Because we do not have any figures for the buying and selling of food outside official channels, we have no way of comparing sales of foodstuffs with figures on population and migration. In the end, we simply have to admit that it is very difficult to construct any single narrative for the effect of labor migration on food production in the Northern Province. What we can say, however, is that the old story we are all so familiar with is not supported by our reanalysis of the data. We may not be able to offer a comforting alternative, but we can at least shed new light on an old problem.

The Changing Meanings of Marriage, Service, and Labor

The social and economic effects of labor migration and general population movement were not confined simply to food security and agricultural production. Labor migration was itself one of the many effects of capitalization and commoditization, therefore, it was part of a wider set of processes that not only transformed ways of working, but fundamentally altered concepts of wealth and status, notions of time, and the comprehension of social relations. In the process, transmutations occured in a very diverse set of human experiences, everything from sexual relations to the performance of ritual were caught up in what was to become a profoundly altered world view. This change in perspective, if that is what it was, gained momentum from, and was sustained by, people's practical engagement with the changed circumstances of day-to-day living. If the transformation we seek to describe was conceptual and symbolic, it was also preeminently practical, embedded in sets of quotidian activities. One area of peoples' lives in which this transformation was wrought most powerfully and was thus very keenly felt, was in the domain of intimate relations, given substance by the sexual division of labor and the institution of marriage.

For the Bemba, however, and indeed for many people in the Northern Province, marriage was not an institution, but a set of processes. Audrey Richards attempted to describe both these processes and what she saw as the changes taking place in the institution of marriage as a result of labor migration and the money economy (Richards 1940b). Her general view was that the monetization of the betrothal present (nsalamu), the main marriage gift (mpango), and the puberty or initiation payment (ndalama shya cisungu), coupled with the decline in brideservice performed by the young husband for his father-in-law, was undermining an institution that had previously depended on the exchange of services between a husband's and wife's kin over many years. In an ideal situation, the continuous flow of gifts and services, through which the social relations inaugurated at marriage were given substance and meaning, should give rise over time to a high degree of mutual dependency and respect between the kin groups involved. The strong interdependency of the parties concerned was reinforced further by the relatively high rate of cross-cousin marriage, which served to sediment ties between brothers and sisters and to reduce the tensions produced by matrilocal marriage.[19]

Overall, Richards saw Bemba marriage as dependent on a very precarious balance that had to be maintained between the husband's relatives and the wife's male relatives. In part, this was simply a way of pointing to the tensions that are supposed to exist in matrilineal systems between the rights of the father and those of the mother's brother. However, while Richards tended to speak of Bemba marriage as an institution and possibly oversystematized and overlegalized our conceptions of it, she also clearly understood not only its basically processual nature, but the shifting sands of negotiation, compromise, and interpretation on which it was based and through which it was constituted, as the following quote indicates.[20]

> . . . the balance of power between the husband and the wife's male relatives is a very delicate one. To reach a stable Bemba marriage the relationship between the two groups must reach a fine equilibrium based on sentiment, reciprocal legal obligation and an accepted division of rights over the children. It is attained after a succession of ceremonial acts and status changes. It has always something of the trial and error nature about it, at any rate during the first years of married life, and sometimes a successful personal adjustment of the interests and social allegiances of both parties is never reached.
> (Richards 1940b:51).

According to Richards, a Bemba marriage could take from five to twenty-five years to complete. The process began with the presentation of the nsalamu (betrothal gift) and continued through many phases to the ukuingishye shifyala (entering in of the son-in-law). This rite, which was supposed to mark the ending of taboos between son-in-law and parents-in-law, and the former's incorporation into the kin group of the latter, was also the moment when a man could be permitted to take his wife back to his own village if he wished (Richards 1940b:72). Bemba marriage was an integral part of creating and maintaining a political following; the crucial node in a wealth-in-people system that depended on rights over the services of others. This much is clear, but the question we have to ask ourselves is whether labor migration and the money economy brought about a breakdown in marriage and in the kinship relations upon which it depended.

In attempting to answer this question, we have to begin with Richard's contention that the monetization of marriage payments, coupled with an increasing tendency to substitute a cash payment for brideservice, was undermining the stability of Bemba marriage. According to Richards, what had once been sets of social relations made concrete through the exchange of services had now been replaced by cash transactions and thus social relations were in danger of breaking down.[21]

The idea that money causes a breakdown in social relations is a common theme in much anthropological and historical writing. Money, in these accounts, corrupts social values, while inaugurating new ways of giving value to identity, status, labor, and obligation.[22]

The first point to be made here is that cash transactions were a standard part of Bemba marriage arrangements by the time Richards came into the area for the first time in 1931. She herself never saw or recorded a marriage that did not involve cash. She had to assume the existence of a past in which marriage agreements had not involved monetary transactions of various sorts. In fact, much of what she reported as typical of traditional Bemba marriage was never observed by her. If these activities had ever taken place, they had done so in a past that was already quite conceptually and symbolically distant by 1931–34, if not actually chronologically so. Richards herself acknowledged this problem in the introduction to her famous discussion of Bemba marriage:

> . . . I have selected for description the type of marriage practised some fifteen or twenty years ago, that is to say the institution as outlined by the old and middle-aged men in accounts of their own life histories and as practised with a few modifications in outlying parts of the country. I have done so because this conception of marriage appeared to me to embody the ideals to which the majority of people are still trying, however ineffectively, to conform . . .
>
> (Richards 1940b:12).

Traditional Bemba marriage, when the nsalamu was no more than a token gift such as a copper wire bracelet, and the mpango was only two or three barkcloths was not in existence even in the 1930s. Richards seems to imply that it did exist around 1910–15, but we should remember that the late nineteenth century had been a period of significant disruption to marriage systems in this area (Wright 1993), that Bemba men had long been involved in migrant labor in Katanga, Lupa, and the sisal plantations of Tanganyika, and that the administration had encouraged the payment of taxes and wages in cash from the earliest days. Richards asserted that the Bemba did not understand the role of money as a medium of exchange and did not use it regularly in everyday transactions (Richards 1939:216–221), but by the time Richards began her fieldwork, the exchange of labor for money was an established fact of life in the Northern Province. When Richards argued that the subsitution of brideservice by a payment of 10/s was an indication of the monetisation of social relations and an example of their breakdown, she was, we would argue, both recording and misunderstanding the crucial role of money in Bemba marriage and in Bemba life in general.

This becomes apparent if we look in detail at what Richards actually says. She notes, for example, that the Bemba regarded money as a medium to be used in specific transactions only. Aside, from the purchase of European goods and the payment of taxes, where there was no choice in the matter of whether money was used or not, these transactions included marriage payments, native crafts, the fulfillment of ceremonial obligations and debts of honor (Richards 1939:218). At first sight, this might seem a rather surprising list, but one could argue that this indicated that the Bemba substituted money in all instances where they had previously made good an obligation in services or in kind. This included such items as baskets, pots, and axes, which had always been paid for by goods in kind or by service. The substitution of money for barkcloths in the case of marriage payments was noted by Richards to have been "an easy transition" (Richards 1939:218–219). This transition was easy precisely because what money represented in this case was labor, and more specifically labor conceived of as service. In the past, barkcloths presented at marriage were a sign of respect and symbolic of the young man's own labor—he made them himself—and of his offer of labor services. They were not wealth objects in themselves, but rather indicative of labor to be offered in the future. The provision of labor to another was an indication of hierarchy and thus of respect in Bemba society (see Chapters 3 and 8). The exchange of goods at marriage then was not an exchange of wealth objects, but a promise of service to come. What was distinctive about this situation was that goods and service were already symbolically and practically bound up with each other. Labor as service could be made concrete in the form of a good, not in Marx's sense of its value being equal to the amount of congealed labor embodied in it, but in the sense that labor as service was itself a type of good. The control of others' labor in this wealth-in-people system was a form of wealth, as well as an indication of status. Even the more notable chiefs did not attempt to accumulate great stores of food or wealth items, for a wealthy man did not collect material wealth in the form of goods or property, but rather reckoned his wealth in the amount of service, and thus respect, he could command (Richards 1939:213–216). In this situation, the very act of service was itself a form of respect. It is for this reason that the Bemba chiefs of the nineteenth century demanded services from their slaves and tribute from the people they conquered (Richards 1939:400–401).

Richards may have missed this point because she saw the use of money in marriage and other ceremonial transactions in the 1930s as an indication that the Bemba regarded money as a kind of prestige object or valuable.[23] We would argue that this view of money does not help us to make sense of the use made of it in rituals, in settlement of debts of honor, or indeed in the case of the chief who was reported to be more "flattered on receiving a coin, however small, than anything else," because, as he said, "money is respect (umucinshi)" (Richards 1939:220). The respect embodied by the gift of money came, perhaps, not from its value as a wealth object, but from its status as a good that embodied labor as service.

The notion of labor both as a form of good, and as the basis for wealth, explains why goods could substitute for labor in certain ritual transactions, particularly those involving asymmetrical social relations, and why money should have come to substitute for labor in the very same way. European goods and waged labor, we suggest, found easy acceptance in a system that already emphasized the fact that labor was a form of good, and thus that goods were substitutable for services precisely because they were not wholly distinguishable from them.

The practices of missionaries and traders in the Northern Province in the first half of the century did nothing to reinforce a rigid distinction between goods and services, because they so often remunerated their labor in kind, and engaged in various kinds of barter transactions.[24] European goods, like calico, beads, and salt, as well as a variety of other items, became part of a cycle of services and goods embedded in a set of asymmetrical social relations in a manner that was probably only too familiar to the people of Northern Province.

When Richards discussed the use of money in ritual transactions, she implied that it was used in this way because the Bemba did not understand the real function of money. However, she also stated that young people and those living near urban centers did use money to purchase the necessities of life, including food, and so presumably must have understood its real function (Richards 1939:220–221). Certainly, we do not wish to imply that individuals in the Northern Province in the 1930s did not comprehend the uses and meanings of money as a generalised medium of exchange. It seems very likely that they did. After all many of them had already spent long periods as industrial wage laborers. However, these understandings of money, we suggest, coexisted with other views that allowed money as a good to be integrated into a conceptual system based on labor as a form of good. In fact, it seems quite likely that by the 1930s, these different views of money were indispensable to each other, as if the fact of wage labor had solidified or made concrete a symbolism that prior to the advent of the money economy had been partial and implicit. It was perhaps only with the particular experience of the processes of capitalism and commoditization in the form of wage labor that certain connections, identifications, and substitutions became explicit.

Marriage Payments

We cannot be certain at this distance of time, what all the associations of money and labor might have been. However, we can look at some of the changes we know to have been taking place and try to rethink the classic interpretations that have been made of them. In many parts of Africa, the advent of wage labor and a cash economy brought about a massive inflation in marriage payments. Such payments became a method of accumulating wealth, and the resultant difficulty in repaying the huge sums involved had a noticeable impact on marriage stability. In addition, the question of the legal status of the children of the marriage became ever more firmly tied to the formal completion of payments (see Mann and Roberts 1991). However, when we look at what was happening to Bemba marriage payments, we can see from the small sample covered in our own research that although they did increase slightly over the period 1925–1960, they were never large, and they could never have functioned as a method of capital accumulation (Table 6.3). This does not mean that the mpango payment was insignificant, but if we take the average unskilled rural wage in 1934 in the province of 7/6 per month, and the average mine wage for surface work of 22/s per month, and compare them with Richards's figure of an average mpango payment of 8/s, we can see that it would not have taken a young man long to raise the sum of money involved (Richards 1940:53).

In addition to the relatively slow rate of increase in the monetary value of marriage payments, the results of our research show (and this is a point substantiated by Richards's findings) that there was little uniformity in the size of payments (see Table 6.3.). Richards noted in the 1930s that marriage payments did not appear to be fixed, with the son-in-law of a rich clerk giving 10/s nsalamu and £2.10s mpango, and a man in a neighboring village giving his father-in-law 2/6. She also remarked that there was little regularity in the manner and timing of payments (Richards 1940b:54). Our own research found great variation in levels of marriage payments. In addition, we could find no fixed relationship between the amount of nsalamu and the amount of mpango paid. It is true that the overall level of payment seemed

TABLE 6.3   Marriage Payments and Services 1925–1959

Year of Marriage	Nsalamu £ s. d	Mpango £. s. d.	Brideservice
1925	£1	£5	Y
1926	2/s	£1	Y
1928	5/s	£3	Y
1933	5d	£1	Y
1933	1/s	£1	Y
1934	3d	2/s	Y
1939	5/s	£2	N
1941	3/s	10/s	Y
1943	2/s	£8	Y
1943	£3	£4	Y
1943	5/s	£2	Y
1944	5/s	£4	Y
1946	2/s	7/s	10/s
1946	4/s	£2	£4
1946	5/s	£2	N
1946	5/s	£4	Y
1948	£2	£10	Money
1948	5/s	£2	Y
1948	1/s	£6	Y
1949	10/s	£10	£5
1950	2/s	£1	Y
1950	5/s	£2	Y
1951	£2	£5	Y
1951	£1	£4	N
1952	2/s	£2	Y
1952	3/s	£1	N
1953	£1	£4	Money
1955	£2	£10	N
1958	1/s	£8	Y
1959	2/s	£7	Y

(*Source*: Interviews conducted in 1988, based on first marriages only.)

to be loosely associated with the wage level and the professional status of the son-in-law. However, informants stressed that each man paid what he was able to offer. Richards implied, however, that increases in marriage payments and their overall level were associated with the status of the father-in-law, and with the desire of "better-off natives" to gain more control over their children (Richards 1940b:54). These views were not substantiated by our own research, informants having made no connections between the size of marriage payments and increasing control over off-spring, although they did stress that proper marriages must involve marriage payments. Our own interpretation, based on our findings, would rather stress the importance of the economic and social status of the son-in-law rather than that of the father-in-law as being relevant to the determination of the level of marriage payments. What was at stake here, we believe, was an equality of respect between the parties involved, with educated or well-employed men not wishing to put themselves in a debt of service to their father-in-law—a situation that would imply asymmetrical relations—and so they paid more in marriage payments as a statement of the equality of relations between themselves and their in-laws.

While research on the Copperbelt has identified marriage as a route to social mobility for women, there is no evidence from our own research to support this view in rural areas (Parpart 1986:153–154; Epstein 1981:317–319). None of the migrant men we interviewed gave any importance to the notion of marrying into a good family or indeed into a rich one. The women interviewed did not express any views of this kind either, and we recorded no instances of fathers rejecting prospective sons-in-law on the grounds that they were not wealthy enough. In fact, we only recorded one example of a father removing his daughter from a suitor and that was for failure to pay any mpango at all. If he had paid even a small amount then the marriage would have gone ahead. In this wealth-in-people system, marriage would appear to be a domain in which the older concepts of wealth, as embodied both in the service of respect and in respectful service, seem to have endured. We would want, therefore, to take issue with Richard's contention that the monetization of marriage payments was one of the factors leading to a breakdown in marriage and in the relations of service on which marriage depended.

We can pursue this argument by looking at another development that Richards regarded as evidence of a breakdown in social relations: the subsitution of a cash payment for brideservice. Richards reported that under ideal conditions a young man would move to the village of his bride and work for his father-in-law over a number of years, during which time he was fed by his mother-in-law and only gradually assumed the responsibilities of an independent head of household. His young bride continued to work and cook with her mother for some considerable time, and her assumption of full domestic responsibilities was equally gradual. As time passed, ties of respect and dependency between the two kin groups involved in the marriage would develop and grow (Richards 1939:112–113, 124–127; 1940b: 61–65). However, Richards argued that with the impact of the money economy and with so many young men away at the mines, many used this opportunity to shirk their responsibilities to their fathers-in-law (Richards 1939:172). In addition, many young men preferred to pay their fathers-in-law in cash rather than work brideservice, and this substitution of cash for labor, according to Richards, did away with the enduring relations of service on which the social relations of marriage depended (Richards 1940b:74). Richards contended that the payment substituted for the period of brideservice was regarded as separate from all other marriage dues, and she noted that it was set at a fixed rate of 10/s during her period of research (Richards 1939:133; 1940b:78).[25]

In our sample of women and men migrants, we found only five people married in the period 1925–1960 who explicitly spoke of the groom as having offered a separate sum in lieu of brideservice. However, a number of informants linked the total sum paid as mpango to the amount of brideservice the prospective groom was able to perform. One woman informant who had married in the 1950s actually said: "The work done by the fiancé determined the amount to be requested for the mpango. If he worked hard the demand would be very little." It is not possible to provide a numerical analysis of the relationship between groom's salary, size of mpango, and length of brideservice because the figures for salary and length of brideservice are not reliable. The problem in terms of brideservice is that some informants, for example, might speak of working for three years for their fathers-in-law, and this may well mean that they cut citemene on three occasions rather than that they worked for three agricultural seasons. Richards found the same problem when she noted that

> *It is difficult to know how long this period of service lasted in the old days. Some elderly Natives said they worked "continually" for their fathers-in-law, but this probably means a general readiness to co-operate with the wife's people after the period of service is over*
> (Richards 1940b:52).

Richards' comment makes it clear that one of the problems we face when trying to ascertain changes in the practice and meaning of brideservice within marriage is that we have little idea of how much brideservice was performed in the more distant past. As with so much we have been investigating in this book, it seems to have varied a great deal. Richards seems to have assumed that the length of brideservice performed in the past would have been much longer. To some extent, an interpretation rests on the definition of brideservice. According to Richards, many men married very young at about fifteen years of age, although her informants apparently insisted that in earlier times age at marriage had been higher (Richards 1940b:61–62). There is no way now of assessing this information, and we could obtain no reliable information on this point. However, if Richards was right about the early age at marriage for men, then we must envisage that some form of brideservice or other could well have gone on for a long time, given the extreme youth of the parties to most marriages. Even once the young people were old enough to manage their agricultural and domestic responsibilities, it seems that matrilocal marriage would have entailed years of cooperation between households of different generations. The difficulty then, as Richards suggests, is in deciding when the period of brideservice definitely had come to an end. It seems likely that this would have been a matter for contestation and interpretation, rather than a matter of the application of a rule. What would have been at stake in the final analysis would have been the control of labor on which political structures depended.

It seems clear from our data that the substitution of cash for bridewealth service cannot be taken as unequivocal evidence of the breakdown of marriage or of social relations. Out of the thirty-one marriages contracted before 1960 for which we have reliable data, only four sons-in-law performed no brideservice, and only five substituted money. However, the amount of brideservice performed varied from two days to three years. Two of the men interviewed, for example, worked only very briefly for their in-laws while on a visit home and this was counted as legitimate brideservice. What was thought to be much more important by informants were the small gifts that flowed between the rural and the urban areas, often carried during visits. These gifts were of many different kinds, some made them regularly and others did not. Out of sixty migrants interviewed, thirty-two said they sent money home to parents and in-laws, twenty-seven sent clothes, fourteen took clothes or blankets whenever they visited, five sent tools, axes or hoes, six sent household goods, one sent seeds, and ten claimed to have sent nothing at all until such time as they returned to the rural area to settle.

What our data suggest is in a way quite opposite to Richards's contention. Despite the fact of migration and of the distances between rural and urban households, people seem to have continued to send money and goods home in lieu of service. They continued quite literally to service their kinship links, and most of our informants claim they did so because they knew one day that they would want to come home. This is itself instructive because in many parts of Africa, migrant remittances are high in situations where there are opportunities for investment and capital accumulation in the rural area (in the form of livestock, land, and other property). However, there were almost no such opportunities in the Bemba areas of the Northern Province in the past, and there remain few at the present time (see Chapter 8). Land was not short and there was no market in it, residence shifts were frequent, and livestock existed only in small numbers. In other words, there was little to invest in other than social relationships, and the migrants tended to do this symbolically rather than practically. The total cash value of remittances was, as we shall see, very small.

In our view, the substitution of goods and cash for brideservice did not necessarily change the meaning of marriage and social relations as radically as Richards suggested. In fact, for most people cash and goods were perfectly comprehensible and acceptable substitutions for brideservice. Although practically they were certainly not the same thing—a point that anthropologists and colonial administrators made endlessly—they were conceptually and symbolically of the same order. It was when the goods, no matter how small in value, stopped coming that people felt that social relations had broken down. It is for this reason that young women waiting in the rural areas for their husbands to come home had to refuse the clothes sent to them, or they would stand no chance of receiving a divorce in the Native Court. Goods were capable of servicing social relationships, and money—although hardly ever used for agricultural purposes—was a perfectly acceptable substitute for labor.

When looking at more recent marriages (Table 6.4), we can see that continuing capitalization and commodification has not led to a massive inflation in marriage payments or to a wholesale abandonment of brideservice. The data suggest that although there have been increases in marriage payments and a decline in reported brideservice over time, these changes are nowhere near as large as those that Richards might have predicted. In fact, given the levels of inflation in the Zambian economy since Independence, marriage payments, relatively speaking, are very low. No really firm conclusions can be drawn from the data we have available because of the small size of the sample, but it is clear that marriage payments and brideservice have persisted. From discussions with informants, it seems that brideservice in the formal sense is now a matter of a few weeks' work or in some cases only a few days, although there are still some individuals who work for long periods for their fathers-in-law. The more interesting question is why brideservice has persisted at all, and this, we suggest, has much to do with the continuing importance of labor as a form of a material good, especially in those households engaging in maize cash-cropping (see Chapter 8). If the rise of waged labor served to make explicit certain links between money, goods, and services in the period between the wars, the rise of cash-cropping in the Northern Province in the late twentieth century seems to have reinforced those links once more, but not in entirely unexpected ways.

Married Men and Abandoned Women

The fact that marriage and social relations cannot be said in any straightforward sense to have broken down either in the 1930s or at the present time, does not im-

TABLE 6.4 Marriage Payments and Services
1965–1984

Year of Marriage	Nsalamu (kwacha)	Mpango (kwacha)	Brideservice
1965	K0.5	K25	Y
1973	K1	K50	N
1976	K2	K67	N
1976	K5	K70	Y
1977	K2	K80	MONEY
1977	K10	K200	Y
1978	K10	K100	Y
1979	K5	K200	N
1979	K20	K100	GOODS
1980	K5	K60	Y
1980	K20	K300	Y
1982	K15	K75	MONEY
1982	K20	K170	Y
1982	K50	K350	N
1983	K2	K60	Y
1984	K10	K150	Y

(*Source*: Interviews conducted in 1988, based on first marriages only.)

ply that the meanings and experiences associated with them have not changed. The point we would wish to stress is that there have been changes, but these changes have taken place in the context of a set of recurring problems—how to control labor, how to maintain social networks, how to reproduce households— that appeared at different times in new forms, both conceptually and materially, but which required people to draw on old repertoires in order to develop new ones. The old and the new copied, replaced, and substituted for each other, but did so within a long historical process that was never complete or completed. There are, therefore, many different stories to tell about marriage and about what it meant for particular individuals in the Northern Province in the 1930s, just as there are today. The experience of marriage differed and differs greatly for different sections of the population, for rural as opposed to urban dwellers, and for men when compared to women. It is very difficult at the present time, and doubly difficult for the past, to hear the various voices that proclaim these different experiences and meanings. The process of interpretation and contestation is often muted by the material available and by the research methodologies employed. This problem can be illustrated through an examination of the phenomenon of "unattached women," who posed an administrative problem for colonial officials and who constitute an analytical problem for us.

From the 1920s onwards, there were frequent complaints about "unattached women," sometimes divorced, sometimes unmarried, but always "unattached." It is clear that a certain number of such women left the rural areas and moved to urban centers. They did not go in search of formal employment, but rather in search of marriage, adventure, and the benefits of an urban lifestyle. These women, unlike the women migrants we interviewed, did not migrate at the request of their husbands. An officer touring Chief Mwamba's and Chief Nkolemfumu's areas in 1929 noted the following:

> . . . *a large number of unmarried women appear to have gone to Broken Hill and Ndola of their own accord . . . gone to find husbands. This indiscriminate migration of young women to the mining areas is a thing to be deprecated not only from the moral point of view, but from the point of view of the effect it has on village life . . . It is however extremely difficult, if not impossible to prevent*
> (ZA 2/4/1 Awemba: TR Nov. 1929).

We have no idea how many women left of their own accord either in total or as a proportion of the total number of women who migrated. It seems likely that their numbers increased over time, and we certainly know that Native Authorities tried to put various measures in place to prevent unmarried women leaving the rural areas, including the rather crude one of installing road blocks along the Great North Road. By all accounts, these measures were very largely ineffective. In 1935, some Native Authorities introduced a law that gave grounds for divorce if a woman had been abandoned for more than thirty months.[26] This law was apparently designed to encourage men to return home quickly to prevent their wives from divorcing them. However, as time went on Native Authorities tended to increase the period of time that had to elapse before a case could be made for abandonment in an effort to keep women married and in the rural areas. The same report noted that steps were being taken by the Bemba "to try and prevent the loss of women" (SEC 2/1297:1935–37). There seems to have been very little coordination between the various Native Authorities, whose circumstances in any case varied a great deal, and laws came and went without much success. However, these measures represent attempts on the part of colonial authorities and chiefly elites to stabilize marriage and residence patterns.

Anxieties about marriage stability were constantly expressed by officials and by missionaries, and the breakdown of family life became synonymous with the breakdown of society. Divorce was reported to be on the increase by almost every commentator from the 1920s onward. There are, however, no reliable figures on increases in the divorce rate, the definition of divorce being partly dependent on the equally problematic definition of marriage. Richards noted in the 1930s that young men spoke of having "married" two or three girls, by which they meant they had provided a betrothal gift (nsalamu) (Richards 1940b:61). A breakdown of relations at this very early stage would have been most unlikely to have been considered a divorce by the parties involved. Similar difficulties could also be encountered at later stages in the long process that constituted Bemba marriage. It was often the case that young men did not pay all the mpango payment at once, and thus it could be difficult to decide at any particular moment how far the marriage had progressed. This gave rise to a great deal of negotiation over the question of when a marriage was really a marriage, and when any breakdown in those relations could be considered a divorce.[27] As Richards argued, Bemba marriage was very unstable, especially in its early years, and had likely always been so; thus, informants could and did have very different ideas about when a marriage could be considered a proper and secure one. In many cases, marriages were not considered to be final until more than one child had been born (Richards 1940b:52–58, 72). Into this situation of process and negotiation came Indirect Rule and the codification of custom. Under colonial law, a marriage was a marriage if the consent of the parents had been given and if an mpango had been paid. The formalization of marriage as an institution, and of the responsibilities entailed in that institution—such as the fact that a man had to pay tax for his wife—inevitably lead to a corresponding formalization of divorce, as well as an increase in its reported incidence.[28]

A further complication involved the return of the mpango at divorce. This seems to have occurred from time to time prior to the establishment of British rule, but it is not clear under what circumstances (Richards 1940b:104–111; Gouldsbury 1915). However, with the codification of customary law, it became possible to claim the return of mpango at divorce, and this gave rise to a situation in which formal divorce rates increased not only because it was necessary to be divorced, but because it became possible to make additional claims for money and for children under the new system. The bilateral kinship system of the Bemba had always provided for the possibility of children going with either parent at divorce, if the child were old enough. This situation had no effect on the matrilineal affiliation of the child. However, the new marital regulations provided for a situation in which the rights of parents had to be formalized, and what had once been open to interpretation and negotiation became much more rigid. As a result, tensions between the rights of the father and those of the matrilineal descent group represented by the mother became severely exacerbated.[29] The mpango had formerly conferred no rights on the father for the control and custody of his children, but under the impact of European law, it came to signify both the legal status of the marriage and the rights of the father over the off-spring of that marriage. Thus, the legalization of marriage brought about far-reaching changes in the meaning of marriage itself, in the symbolism and practical significance of marriage payments, and in the rights and duties of parents.

It is in the context of these changes that we have to understand claims about the rising incidence of divorce. Richards recorded divorce rates of 20.5%–44% in different areas of the Northern Province in the 1930s, with 6% having been divorced twice, 4% three times, and 1% four times (Richards 1940b:99). However, she also pointed out that figures of this kind gathered from informants are potentially misleading because the use of the term divorce covered a number of different eventualities, as discussed above.[30]

> There are a number of forms of separation between a man and wife, from a boy and girl quarrel during the betrothal period, a separation by family arrangement during the early years of marriage, a return of the wife to her own village which can be final or provisional, to the divorce by mutual consent or at appeal of one party obtained before the Native Courts. The latter will probably be the final outcome of a broken union but to any of the intermediate types of separation the Bemba apply the term ukuleka ("to leave") or ukulekana ("to leave one another") which we usually translate by "divorce"
>
> (Richards 1940b:100).

It seems likely that divorce rates were high in the period 1930–1960, but because we have little idea of divorce rates before 1930, it is hard to know if such rates were evidence of breakdown.[31] What is clear, however, is that the experience of marriage had changed for many men and women as a result of labour migration. Evidence from the Copperbelt suggests that men were often involved in serial marriages, although the legal and temporal status of many of these unions was variable. Marriage, whether formal or not, was a way for women to gain access to resources (such as cash, food, and clothes) and for men to gain access to domestic labor. The provision of housing to married couples was a particularly crucial factor in promoting the benefits of marriage (Wilson 1942; Mitchell 1957, 1987; Epstein 1981; Parpart 1986, Chauncey 1981).[32]

There is further evidence for the fact that the meaning of marriage differed for men and women. Despite a high turnover of partners in some situations, it is clear that all men were able to marry if they wished. Many of the informants we interviewed returned from work to the rural areas to marry a woman who had already been chosen for them by their families. Of course this did not preclude a man from making a temporary union in town both before and/or after this event. Labor migration almost certainly produced a situation in which young men had much more autonomy with regard to marriage than they had had previously. However, as we have already argued, many chose to maintain strong kinship links in their home areas, and most had cemented their rural marriages and relations with in-laws through marriage payments. Neverthless, the balance of power between the genders and between the generations had shifted. In the past a young man had had to marry in order to gain independence and to attain adult status through running a household and producing children. Participation in a kin network was thus essential for him if he was to build up the links of dependency and respect that would enable him to become a man of influence in his turn and to draw other men around him. Labor migration provided alternative routes to independence and status. These new routes did not entirely supercede the old ones, but rather complemented them. The fact that residence with a woman's family was no longer really possible under conditions of labor migration meant that a young man who could supply his in-laws with gifts was actually in a stronger position within the marriage than he might have been in the past. He was both more secure and more independent.

The situation for women appears to have been rather different. From the late 1940s onward, the colonial records, as we have indicated, report that villages were full of "unattached women." These reports may, of course, provide a distorted picture, less reflective of reality than of the prevailing discourse on labor migration and social disintegration. They claim, however, that many of the "unattached women" found in villages were neither divorcees nor abandoned wives, but rather unmarried girls and young widows, and it is this claim that we attempt to investigate here. An officer touring Chief Nkolemfumu's area in 1959, commented that the very low frequency of Copperbelt "grass widows" was unexpected. "In this area the married man who goes away in search of work almost invariably takes his wife. The drain of single men, however, inevitably leaves large numbers of unmarried girls in the villages (SEC 2/800 Kasama: TR Sept. 1959). Another officer touring Mpepo's area in 1953 remarked that there were a ". . . large number of women in villages not attached to any registered male in the village . . . most of these women

are unmarried wives and have quite a number of children living in the village"
(SEC 2/844 Mpika: TR April 1953). It is hard to know what this officer meant by an
"unmarried wife." Perhaps, these women were simply divorcees, but then we
might have expected the officer to simply say so. The point he was concerned to
make was in fact one closely connected to his tax collection duties. Here were a
number of women, most of whom had children (hence the use of the term "wife")
who claimed not to be attached to any registered taxpaying man, and whose links
to the villages in which they were resident were far from clear.

Reports of this kind from the 1950s are supported by Ann Tweedie's extensive
ethnographic research. Tweedie studied five villages in the Mpika District in 1958–
1960, and she commented on differences in the marital status of men and women
within these villages. All the men older than twenty years of age, with the exception of two, were married. However, 35.8% of women over the age of fifteen years
were unmarried, widowed, or divorced, or living in the villages when their husbands were elsewhere. What concerned Tweedie was not so much the level of desertion or abandonment of women, something that all commentators noted, but
the fact that there were significant numbers of young unmarried girls and of young
women who had been widowed, divorced, or abandoned, who appeared to be
finding it difficult either to marry or to remarry. Tweedie was unable to collect data
on the length of time widowed and or divorced women had to wait for remarriage,
although she did note that one woman who had been divorced at age twenty-two
had had to wait seven years for remarriage, and three others between seventeen
and thirty years of age had had to wait four years (ATW: Private papers). Tweedie
suggested that the difficulty of finding a husband, and in particular the problem of
remarriage, was causing a rise in the incidence of polygyny.[33] Unfortunately, there
is absolutely no data with which to check this assertion. Discussions with informants produced no evidence of any concern about rising polygyny in the 1950s,
and the situation is further complicated by the fact that widowed women were inherited traditionally by one of their husband's male relatives, and most usually, as
a second wife.[34]

Although Tweedie saw that it was a problem for young women to find a husband or to remarry, her data also showed that there was a great deal of flexibility
in residence rules, which allowed widowed, divorced, or abandoned women to return to the rural area to live with their classificatory brothers or with the children
of one of their sisters. The flexibility of residence rules and the strength of sibling
ties meant that women who were no longer being supported by husbands had
a number of options open to them. The mobility of women within the rural area
had probably increased, giving rise to a situation in which colonial officers found
it difficult to "pin them down." The "unattached wife" then, is probably best
viewed as a woman who has the possibility of "attachment" to a wide range of near
or distant kin, possibilities that she must exploit because she is temporarily without a husband.

It is difficult to know what to make of the observations of colonial officials and
of Tweedie's data. We have argued earlier in the chapter that the ways in which
population and labor data were collected in the colonial period make any analysis
of the effects of labor migration on sex ratios very tentative. We have suggested that
such figures may sometimes overstate the distortion in sex ratios because they
focus exclusively on the number of men absent from any given area, without si-

multaneously providing us with information on the numbers of absent women. There were, nevertheless, significant changes taking place in marriage that may have contributed to the perception of a problem of "surplus women" or "unattached wives."

The age at marriage for men had risen as a result of labor migration. Many men now left the rural area in their teens, many in their early teens, and stayed at work for a number of years before they returned to arrange a rural marriage. This may well account for the distorted sex ratio in the age cohort 15 to 24 years. Thus, there may well have been numbers of young, eligible, but unmarried women in some villages. What seems likely is that most of these women did eventually marry, but that it was the rising age of betrothal and of marriage imposed by the conditions of labor migration that caused parents so much concern, as reported by Richards (1940b:114) and Tweedie. This interpretation is given some support by the fact that we know men were marrying and that many of them were maintaining links with the rural areas, as well as taking their wives with them. There is, therefore, no evidence to suggest an overall decline in marriage for women.

Distinct from this group of young unmarried women, however, was the question of "unattached wives." Some of these were young women, most with one or more children, who had been betrothed and later abandoned by their prospective husbands. They had, in other words, been victims of the very processual nature of marriage in a changing context. Others were likely to have been somewhat older women, many of whom had spent some time in urban areas with their husbands, but whose marriages had broken down or who had been widowed. Upon divorce, widowhood, or abandonment, a woman often returned to the rural area, but given the flexibility of residence amongst most groups in the Northern Province, her exact destination would have been a matter of negotiation, just as it was for returning male migrants. Richards saw this residential indeterminacy as a weakpoint of Bemba society which, in the context of labor migration, would further contribute to social breakdown. Such flexibility might, however, have been a real advantage to divorced or abandoned women who could exploit the possibilities of rural-rural migration and mobility. Pottier has documented a parallel process for Mambwe women who, in the 1970s, as a result of divorce, widowhood, and recession in the urban economy, could be observed returning to villages in which they had male kin (Pottier 1988:13;Ch. 7).[35] In the Mambwe case, the combined effects of land shortage and patrilineal inheritance meant that women without husbands had no other way of gaining access to the land resources necessary for survival, except by turning to their fathers and brothers. The situation for Bemba women in the 1950s and 1960s would have been quite different. There was no land shortage, and the bilateral kinship system gave them the freedom to operationalize many different kinship links. These women were usually welcomed in the communities into which they came because most villages suffered from a generalized labor shortage, rather than simply the absence of men. Tweedie noted in the late 1950s that female-headed households had smaller millet acreages, but she also recorded that all these households had direct rights to millet by one means or another, either through kinship links or through having provided labor in stacking branches or mounding, or through having purchased labor (ATW: Private papers). A critical reading of her data suggests (as we have already suggested) that the crucial factor was not necessarily male labor for cutting citemene because that could be provided through one means or another, but the overall labor supply.

There was a more general shift taking place in marriage by the 1950s, however, which cannot be contained within a discussion of unattached women. This was a shift in the balance of power between partners of a marriage. At an earlier period, when age at first marriage for men was very low and uxorilocal residence was necessary for a young man to build up links and start to acquire a political following, women's position within marriage had almost certainly been much more secure. It was men rather than women who needed to be married. One of the major changes that began with labor migration, and which has continued to the present day, is that it is women rather than men who need to be married (see Chapter 8). Their economic and social security is now much more dependent on the marital tie than it ever was before. This perception of dependence is one that many women who were interviewed seemed to hold simultaneously with views about women's independence and autonomy. A number of informants, when asked about the changing nature and experience of marriage, said much the same thing: "nowadays marriage is too commercial;" "the money economy has led to changes in the approach to marriage;" and in one very memorable phrase, "traditional marriage has been profaned by the money economy."

One very important task in trying to assess the changing meanings and experiences of marriage would be to investigate questions of sexuality, marital relations, and ideas about fertility. This is particularly crucial in the case of the Bemba-speaking peoples of the province because Richards demonstrated so well in the 1930s how ideas about chiefly power were linked to fertility of the land and the people, the latter requiring careful handling of sexual relations. Similar ideas about fecundity and the fertility of individual households meant that husbands and wives had to be careful to separate sexual relations from other kinds of productive and transformative tasks, notably cooking and eating (Richards 1939). Richards also addressed issues of sexuality and marital relations in her book on chisungu (the girl's puberty initiation ritual), where she describes the rites and the teachings that make up the ritual (Richards 1982). The women we interviewed all insisted that chisungu is now rarely performed. In fact, we know that even in the 1930s there were many girls who were married without having the chisungu "danced." The reasons for this decline are not particularly apparent and we cannot assess what it means in terms of changing ideas about sexuality and fertility for women and men in contemporary marriages. The women we interviewed all linked the decline in chisungu to an increase in premarital sexual relations and sexual immorality in general. This suggests that there have been significant changes in ideas about sexuality and marriage relations, but much more work would need to be done to determine this.

Most of the women we interviewed were reluctant to discuss the details of their own chisungu rites, but all of the women married before the 1960s had chisungu "danced" for them. Those women we interviewed who had married since that time deny having gone through chisungu, but it is difficult to assess the validity of this data. One old woman said that "it is still danced in one or two isolated villages." One very general appreciation on the part of women informants, however, was that the decline in chisungu was part of a more general decline in "respect" related to marriage. Women were particularly anxious about grooms taking their brides before the mpango had been paid in full, the decreasing expenditure on marriage ceremonies (this does not apply to church ceremonies), the fact that young men interact directly with their bride-to-be rather than sending a go-between, and the failure to make ceremonial offerings to the son-in-law to allow him to eat in his new home. All these things were identified by the women we interviewed as being the major changes in Bemba marriage. They did not directly link these changes to the decline in chisungu, and we understood that women felt these changes to be the more important. However, overall what many women informants expressed was an anxiety about a lack of respect for marriage and a lack of respect for women in marriage. This view would have to be backed up by much more extensive work with women informants, but it would accord well with the idea that women's position within marriage has become more insecure over time. Women explicitly linked what they saw as a decline in respect to the increasing dominance of the money economy.

Looking back, it seems likely that colonial commentators, including Richards, were undoubtedly correct in their observation that the commercialization of the economy and the high incidence of labor migration were causing difficulties over marriage. These difficulties, we would argue, were felt more acutely by women than they were by men. The meaning and experience of marriage have undoubtedly changed for both women and men, but we can make little sense of this by subscribing to a simple notion of breakdown. The changes in the institution of marriage were, and are, far more complex, as we hope we have shown, than can be contained within such a model.

Residence and Remittances: The End Products of Labor Migration

The final question that has to be asked about labor migration concerns its contribution to the rural economy. In many parts of Africa, and elsewhere in Zambia, remittances from labor migrants have been used to invest in agriculture. This was indeed what Watson argued for the Mambwe, and he linked the success of this investment to the fact that under the patrilineal system of the Mambwe, men had a fixed allegiance to a particular village (Watson 1958:226–227). In short, their long-term interest in land and resources located in a specific village encouraged high levels of remittances, as well as a form of circulatory migration that would end with permanent residence in the natal village. Pottier has recently suggested that the Mambwe no longer hold fixed allegiances to specific villages, and that they return to a general location, making their final residence choice according to a number of considerations, including affinity and links of friendship. He also argues that circulatory rural-urban migration has come to an end and has been replaced by rural-rural migration (Pottier 1983; 1988: Ch. 4).

In this chapter, we have shown that rural-rural migration was an option for people in the Northern Province from the 1920s onward and that it grew over time. Circulatory rural-urban migration was just one of many strategies open to individuals, although there is no doubt that the numbers of people going to the Copperbelt increased steadily, reaching a peak in the 1950s and 1960s. From the 1930s onward, many men stayed away for long periods with their wives, but ultimately returned to the rural areas to retire. This pattern is also typical of the men and women we interviewed in 1988. It is true that some had migrated over long periods of time and others had stayed away only for short periods, some had gone to the

Copperbelt, and others had resided in small towns within the province. However, all 60 informants declared that they had always intended to return to the rural areas to retire, and the chief reason given was that it was impossible to live on a pension in an urban area. We asked our informants two important sets of questions. The first set concerned their village of origin, that of their partner, and their decisions regarding marital residence as they related to their final decisions about resettlement in the rural area. The second set of questions focused on remittances, their frequency and quantity, and their relationship, if any, to a future return to the rural area.

The answers to the first set of questions are not straightforward. What is evident is that returning migrants do not return to a specific village, unless they are returning to take up a headmanship, as in the case of two of our sample, or unless they are returning to reside with a specific headman, as in the case of two women informants, both of whom went to live (with their husbands and children) in villages where their brothers had become headmen. Most migrants speak of returning to a home area, and this is in fact what happens. However, this notion of home area can be very broadly defined. For some, it means anywhere within a particular chief's area, while for others it can mean anywhere within the district. Out of thirty male migrants interviewed, we found only five men who were returning to their natal village. None of the returnees were farming in an area where they had no kin, but the definition of kin in this instance could be very broad, and informants who refer to "cousins" or "nephews" are often unable to trace the exact genealogical connection, usually because the relevant link is thought to be two or three generations back. A further complication results from the high levels of residential mobility and choice that exist in the area. Many men who referred to returning to their father's village or to the village of their parents, when pressed, were actually returning to villages to which their relatives had moved very recently, and these villages were often at a considerable distance from their original natal village. Once again, links between siblings appeared strong and many women reported settling (with their husbands and children) in the villages of their brothers, but in some instances these were classificatory brothers.

Residential choice is discussed in detail in Chapter 8, but one factor that makes data collection difficult, is that returnees often have a number of options as to how they will describe their residential choices. The choice they make is usually a strategic one. For example, many men who describe themselves as returning to their father's village could just as well describe themselves as returning to their mother's village, but they don't. In addition, connections persist over generations. Thus, women often refer to moving to a village of their sister's children, and men born outside the district talk of returning to their father's village when they mean the one in which their father was born. Residential choices are thus very fluid, and open to constant negotiation and renegotiation. The consequences of this for land holding and resource allocation are discussed in Chapter 8.

However, what our data does show is that most migrants return to a home area rather than a specific village. This is regarded by the returnees themselves as a strength, rather than as an indication of the breakdown of social relations. As far as they are concerned, kinship and social relations are alive and well, and this is not surprising given that it was the anthropologists, and not the people themselves, who saw central African matrilineal systems as constantly on the brink of social dissolution. What is interesting in this context is that there is some evidence to suggest that married couples return slightly more frequently to a village where the husband has kin links of some kind, rather than to a village where the wife has such links. But, nothing definite can be said on this point because the size of the sample is too small, and because in many instances both husband and wife have kin links of some sort in the village or area to which they return. This is partly a result of cross-cousin marriage. Additionally, the whole concept of returning to a village is itself open to question, because many returnees, although speaking of having returned to a village actually reside outside the village. The pattern of residence on individual farms started by peasant farmers in the 1940s and 1950s continues to the present day. This residence pattern has consequences for agricultural labor and child nutrition, which are discussed in Chapter 7.

There is some evidence from the 1980 census (Government of Zambia 1980) that there are higher numbers of divorcees and widows in rural areas than there are in urban areas. This situation, which is clearly analogous to the "unattached wives" of colonial days, is related to the fact that when women are divorced or widowed they return to live with kin in rural areas because they can no longer support themselves and their children in urban areas. This situation has become more acute in the last fifteen years, because employment levels and real urban incomes have declined. Women in this situation do not necessarily return to their natal villages, but there is some evidence from case histories and interviews that they often return to villages where their sister, sister's children, or their own children are resident, although there are also examples of women moving to live with their brothers.

Agricultural Investment

Migrant labor and the recent depredations of recession may have transformed social relations, but we cannot understand this transformation simply by characterizing it as a breakdown in kinship or in marital relations. This argument is pursued and developed in Chapters 7 and 8. However, the facts of residential mobility and choice for returning migrants have to be set in the context of data on remittances.

We know very little about the level of remittances coming in to the Northern Province from labor migration. From the beginning of the century, a number of recruitment agencies operated in the province, including the Rhodesia Native Labour Board. Oral histories and colonial archives suggest that most migrants preferred to seek work independently, and that they disliked recruitment agencies because of their compulsory schemes for deferred pay. However, returns from these schemes are our only source of data on remittances for the early period. In 1926, the amount of deferred pay sent back to the province was £54,295, in 1927, it was £45,091, in 1928 it was £52,823, in 1933 it was £8,586, and in 1939 it was £3,500 (Northern Rhodesia Annual Reports for Native Affairs, 1911–1939). These figures do not cover anything other than cash wages, and they do not include workers who went to the Copperbelt. This means that they are underestimates but, in the view of colonial officials, not great underestimates. The Annual Report for Native Affairs for 1933 declared that Africans outside deferred pay schemes were most unlikely to bring back anything other than a very small proportion of what they would have earned in deferred pay. We do know that as the Copperbelt grew in popularity as a destination, so the levels of remittances apparently declined. Men brought back clothes, cloth, and hoes and axes, and they used cash to pay their taxes. However, there is little evidence that workers who went independently to the Copperbelt remitted large sums of money.

In 1940, the Provincial Commissioner wrote to the Chief Secretary saying "While it may be too early to say how the savings of Africans are spent it is quite definite that none of them find their way to Northern Province. All the savings of a Copperbelt worker are spent before he leaves the 'line' and many walk home because they have not money for lorry fares. The number of native mine employees was at 30 April 23,548, their joint savings of £7000 represents just under 5/11.25 each." The Commissioner pointed out that this was a trivial amount when compared with £10–£12 for each Rand worker and several pounds for each Tanganyika sisal worker (SEC 2/788 Kasama: TR March 1940). Godfrey Wilson calculated that in 1940, an average 10.5% of the annual cash wage at Broken Hill was given as gifts to rural relatives, and a further 7.2% was taken to the country by the worker or his wife for their own use. However, most of the wealth transferred to the rural areas was in the form of goods not cash: 10.4% in cloth, 5.8% in cash, 0.4% in pots, and 1.1% in bicycles, guns, sewing machines, and hoes. Wilson estimated that no more than £2,000 of the wealth personally transferred from Broken Hill (i.e., about 3% of cash wages) was ever used to pay rural taxes or buy food. He also suggested that the largest proportion of gifts were taken in person to the rural areas, with a certain amount being sent by other people or being transferred when rural visitors came to town (Wilson 1941:43–44).

Wilson's findings are remarkably similar to our own in some respects. We found it impossible to calculate the total amount of remittances made by any individual over a working life. We also found it difficult to find informants who could tell us what proportion of their income had been remitted in one form or another. However, what we did find is that most remittances, whether of goods or money, were transferred in the course of personal vists, either when the worker or his wife returned home or when rural relatives visited the urban areas. Goods were also sent with friends or relatives when appropriate. Among our informants, there was little expectation that what was required was a regular sum of money. This would no doubt have been welcome, but the fact of the matter is that most people made remittances either when they were visiting or being visited or in response to a particular request. A small number of professional men (police officers, agricultural assistants, and teachers) claimed to have sent money regularly, but the sums involved were very small; although thirty-two out of sixty informants did send some money at one time or another. In general, remittances then were linked to moments of personal contact. This had the advantage of allowing urban dwellers to maintain a wide range of contacts with people from rural areas. Instead of making specific and regular payments to one household, they tended to disburse gifts to a wider range of people whom they might visit or who might visit them. This made vists home and/or unwelcome vists in town from rural relatives very expensive. However, it did mean that when the time came to return home, a wide range of contacts and networks remained open to the returnee. Once again, the data suggests that people from this area were truly investing in kin, rather than in land or capital, or indeed a natal household. Thus it is not surprising that out of the thirty women migrants interviewed, twelve said that they had sent children home to live with relatives, either because they were worried about morals, about Bemba language and traditions, and/or because people in the rural areas wanted help with domestic labor. Two others sent children home on long visits. The children were in these instances, like the goods and the money sent as remittances, an indication of social commitment and respect. In this sense, the notion of labor as service being both a form of a material good and the basis for social relations has remained right up to the present day.

The fact that returning migrants were able to retain social networks while making intermittent remittances is supported by the data on the relationship between remittances and agricultural investment. Only four out of sixty people interviewed (thirty women and thirty men) said that the money they had sent home had been used for agricultural purposes. One man said that when his wife had "been at home" she used to pay people to cut the citemene, and another said that his mother and brothers had asked for "clothes and money to help prepare the land." All the rest said categorically that they knew it had not been used for such purposes because, as one man said, "nobody thought of farming in those days." A further five people claimed to have sent hoes or axes, and one person sent seeds. No one claimed to have sent money home for the purchase of land or capital equipment. What was most notable about these interviews was the certainty with which people dismissed any idea that remittances were used for agricultural investment. Some people, however, did say that they knew money had been used to purchase food and household items, and there was an almost unanimous understanding that the purpose of remittances was to assist in household reproduction and to maintain social relationships. This understanding persists, in part, because many of these people now have married children and they expect their children to contribute to household reproduction, and increasingly to agricultural projects, especially the purchase of inputs (see Chapter 8). The fact that remittances were not used for agricultural investment, at least until the end of the 1970s, underlies the fact that by making remittances, migrants were investing not in land or in capital projects, but in people.

However, since the development of maize cash-cropping, the perception that remittances might usefully be invested in agricultural production has been growing. Several informants bemoaned the fact that they had not saved enough to make agricultural investments, and about 25% said they could not invest in maize-cash cropping because the price of inputs was too high and they didn't have enough money. There was a very clear realization on the part of returned migrants that agriculture could provide a living, but that it was risky and it required high levels of capital input. Of the sixty migrants we interviewed, ten retired before 1950, twenty-two retired before 1970, and twenty-eight retired before 1990. When asked what articles they had brought home with them from the urban areas, five said they had brought clothes, eleven had brought pots and plates, six had brought hoes and axes, seven had brought bicycles, five had brought sewing machines, three had brought vehicles, one had brought items of furniture, and three had brought radios. This list is not a complete inventory of everything people brought with them because it is obvious that a man who came home with a car also came home with clothes. This is, in fact, a list of what people said they had brought home by way of investment. It differs very little from the list that Godfrey Wilson collected, although vehicles are a new addition.

In the 1950s, when migrants returned to become peasant farmers, they did so to show that they had made money and that they had successfuly invested in the project of modernity, which would later, as we indicated in Chapter 5, include nationalist aspirations and hopes. In the late 1980s, people were genuinely returning with the intention of investing in agriculture to make a living. Young men were taken up with the problem of how to make a success of agriculture in the rural areas, and they seemed much less oriented to a working life in the city because the problems of unemployment and high food prices there are well known. The dream of self-sufficiency through working the land began to replace the dream of self-sufficiency through urban employment. Time and time again, we heard people saying "I want to work for myself," "I want to be my own boss."

Despite the fact that agriculture is now a genuine investment possibility, we found that about 25% of the migrants we interviewed made use of skills or resources that they had acquired during their time as migrants to gain access to an off-farm income. Nonagricultural income remains centrally important to many households' economic strategies. One woman who had returned from the Copperbelt with four bicycles began fish trading, and men who had been bricklayers or carpenters continued after a short break to work, albeit irregularly, for the Catholic mission or the Zambia National Service Camp. The informants themselves related these employment decisions both to the length of time it takes to set up a farming enterprise that will bring real returns, and to the risks involved in making a living out of cash-cropping. They also argued strongly that, because of the cost of inputs, cash-cropping was not a viable option without recourse to another source of income. (The relationship between on-farm and off-farm income is discussed in Chapter 8). It was also clear that some returning migrants found farming extremely difficult and that they lacked the necessary skills and knowledge to create successful farming enterprises. Several men had lost all their savings in one or two years of ill-judged agricultural investment, and subsequently found themselves very short of money. Very often, women and children also found it difficult to adjust to this new way of life, and not a few men had retired to the land only to go back into waged employment.

The ideas of home and of returning to the land have been partly created by recent government propaganda and have been fueled by stories of a maize boom in the Northern Province. However, although returning to the land may sometimes seem to be an unrealistic vision, for many it is more realistic than the idea of supporting oneself through old age in town. Unemployment and recession, the rising cost of urban living, and an increase in the outlay necessary to maintain education and housing provision have made it impossible for many to contemplate maintaining a family in town. The stabilization of the labor force, seen by some as the end point of labor migration, may have gone into reverse. For the moment young men in the Northern Province stay at home, and the story of labor migration has come full circle.

7

Working for Salt:
Nutrition in the 1980s

If the food systems of the Northern Province had seemed to a succession of researchers to be potentially vulnerable to the migration of men and the commoditization of the economy in the past, it is not surprising that this issue should be prominent once again in the 1970s and 1980s, when strenuous attempts were being made to commercialize the agricultural economy of the region and, in particular, to introduce maize cash-cropping. During this period, a large number of studies were conducted on the food security and nutrition of households in the Northern Province, as well as in other parts of Zambia.[1] In one particular study carried out by the Integrated Rural Development Programme (IRDP) (operating in the Serenje, Mpika, and Chinsali districts), it was suggested that a correlation existed between increasing farmer commercialization and the declining nutritional status of children under the age of five in farming households (IRDP: 1986). There is more than an echo of Audrey Richards' arguments here.

Anthropometric measurements (weight, height, and mid-upper arm circumference) were collected on a sample of children in different household types. The initial results appeared to suggest that children of farmers growing maize for sale were more likely to suffer from nutritional stress (Table 7.1). The study also emphasized that the problem was more acute in the predominantly cassava-eating areas of the province among households that were most fully integrated into the cash economy. One possible explanation for the association between commercialization and poor nutrition, as noted for other parts of Africa, was that households were selling too much of the food they produced or that their commitment to cash-cropping had the consequence that they grew insufficient quantities of food crops. This hypothesis could be almost immediately discounted for the IRDP survey because the so-called "commercial households" had larger acreages under food crops than other household types, and retained more of the food they produced (Table 7.2). Instead, the tentative hypothesis put forward by the IRDP was that the increasing demand for labor in more commercialized households was restricting the time available to women to prepare food (especially weaning food), with important consequences for the nutrition of small children. This hypothesis was

178

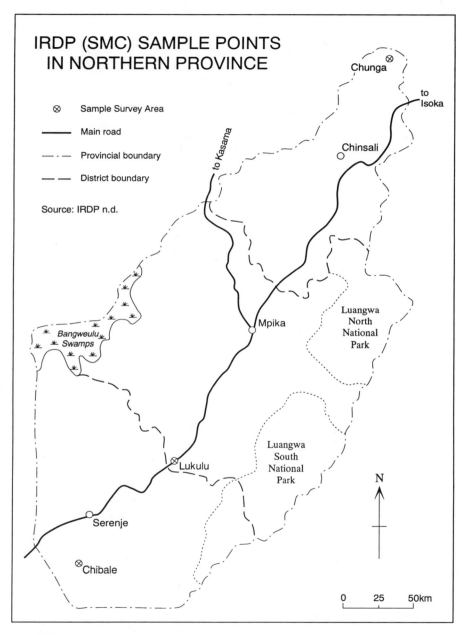

IRDP (SMC) SAMPLE POINTS
IN NORTHERN PROVINCE

⊗   Sample Survey Area

━━   Main road

━·━   Provincial boundary

━ ━   District boundary

Source: IRDP n.d.

Chunga

to Isoka

to Kasama

Chinsali

Luangwa North National Park

Mpika

Luangwa South National Park

Lukulu

N

Serenje

Bangweulu Swamps

⊗ Chibale

0   25   50km

Map 7.1

TABLE 7.1. Nutritional Status of Children Under Five Years of Age by Farmer Category

Farmer Category	No. of Bags Maize Sold	Adequate Nutrition (%)	Mild Malnutrition (%)
1 "subsistence"	0	70	26
2 "emergent"	1–30	52	41
3 "commercial"	30v	50	44

Sample consisted of 205 households containing 166 children between the ages of 6 months to 60 months in areas where hybrid maize production has significantly increased.

(*Source:* IRDP (Serenje, Mpika, and Chinsali.)

TABLE 7.2. Household Size, Landholdings and Food Production by

	Farmer Category		
	Subsistence	Emergent	Commercial
Average family size	6	6	6
Average no. of dependents	2	2	2
Farm area (hectares)	1.94	2.72	4.51
Farm area (hectares) devoted to food crops	1.57	1.66	1.80
Average quantities of food retained by households			
Bags of maize	4	7	11
Finger millet (kg)	13	17	18
Beans (kg)	106	279	420

(*Source:* IRDP (Serenje, Mpika, and Chinsali.)

supported by independent data from both the Western and Northern Provinces, which suggested that maintaining the frequency of feeding of young children was a serious problem for some households, particularly in the busiest agricultural seasons (Freund and Kalumba 1984).[2] A survey in the Eastern Province suggested that levels of food retention and levels of cash income, as well as intrahousehold resource allocation mechanisms all affected the nutritional status of household members (IFPRI 1985). The findings from this last survey supported the well-known contention that income levels are a significant determinant of nutritional adequacy (Kwofie 1979; Sen 1981; Lipton 1983), but it also stressed that they are not always a sufficient condition and that household decision-making processes and women's access to income and resources are also crucial factors.[3]

The ensuing discussion on the constraints affecting women, and especially those impinging on their labor time and control of resources, was given further weight by the renewed debate concerning male absenteeism. The contemporary guise in which this debate is currently presented is in the form of discussions about female-headed households (Table 7.3).[4] One study of female-headed households in the Mpika District observed that these households were characterized by inadequate equipment for food production and processing, as well as by an inadequate

'Marriage and material culture'
Photo: Audrey Richards, 1930s

TABLE 7.3. Percentages of Female/Male-Headed
Households by District for Northern Province

District	Female-Headed	Male-Headed
Chinsali	37	63
Isoka	35	65
Kaputa	34.5	64.5
Kasama	37	63
Luwingu	38	62
Mbala	35	65
Mpika	35	65
Mporokoso	37	63
Chilubi	52	48

(*Source:* Government of Zambia, Central Statistical Office,
1980 Census.)

supply of adult labor. These two factors taken together—although the latter was
the more significant—accounted for the smaller acreages cultivated by these
households and their insufficient levels of food production (Jiggins 1980; see also
Geisler et al. 1985:22). The correlation between constraints on labor supply and

small acreages for female-headed households was confirmed by a survey in the Luapula Province (ARPT 1984).

Female-headed households, it was asserted in a familiar line of argument, could not cultivate sufficient citemene fields because women were unable to cut the trees. Once again, male absenteeism was brought forward to explain nutritional inadequacy. The analysis of one survey conducted in the Kasama area claimed that in the absence of men, female household heads moved to growing cassava on permanent fields, and were unable to meet their basic nutritional requirements, partly because cassava is less nutritious than millet, and partly because it has little market value (Stolen 1983a).[5] In a further study, female-headed households were found to comprise the poorest stratum of the population sampled. These women carried out ukupula (food for work), brewed beer, and collected bush foods in order to gain access to cash. However, the same study showed that out of fourteen female-headed households, only two had no citemene fields, eight had had citemene fields cut for them by their relatives, and four had organized a beer party (ukutumya) in order to get their fields cut (Vedeld and Oygard 1982). Although there is no doubt about the poverty of these households, we would argue that there is little direct evidence to suggest that this poverty is straightforwardly attributable to male absenteeism or to an inability to cultivate citemene fields (Moore and Vaughan 1987). The questions we have to pose are what has been happening to diet and nutrition in the 1980s, and what is the connection between declining nutritional status in some households, women's workload, and women's access to resources?

Rather than begin this discussion with the "maize boom" of the 1980s, we prefer to take a longer historical approach and, drawing on the research summarized in Chapter 3, consider the absolute and relative changes that have taken place in the diet in the Northern Province since the 1930s. This is not an easy task because the sets of data available were collected in very different ways, and as with much nutritional research, are likely to contain a significant percentage of error. Nevertheless it seems to us a task worth attempting, and we summarize the main results of this exercise in the Tables 7.4 and 7.5.

If we look at aggregate figures for food consumption in the Northern Province from the 1930s–1980s, there appears to be very little significant difference over time (see Tables 7.4. and 7.5.).[6] The amount of staple consumed daily per capita has remained remarkably constant, although there seems to have been slight increases in the amount of animal protein, fat, and iron in the diet. However, these increases, like the basic figures themselves, are difficult to substantiate or assess given the problematic nature of the data, its inherent inaccuracies and, in many cases, its highly aggregated nature (especially the Ministry of Agriculture (MAWD) figures in Table 7.4). Nevertheless, the apparent continuities in the character of the diet in the Northern Province are surprising given the changes that we know to have taken place over the last sixty years or so. Chief among these changes are the spread of cassava cultivation, the introduction of hybrid maize both as a cash-crop and a food crop, and the decline in finger millet production in some areas. In terms of cultivation patterns there has been also a general decrease in crop diversity, as evidenced in the almost total absence of sorghum and early maturing varieties of millet in some areas, and the significant decrease in the number and quantity of relish crops grown.[7]

TABLE 7.4. Diet: Average Per Capita Daily Intake Compared Over Time

Source	Calories	Total Protein (gm)	Animal Protein (gm)	Fat (gm)	Calcium (mg)	Iron (mg)	Carotene (mmg)
Richards 1933–34							
1	2,061	63.5	19.0	17.2	1,640	19.4	
2	1,888	47.9	0	14.0	1,610	16.4	
3	816	23.3	2.8	2.9	890	9.4	
Preston Thomson							
1947	1,392	42.1	9.3	9.9	1,202	22.7	2,337
MAWD							
1984	1,700	51.0	21.0	18.2	1,210	38.0	2,290

1: hot season; 2: wet season; 3: hunger months. Figures from Richards and Preston Thomson are based on village level studies. Figures from MAWD (Ministry of Agriculture and Water Development) are aggregates at the provincial level.

(*Source:* Richards 1939; Preston Thomson 1954; Hurlich 1986:67.)

TABLE 7.5. Food Availability in kg per Person per Day Over Time

Source	Date	Average	Range
Richards	1933–34	0.3	0.2–0.9
Gore-Browne	1936	0.4	—
Moffat	1936	—	0.4–0.5
Preston Thomson	1947	—	0.25–0.3
Bookers/RDSB	1979	0.4	—
IRDP (SMC)	1984–5	0.4	0.3–0.6

All figures rounded up to nearest decimal point where possible.

Source: Richards and Widdowson 1936; Gore-Browne 1938; SEC 1/1039; and Sharpe 1987; 1990.

Major changes in the amount of maize, cassava and millet grown have occurred across the province, but they have been variable in their scale and in their effects. The IRDP (SMC) sample points are atypical of the province as a whole because they were initially selected on the grounds that these were areas already demonstrating a commitment to growing hybrid maize. Although it is true that in the years 1975–1988, the production of hybrid maize in the province increased dramatically, it is not true that all areas of the province or all farmers in the province were equally affected, as we shall show in Chapter 8. One result of this is that considerable variability in cropping patterns exists across the province, and indeed within a single location. For example, within the IRDP samples, average sales of hybrid maize per farmer increased from 11.5 90 kg bags in the 1980–81 season to 30.9 bags in the 1986 season, and the percentage of farmers growing maize for sale within the sample rose from 60% to 81% (Sharpe 1987:Section 6). However, we conducted field work in a small area of the Kasama district where the overall commitment to growing hybrid maize was not nearly as marked, and where considerable

TABLE 7.6. Mean Cropped Area Changes 1980–1986

	1980–81 (hectares)	Total Area (%)	1985–86 (hectares)	Total Area (%)
Hybrid Maize	0.76	28	1.11	48
Local Maize	0.37	14	0.10	4
Finger Millet	0.58	22	0.35	15
Others	0.92	36	0.74	33

Others: beans, cassava, groundnuts, sweet potato, ntoyo (voandzia subterranea), pumpkin, vegetables, and intercrops.
(*Source:* Sharpe 1987:Section 6; 1990.)

variability in cropping patterns existed between farmers (see Figure 7.1). In our nutrition and household consumption survey of twenty households conducted in 1988, thirteen were growing hybrid maize for sale, but the amount of maize sold by any one household in the 1987–88 season ranged from 1–123 × 90 kg bags. Of the seven remaining households, four reported growing local maize, while three reported growing no maize at all (although it is likely that the latter were producing small amounts of local maize in their village gardens for consumption when green). In general, data from the province suggests that the introduction of hybrid maize leads to a reduction in the area of land devoted to other crops, notably local maize and finger millet (Table 7.6). However, in maize cash-cropping households in the northeast of the province there was also an expansion of finger millet production on permanent fields (ibala). The explanation for this phenomenon may be connected to decreasing opportunities for citemene production and the increasing need for millet to produce beer as payment for labor (Bolt and Holdsworth 1987; Bolt and Silavwe 1989).

In his analysis of the IRDP (SMC) data, Sharpe notes that figures for mean changes in cropped areas are indicative of what is taking place overall at the sample points, but that wide variations exist in the actual cropped areas of individual farmers and this reflects their individual circumstances (Sharpe 1987:Section 6; see also Bolt and Holdsworth 1987). Once again the available data, including our own fieldwork, demonstrates that there is enormous variability in cropping patterns across the province, just as there was in earlier periods (Chapter 2), and that these microadaptations reflect not only local ecological variables, and differences in infrastructural and marketing support, but also the specific circumstances of individual farmers.

Women's Views on Diet and Nutrition

The variable nature of cropping patterns across the province and of farmer circumstances means that that there are no easy generalizations to be made about changes in diet and household consumption patterns. Much of the information available is at the aggregate level and is likely to mask rather than to reveal the salient differences. The nutrition and household consumption survey of twenty households that we conducted in 1988 sought, among other things, to solicit the views of women on dietary sufficiencies and insufficiencies, as well as on dietary change. We were particularly concerned to investigate the links between nutritional inadequacy and women's workload (suggested by the IRDP investigations), but to do so from the point of view of the perceptions of individual women.[8] Sixteen out of twenty women said, in open-ended interviews, that they felt that the diet was poor because of the high price of essential commodities, a clear statement of the important role of cash in household reproduction. The items listed by the women as essential were salt, sugar, soap, clothes, parafin, cooking oil, mealie meal, relishes (primarily meat, fish, kapenta (small dried fish), vegetables), and school fees. There was a straightforward perception that the increasing monetization of the economy had made life much harder. As one woman said: "At present the diet has become very poor because prices are too high. In the past, little money could buy alot of food." The local term for this economic situation of inflation and high prices leading to grave difficulties for people is "mitengo." A number of informants mentioned that either they had been involved in, or they knew other people who had been involved in the vicious circle of having to sell food to gain cash, and then finding themselves without enough to eat. One informant explained: ". . . food is not enough. Also, most people have become business-minded such that even the little that is grown is used in the brewing of beers. From beers not enough money is realized to combat hunger." The phrase "business-minded" refers to the fact that people have to seek any opportunity they can to generate cash. Beer brewing is the most reliable way of turning millet into cash, and it is a source of income over which women retain a significant degree of control, but, as our informant stated, it rarely brings in sufficient money for household needs. Additionally, some informants felt that people had grown enough to eat in the past, and that the food had been sufficient because people had helped each other and because there had not been the same need for cash: "In the past food was plentiful because people were not business-minded." Others, however, thought that more food was being grown now than in the past, but that families still ran out of food because of the way in which food was needed to generate cash. None of the women interviewed felt that they had ever experienced famine in their lives, although some reported experiencing shortages or hunger (nsala).

When the nutrition and household survey was completed, we asked the two research assistants who had worked on the survey to analyze the data and give their interpretations. One of them clearly identified "budgeting" as a problem.

> *Budgeting of foods and income is not so good in certain families. Food shortages in these homes happen not because they don't grow enough but due to poor budgeting. People want MONEY first and then other things such as food are bought using the same money from the sales of farm produce. All the food produced is sold especially maize a major source of mealiemeal and part of it—plus millet—are used to brew beers. From these businesses sometimes very little income is realised, such that to buy similar foods is quite expensive. So shortages begin to occur in these ways.[9]*

This analysis, like those provided by the women interviewed, suggests that some households are engaged in overselling of staples, contrary to the IRDP (SMC)

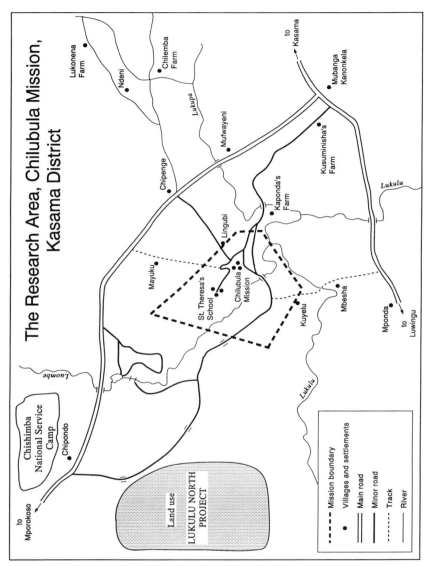

The Research Area, Chilubula Mission, Kasama District

Map 7.2

findings (see above). Overall, however, we would argue that this problem, which is much more acute for some households than for others, is not simply a matter of overselling. Rather it is a question of how to manipulate exchange cycles in order to ensure household reproduction. This argument is developed further in Chapter 8.

The Question of Relish

> *". . . to improve the food-supply in this area by increasing the production of millet would be useless without at the same time ensuring an adequate supply of vegetable crops for relish"*
>
> (Richards 1939:48).

Nine out of twenty women declared that shortage of relish foods was a major problem, and that this affected the quality of the diet. Richards had recorded one of her informants as saying in the 1930s: "when we have little umunani (relish), then we cook little porridge" (Richards 1939:48). This view was widely held by women in the 1980s, who definitely thought that the supply of relish affected both the quantity of food eaten and the number of meals consumed per day. One informant straightforwardly said: "lack of relish nowadays contributes to the many cases of malnutrition" (see later for further discussion on this point). Sharpe noted that for IRDP (SMC) farmers, the area devoted to relish crops declined by about 50% between 1980–86. This decline might not have been particularly large in terms of overall hectarage, but its affects on household nutrition were disproportionately large because of the link between relish supply and the size and frequency of meals (Sharpe 1987: Section 6). The overall quantity of relish grown would appear to have declined because of the dramatic increase in women's on-farm labor consequent on the introduction of hybrid maize resulting in the withdrawal of labor investment in relish crops on village gardens; an argument we explore in Chapter 8. At the same time, there has been a decrease in the overall level of intercropping in the agricultural production system, which has had a negative affect on the quantities of relish crops produced because many were previously intercropped with staples, as well as with each other. The available data shows that with commercialization of agriculture there is a marked decrease in levels of intercropping on both citemene and ibala (permanent) fields, and a consequent decline in crop diversity (Bolt and Holdsworth 1987:Section 4; Sharpe 1987:Section 6; Evans and Young 1988:Section 4, see also Chapter 2 this text). The Adaptive Research Planning Team's (ARPT) results suggest that it may be farmers in the initial stages of commercialization whose food supplies and general well-being are most vulnerable to the rapid substitution of diverse crop production under citemene systems for specialized production on permanent fields (Bolt and Holdsworth 1987; Bolt et al. 1989). Women informants who were interviewed on this point expressed anxiety not only about the overall decline in the availability of relish crops, but also about a decrease in the variety of relishes available.

There is a series of further identifiable factors that tend toward reductions in crop diversity and the number and amount of relish crops grown. One is that in some circumstances women's labor is now more directly controlled by men because of the labor pressures associated with hybrid maize production. Thus, women in some households are less able to control the allocation of their own labor than they were previously. Under such circumstances, women find it increasingly difficult to maintain the flexible production strategies that had earlier characterized their decisions concerning cropping patterns, both on the citemene fields and on the village gardens. These issues are discussed in more detail in Chapter 8.

A second set of factors that affect the availability of relish crops and the range of foodstuffs grown are those associated with changes in residence patterns. The government policy of village regrouping after Independence and the consistent drive to persuade farmers to abandon citemene in favor of permanent cultivation has had some moderate success, producing a situation in which villages move very much less than they did in the past. The result is that wild relish resources surrounding villages have long since been depleted, and the same is true of game. One woman interviewed in the 1988 survey said that she had abandoned the consumption of certain bush products because they were no longer to be found locally, and she did not have the time to go and look for them further afield. The crucial factor here of pressure on women's time was recognized by Richards even in the 1930s, when she commented that households "which have not got young girls to help scour the bush for relish go short of food when the agricultural day has been heavy" (Richards 1939:104).

The general increase in population levels and the concentration of settlements close to roads and market centers have contributed to this problem. Bush meat and fish are now very rare in certain areas, and extremely expensive in others. Regulations governing hunting also restrict the exploitation of game resources.[10] Kapenta (dried fish) and butchered meat are on sale in some places, but they are also very expensive. One divorced woman interviewed in the 1988 survey said that she could not remember the last time she had eaten meat. There may be an element of anxious overemphasis in some informants' accounts of the absolute scarcity of meat and fish, but there is no doubt about the overall decline in the availability of relish crops in some areas.

Increases in population and changes in residence patterns do not entirely account for the fact that some households suffer from an acute shortage of relish crops. Because of their general scarcity, these crops (both those that are cultivated and those that are gathered) have very high barter and cash value. Beans, groundnuts, and sweet potatoes are regularly sold for money or bartered for scarce commodities. One woman interviewed in the 1988 survey said that she bartered groundnuts for salt; and traders from the Copperbelt frequently visited the rural areas around Kasama and Mporokoso where they would barter clothes and other commodities with local women in order to obtain beans for resale to urban populations. The bursar of one girl's secondary school in the late 1980s found it so difficult to get sufficient relish to feed the pupils that she would drive out to rural areas with her van loaded high with blankets and clothes to barter for beans.[11] Evans and Young (1988) and Sharpe (1987) record the extensive use of beans, groundnuts, and sweet potatoes in barter exchanges in the IRDP (SMC) areas in Northern Province. In 1982, Vedeld and Oygard recorded that one tin of beans, in a rural area, could be exchanged for twelve bars of soap (which was held to be equivalent to K12), though if the same amount of soap had been purchased in Kasama it would only have cost K6 (Vedeld and Oygard 1982:36). The poor development of markets, the poor infrastructure in many parts of the province,

'Girls collecting mushrooms'
Photo: Audrey Richards, 1930s

and the declining availability of any form of public transportation clearly produce this situation.

Gathered bush products, in areas where they are still readily available, have equally high barter and cash values. These foodstuffs are particularly important for women because they are a resource that they can control. Eighteen out of the twenty women interviewed in 1988 said that they regularly collected foods from the bush, and that they relied upon them to supplement the household diet and to raise cash. The role of bush foods in supplementing household diet was perceived as particularly crucial by many informants in situations in which pur-

TABLE 7.7.   Gathered and Hunted Foodstuffs in the 1980s

Item	Seasonal Availability
Boa (mushrooms)	November–April
Masuku (loquots)	October–December
Imfungo (wild plums)	October–December
Caterpillars	November–April
Mpundu (wild fruit)	June–August
Mangoes	November–April (from old village sites)
Guavas	May–July (from old village sites)
Fish	All Year
Bush Meat	All Year

(*Source:* Nutrition Survey 1988.)

chased dietary items (notably fish, meat, and other relishes) are becoming increasingly expensive, and prohibitively so for some households. As one woman said: "Thanks to God for a free gift because it really helps us to cut down on food expenditure." The major seasonal bush foods gathered by women in the survey are listed in Table 7.7. Other wild foods, such as wild spinach, bitter leaves, and wild orchid are still eaten, but were not mentioned in the survey because they have less potential as cash earners. Thus, the very foods that Richards mentioned as essential to the Bemba diet in the 1930s are still central to the diet of the Northern Province today, but their value is, if anything, enhanced by the fact that they provide women with much needed access to cash or to bartered manufactured items. It is indicative that of the two women in the sample who said that they did not collect bush products, one said that she had no need to do so because she was in business, and the second said that she would not do so because her husband was employed and they were a "modern" family. Thus there are cultural and status considerations that affect the consumption or nonconsumption of bush foods, although income is still the crucial factor.

One women interviewed in 1988 claimed to have realized K500 from the sale of caterpillars. This was a very substantial sum. Both Evans and Young (1988:Section 6) and Sharpe (1987:Section 9) record women selling caterpillars and exchanging them with traders. In such situations, certain bush foods, like some cultivated relish crops, may function as a substitute for money rather than simply as barter items. Thus, although bush foods continue to play a vital role in the diet of populations in the Northern Province, the extent to which individual households actually consume them depends on a number of factors that relate not only to their immediate availability for gathering, but also to household income and to women's needs for cash and commodities.

It is clear that the link identified by the women in the 1988 survey between nutritional inadequacy and the decline in relish crops is an important one. This decline is itself associated both with changes in cropping patterns and women's labor input, as well as with the absolute necessity for households to gain access to cash. This latter point was well recognized by the women in the survey who also saw a clear link between nutritional inadequacy and cash shortages. Once the question of how women get access to cash is identified as central to the issue of

TABLE 7.8. Sources of Household Income 1988

Source	Number of Households Out of a Sample of Twenty
Beer sales	14
Maize sales	12
Full-time employment	8
Gathered foods	7
Vegetable sales	7
Fishing	4
Part-time employment	3
Cigarette sales	2
Business	2
Remittances	2
Pension	1

(*Source:* Nutrition Survey 1988.)

adequate nutrition, we can begin to unravel how wider household consumption needs affect food availability (see footnote 3).

Consumption and Cash

Table 7.8 gives the sources of income of the households in the nutrition survey of 1988.[12] Sales of beer, cigarettes, fish, vegetables, and gathered foods were controlled by women. Men controlled the income from maize sales and from full-time and part-time employment, the exception being the case of one divorced woman who was employed by the local Catholic mission. The two businesses mentioned in Table 7.8 were both run by women. The only pension recorded was controlled by a man, and of the two sets of remittances noted, one was controlled by a man and the other by an elderly woman. What is immediately apparent is that women's sources of income are by and large distinct from those of men, and the strategies they employ are diverse, although all are dependent in one way or another on food.

Another factor that affects the supply of food within the household is the use of food for labor recruitment. There are a number of ways in which food is used to procure labor. There are the communal working parties—usually of men for the cutting of citemene and/or land preparation tasks, but sometimes of women for the piling and stacking of branches—in which beer is given in return for work (ukutumya). These working parties were recorded by Richards in the 1930s (Richards 1939:145–147), and they have remained an important tool for mobilizing labor, particularly for land preparation, up to the present day. There is some evidence from the IRDP (SMC) sample points that the more commercialized farmers dislike this method of labor recruitment, partly because it is difficult to supervise the quality and the quantity of the work done, and partly because it carries with it an expectation of reciprocity (Sharpe 1987: Section 10; Evans and Young 1988: Section 6; see also Chapter 8, this text).[13] However, as we have already indicated, ukutumya parties are a crucial means of labor recruitment for women, enabling them to procure male labor for the cutting of citemene. Finger millet is essential for beer brewing and the practice of using beer to recruit and remunerate labor acts as a considerable drain on stocks of this staple.[14] The women interviewed in the nutrition survey in 1988 all mentioned that they engaged in the brewing of beer, some for labor recruitment and some for sale. The money raised from beer sales can be invested in labor in its turn or used for other purposes. The number of times any particular household brews in a given agriculture season is extremely variable and depends on the quantities of finger miller available, the amount of nonhousehold labor required for agricultural activities, and the relationship between household income and expenditure. In a survey of three villages, 40–60% of households brewed beer and had organized at least one ukutumya during the period from March 1981 through March 1982 (Vedeld and Oygard 1982:46–47).

There was a general feeling among women interviewed in the 1988 nutrition survey that beer brewing was on the increase. Five women claimed that diets were poor because so much food was used for brewing and because adults were spending more time drinking than cultivating their fields. The wasteful use of finger millet for brewing was a constant theme of colonial reports from the Northern Province and one with strong moral overtones, although it does seem likely that increased demands for cash and for labor under present farming conditions have led to a genuine increase in the amount of beer brewed. Sharpe notes that beer sales had increased in Chibale in line with the increase in hybrid maize cultivation. He also points out that profits on beer are high (between 60–200% of the cost of the raw grain depending on the proportion of finger millet to maize). Therefore, the returns per hour of labor are higher than any other source of income for women and, indeed, higher than the rates of return on maize farming (Sharpe 1987: Section 11). However, it should be noted here that beer is also a form of food, and that many of the women interviewed stated that when food was short their husbands subsisted almost entirely on beer. Several individuals also indicated that joining work-parties was one way of gaining access to beer and food, which prevented them from having to further deplete their own meager food reserves.[15]

The relationship between food and labor established in the farming systems of the Northern Province means that labor-short households are vulnerable, not only because they may be unable to produce enough in the first place, but because they may have to further deplete their food stocks either to recruit labor at critical moments in the following season and/or to raise cash for household reproduction. The seasonal nature of agricultural production exacerbates this situation because as Richards noted in the 1930s, periods of heavy labor investment coincide with periods of low food stocks or food shortage (see Chapter 8). This gives rise to a further problem for vulnerable households because they may have to reallocate their labor away from their own farms and onto the farms of others at critical moments in the agricultural cycle, either because they are in need of food or because they are in need of cash.

This brings us back to the question of ukupula (food for work), which we discussed in Chapter 3. Many women in the Northern Province appear to be performing increasing amounts of ukupula. Women are particularly keen to work for finger millet or groundnuts, not only because they might be food short, but also because these crops have a market value and can either be eaten or used to engage in further exchanges. Finger millet is particularly important in this regard because of the way it participates in the cycle of labor-food-beer-labor-food. Men may occasionally hire out their labor in this way, but they do not like to admit to it, preferring to participate in the communal work parties. Conversely, women are not only keen to work for food, but for other commodities that they can use themselves or redirect into further sets of exchanges. Two very important items are salt and soap. The popularity of this kind of work for women resides in the fact that they can retain control over commodities much more easily than they can over cash, and that in a situation where commodities are scarce (due to a generalized failure of the retail distribution system in the 1980s), this strategy is a much more effective way of gaining access to essential goods. Women's exchange and barter networks are extensive, and, in some parts of the province, salt is a more acceptable form of currency for women than cash. Evans and Young argue that in Lukulu (a matrilineal Bisa area) and Chunga (a patrilineal Nyamwanga area), women of differing socioeconomic levels have come to rely on casual labor during the period from November through February as a means of acquiring their weekly or monthly supplies of essential items such as salt, soap, and sugar (Evans and Young 1988: Section 6). Employers of female labor find that this system suits them because it is much less costly than hiring labor on an hourly basis at the going cash-wage rate (Vedeld and Oygard 1982:48).[16]

An interesting corollary of this labor-for-commodities system is that women make up the vast majority of hired laborers. As Evans and Young remark (1988:Section 6), this is undoubtedly because there is no developed market for labor (partly due to low levels of land pressure and a very restricted land market). Were there to be such a market, the men would probably predominate in agricultural wage labor. It is also not surprising that most employers of labor are men; although wealthier farmers use their wives to supervise the hired female labor.

There is some evidence to suggest that households that are fortunate enough to be able to balance the input/output demands of labor and food can find aspects of the system of ukupula burdensome. One woman interviewed in the 1988 survey claimed that: "The villages are clustered with widows and divorcees who need help. . . it is these people who contribute to low yields as they come to work for food, if not to beg." Many women who practise ukupula also offer other women food in exchange for labor. Although ukupula is certainly a means by which the more commercialized farmers acquire cheap labor, the less commercialized and the less well-off members of the community have to engage in sets of reciprocal relations involving ukupula. This is essential if households are to have access to networks of support when they are in difficulties, and if they are to get access to nonhousehold labor at peak points in the agricultural cycle. However, if they themselves engage in ukupula, then they are likely to find that at other times they are targeted by women who themselves need to work for food, either because they are food short or because they need access to exchange relations. Ukupula is not an entirely asymmetrical system of exchange because there is an element of long-term reciprocal obligation, as we shall see in Chapter 8. This does not, however, prevent it from sometimes being a burdensome institution.

Community, Kinship, and Sharing

Audrey Richards argued in the 1930s that the sharing of food and domestic labor between households (joint housekeeping) provided for improved food security and also eased the burden of domestic labor. As we have seen in Chapter 3, she was fearful, however, that this cooperative domestic economy would breakdown under the impact of the cash economy, and that redistributive mechanisms based on kinship and residence would give way to redistributive mechanisms based on the market. The question then of the degree to which joint housekeeping and sharing have remained features of village life in the 1980s would seem to be central to any analysis of food security and household reproductive strategies.

In the early 1980s, Stromgaard studied a village to the south of Kasama along the Kasama-Mpika road. He found that the village was made up of a number of uxorilocal extended families intermingled with small matrilineal descent groups and some different single families. Stromgaard argued that ukutumya (beer for work) working parties and extensive sharing were only common within the uxorilocal extended families. In the single-nucleus households, as Stromgaard termed them, production and consumption was organized on more narrow lines, involving only the members of the two generational household (Stromgaard 1985a:46–47). Stromgaard further distinguished four types of household in the village: "progressive," "traditional," "poor," and "dependent." Progressive households had relatively high annual incomes and were most often single-nucleus households. The traditional households, characterized by subsistence-oriented production and a limited cash flow, were dominant within the major uxorilocal groups, and they relied on younger relatives and sons-in-law to meet their labor needs. Poor households were typically "incomplete single nuclear families" with inadequate cash resources and no nearby relatives to provide labor or other help. Dependent households, however, consisted of members of the uxorilocal groups who were resource poor, most likely as a result of changes in the developmental cycle or misfortune. The benefits of sharing, both for production and consumption appeared to be clear. None of the households belonging to the dominant uxorilocal groups (bakalamba) could be classed as poor, although some were dependent, and yet others would have the potential to develop into progressive households (Stromgaard 1985a:50). Stromgaard's four-part typology may be open to question, but he does demonstrate usefully the function of sharing strategies as systems of redistribution within uxorilocal groups that serve to protect the more vulnerable members.

As we have indicated in earlier chapters, the use of the term "matrilineal" to describe the social organization of the peoples of the Northern Province may obscure more than it illuminates (see Chapter 3 and Chapter 6). Village structure varies enormously over the province, even within the boundaries of any one ethnic group. Given this variability (and its instability), it is hard to assess Stromgaard's view that a traditional uxorilocal extended family pattern is breaking down (Stromgaard 1985a:50). What is clear, however, is that the presence in a village of two generational households that apparently do not belong to an uxorilocal group cannot be taken by itself as evidence of the breakdown of a previously rule-bound and stable system. Richards noted in the 1930s that Bemba villages frequently contained households that had migrated in and that had only the most tenuous of kinship ties with the bakalamba (Richards 1939:Ch. 7). In the early 1960s, Harries-Jones and Chiwale made exactly the same argument in stronger terms, on the basis of one of the villages previously studied by Richards (Harries-Jones and Chiwale 1963). The political and economic importance of gaining followers had always meant that the structure of Bemba villages was multiplex and strategic, and such villages often contained households whose kinship ties to the dominant matriline were those of affinity and/or allegiance.

The structure of villages in the Northern Province has been further complicated over many years by colonial and postcolonial attempts to control settlement, as we showed in Chapters 4 and 5. Despite policies of regroupment and containment, there are a number of farmers who do not live in close proximity to villages, although they may be surrounded by the homes of people who work for them. Settlement patterns have been further affected by the necessity for proximity to roads and services for the production and marketing of cash crops. All of these factors taken together mean that there are a number of factors affecting settlement patterns, village locations, and the residence choices of individual farmers, other than the ties of kinship. The net result is enormous variability across the province from villages based almost entirely on identifiable kinship links to those whose metaphor of connection is more likely to be one of community or neighborhood. It is therefore extremely difficult to say in any straightforward way that kinship, as evidenced by sharing and residence strategies, is breaking down.[17]

Just as we have argued in Chapter 6 that the metaphor of breakdown does not capture the changes that have taken place in the institution of marriage in the province over time, we would argue that this applies also to the related question of kinship. In considering the role that kinship has played in people's strategies of household reproduction and survival, we would point, as Pottier does in his restudy of the Mambwe in the northeast of the province, to the importance of links between women (Pottier 1988). Watson had argued that the kinship system of the patrilineal Mambwe had actually been effectively strengthened by the effects of labor migration and a relatively bouyant economy (Watson 1958). Pottier's restudy, carried out in the difficult economic and political conditions of 1978–79, described the Mambwe kinship system as responding to economic recession and increasing pressure on land through the strengthening of ties between agnatically related women and uxorilocal residence for young grooms. What marked a real change in the Mambwe kinship and residence system between the 1950s and the 1980s was the growing tendency for middle-aged women to return to villages where they have male kin. This resulted in the development of villages whose cores consisted of agnatically related women. These changes in the kinship system seem to have provided women with the opportunity to develop collaborative strategies for managing the depredations of the money economy (Pottier 1988:Ch. 7).

There is, thus, a certain amount of evidence from different regions of the province that suggests that women's strategies of collaboration with other women are part of an adaptation to increasing economic pressure and hardship. Women who, for one reason or another, do not participate in systems of collaboration may indeed find themselves in a very vulnerable situation. This point is borne out to some extent by Sharpe's investigation of the causes of poor nutrition among children under the age of five years in the IRDP (SMC) households (Sharpe 1987; 1990). Sharpe's work was commissioned to examine the apparent correlation between increasing farmer commercialization and the declining nutritional status of children, which we mentioned at the beginning of this chapter. He found that within the IRDP (SMC) sample points, local patterns of undernutrition varied. One very important factor was isolation. The evidence, from Chibale (a matrilineal Lala area) in particular, was that foci of undernutrition were explicable in terms of households that had withdrawn from interhousehold exchange relations. These exchange relations, which had involved food, labor, childcare, looking after the elderly, and a variety of other small gifts and services had simply become too burdensome for some to maintain, given the labor demands of maize cash-cropping. For the less successful of the cash-cropping households, the demands on labor and food stocks that reciprocal exchange systems imply, appeared to be preventing them from gaining positive returns on their farming enterprise (Sharpe 1987:Section 8).[18] Having withdrawn from these systems, however, many of such households became particularly vulnerable and demonstrated a higher incidence of child malnutrition. The most successful commercial farming households did not face such problems. Having withdrawn from systems of reciprocity, they were able to replace at least some of their welfare functions through cash exchanges. A similar argument is presented in reports produced by the the the ARPT team based in Kasama, which suggest that poor nutrition and a decrease in food consumption are only features of households in the early stages of commercialization. Once commercialization is well established, food consumption improves (Bolt, 1989; Bolt and Silavwe 1988; 1989; see also Ch. 8, this text).

Poor nutrition, Sharpe found, was also a feature of groups of households that had coalesced around specific commercial farmers. These households had followed the original settler onto new land and into new sets of quasicontractual relations governing labor. In such situations, both the commercial farmer and the "dependent" households had resisted the setting up of reciprocal exchange systems, with the result that children had become much more dependent upon the food and childcare provided within the individual households of which they were members. Such settlements showed severe malnutrition affecting the children of all households, regardless of farmer category (Sharpe 1987:Section 8; 1990:600).

The best nourished children were to be found in settlements where interhousehold exchange systems had been maintained. One focus of adequate nutrition in Chibale was a relatively poor group of female-headed households. In this case, the children benefited from the exchange of gifts and services, and from food-sharing and shared childcare (Sharpe 1987:Section 8). This last finding echoes that made by Richards when she argued that the system of joint-housekeeping improved food security. It also points to the importance of collaborative networks between women for maintaining child welfare under the prevailing conditions in the Northern Province. These female-headed households were definitely poor, as Sharpe makes clear, but their levels of food security and the nutritional status of their children reflected the viability of women's collaborative strategies.[19]

The women interviewed in the 1988 nutrition survey were asked a number of questions about the issues of food sharing and joint housekeeping. Our results, as with much of the material presented above, were ambiguous. Some households were sharing food, both cooked and raw, with others. A few women claimed that food sharing and shared domestic labor between households was a thing of the past; although basic observation certainly contradicted this view because much domestic labor, particularly food processing, was done by groups of collaborating women. The ability of more senior women to command the labor of young, unmarried women was crucial to the continued existence of these collaborative groups. There was, however, a general perception among informants that the sharing of food and domestic labor had become much more calculated than in the past, because all were aware of the exchange and monetary value of both food and labor. Many women reported thinking hard about what they were prepared to do for others, including kin, and what they were not. Several women reported conflicts between using food or commodities to engage in exchange or barter (thus furthering immediate individual or household interests) and using the same products to assist kin (and thereby investing in a local safety network). Nonetheless, twelve out of twenty women reported getting help from their extended family, either in the form of food or in the form of domestic and agricultural labor. When asked about the sharing of food, one women said: "The food produced is sometimes transferred to other members of the extended family who are hungry. Both raw and cooked food—including millet, groundnuts, and relish vegetables." It would thus be wrong on the basis of this material to assert that shared consumption and collaboration were things of the past, but it is clear that there is enormous variability in the extent of interhousehold cooperation. How much this variability is directly attributable to contemporary circumstances in the Northern Province is difficult to ascertain. One is reminded of the young couple whom Richards thought were so selfish in the 1930s because they ate by themselves. There may be many more households in this situation in the province today than there were in the past, but the phenomenon is not by any means a new one.

Women, Work and Weaning

From Richards onward, the causes of poor nutrition in the Northern Province have been assumed to be related to the absence of male labor and a consequent decline in the production of foodstuffs. We have argued against this model in earlier chapters, pointing to the adaptations to the absence of male labor that were successfully made by the agricultural systems of this area. Labor is, however, crucial to an understanding of the causes of child malnutrition in present-day Northern Province, and it is women's labor that is at stake here, not so much in the context of the absence of men, but rather in the context of their presence as heads of cash-cropping households.

In our nutrition survey of 1988, six of the twenty women interviewed made a direct link between child malnutrition and women's workload. In all six cases, the field preparation period and the harvesting period were identified as seasons of the year in which children were likely to suffer because women did not have enough time to prepare food. As one informant said: "Mothers notice these problems (malnutrition and wasting) when it is time for citemene and harvesting because there is a lot of work to be done and less attention is given to children." One important perception was that women were traveling increased distances to their fields, and that children who were left at home were going hungry. The standard pattern during busy periods in the agricultural season is for women to leave home very early in the morning and return in the late afternoon to cook. Children might be given leftovers in the morning or a little porridge or some cassava or bananas, but they are likely to be fed only once a day. The increasing distance to fields, which is the corollary of a more permanent residence pattern, increases the pres-

TABLE 7.9. Mean Labor In-put by Wives In Hours: Domestic and On-Farm Work Compared

| | Farmer Category | | | | |
	A	B/C	D	E	F
Chunga					
Domestic	884	817.5	1131	1142	1115
On-Farm	467.5	202	430	456.5	513
Lukulu					
Domestic	961	860	—	944	841
On-Farm	513	462	—	744	624.5

A: Selling no maize or selling for no more than 1 year out of 6; B: Selling 1–10 bags for less than 4 years out of 6; C: Selling 11–30 bags for less than 4 years out of 6; D: Selling 1–10 bags for 4 or more years out of 6; E: Selling 11–30 bags for 4 or more years out of 6; F: selling over 30 bags for 4 or more years out of 6.

(*Source:* Adapted from Evans and Young 1988: Section 6.)

sure of women's time. During his research in the early 1980s, Stromgaard calculated that women spent 16% of their total time walking to and from the fields (Stromgaard 1985a:42; see also Chapter 2, this text). One woman interviewed in the nutrition survey claimed: "These diseases (malnutrition) very much related to work of mothers especially with working places far off. They starve their children at home and those taken to their working places." This particular informant came from a better-off household and her point about "starving" children is related to the fact that children left at home while their mothers are working must wait until her return before a meal is cooked, while children who go to the fields have to wait until they go home before they can eat. Most women interviewed said that they preferred to leave their children at home so that they could have some food during the day. What is interesting, however, is the fact that several informants, when asked questions about child feeding practices, accused "other women" of "starving" their children. It is hard to know exactly how to interpret this, but it is clear that many women perceive a link between child malnutrition and women's workload, and that there is a degree of anxiety about the stress this places on mothers and children.

The introduction of hybrid maize has increased women's and men's on-farm labor (IRDP 1984; 1985a; Bolt and Holdsworth 1987; Bolt and Silavwe 1988; See chapter 8, this text). However, there is also some evidence to link changes in cropping patterns with increases in women's domestic labor. Table 7.9 shows the real burden of women's domestic labor. Women from all household categories spend many more hours on domestic labor than they do in farm work; and, in the case of the more commercialized households in the Chunga sample, they spend double the amount of time on domestic labor. The figures for the Lukulu sample show that the ratio of domestic labor to on-farm work declines with the degree of commercialization of the household, which suggests that women have to withdraw from domestic labor as the pressure from farm labor increases. It should not be forgotten, however, that the amount of time any individual woman spends on domestic

TABLE 7.10. Mean Cropped Area Cassava and Finger Millet (hectares) in Lukulu and Chunga 1986–1987

	Farmer Category					
	A	B	C	D	E	F
Lukulu						
Cassava	0.02	0.04	0.0	0.0	0.0	0.0
F. Millet	0.21	0.9	0.09	0.9	0.15	0.25
Chunga						
Cassava	0.93	0.42	0.33	0.55	0.65	1.15
F. Millet	0.38	0.32	0.38	0.31	0.35	0.53

(*Source:* Adapted from Evans and Young 1988: Section 4.)

labor will vary considerably with the developmental cycle of the household, and that women with teenage and adult daughters may be able to delegate domestic responsibilities.

The tasks recorded as domestic labor in Table 7.9 include household chores, water and firewood collection, household repairs, and caring for children and the sick. The exact definition of domestic labor and the degree to which it can or cannot be unambiguously distinguished from productive labor is always a contentious issue (See Moore 1988:43; 53). However, the figures given in Table 7.9 include food processing, both for household consumption and for storage and sale, under the category domestic labor. The inclusion of food processing under this category helps account for the enormous amount of time that women spend in domestic labor, although there are differences between the sample points that reflect differences in consumption patterns and in the availability of technology. The main staple in Chunga is ubwali (porridge), made from a mix of cassava and finger millet, two crops that constitute 44% of the total cropped area. In Lukulu, very little cassava and finger millet are grown, except by the poorest households, and hybrid maize is both the main cash-crop and the main staple. This is reflected in the fact that households devote approximately 70% of total cultivated area to hybrid maize as compared to 32% in Chunga (Evans and Young 1988: Section 4). The production of finger millet and cassava flour are long and laborious processes, thus women in Chunga spend many more hours in food preparation than those in Lukulu who have easy access to a mill.

Richards noted in the 1930s that one of the most onerous and time-consuming of women's domestic tasks was food preparation. She calculated that it took 3 hours to prepare an evening meal (including time for fetching wood and water) (Richards 1939:103–104). If anything, the amount of labor required to produce a single meal has increased over the last sixty years. This is partly because of changes in population and settlement patterns that mean that far more time is now required to fetch water and fuel than was the case in the past. However, it is also partly related to the growing dominance of cassava as a staple in certain parts of the province. Cassava is most often eaten mixed with finger millet. The preparation of finger millet flour takes a great deal of time (to produce two or three days worth of flour for a household of six would involve a whole days' grinding) (Jiggins 1981).

But, when the staple porridge is composed of both finger millet and cassava, then the time needed to prepare the cassava flour has to be factored in, and this process is also extremely time consuming and laborious. Cassava tubers must be dug, peeled, soaked, pounded, dried, repounded, and sieved. Jiggins (1981) estimated that a month's supply of cassava flour for a household of six requires approximately four to six days' work. Gobezie's study in Nchelenge District in Luapula Province estimated that cassava and food preparation took women four to five hours daily (Gobezie 1984b).

Apart from the time spent in staple food preparation, the figures in Table 7.9 also show that, in Chunga, the hours of domestic labor rise dramatically with increasing commercialization. This is because hybrid maize creates a great demand for post-harvest labor in the form of shelling, bagging, and transporting all of which add considerably to the domestic labor burden of women in the more commercialized households. The Lukulu figures do not show the same dramatic rise in domestic labor associated with increasing commercialization, and this is most likely related to the introduction of maize as a staple and the use of mechanical grinding mills for its preparation (Evans and Young 1988: Section 6).

The combination of increased on-farm work and increased domestic labor puts severe pressure on women's time. The pressure is not felt equally by all women in all types of households, and not all of it is simply the result of the introduction of hybrid maize. It can be seen instead as the outcome of complex changes in diet, cropping patterns, and technology that interact with the socioeconomic status and the development stage of the household. The pressure on women's time also varies with the agricultural cycle, and it is clear from the women interviewed in the nutrition survey in 1988, that the major way in which this pressure makes itself felt is through women's inability to find enough time to prepare food and complete other domestic tasks. There is, therefore, a link between women's workload, child health, and declining nutritional standards, although it is not always easy to demonstrate it with aggregate data because such data are insensitive to the strategies that women employ to try and ameliorate the effects of time pressure.

For example, Evans and Young (1988: Section 6) found that during the dry season, some women prepared and dried balls of cassava that were then stored for up to six months and pounded into flour when needed. This prepreparation of cassava is important because not only is the wet season the busiest part of the agricultural year, but it is not easy to dry cassava during the rains. Some women also made stacks of firewood during the dry season in order to save time collecting wood during the busy wet season. Over half of the women interviewed in the 1988 nutrition survey turned over some of their domestic work to daughters, granddaughters, and sisters. It was also observed that elderly women were often allocated domestic tasks or asked to supervise younger girls. Evans and Young (1988:Section 6) note that the IRDP (SMC) data that is collected by the half-month is particularly insensitive to the daily compromises that women make.[20] These compromises, which were also mentioned by women interviewed during our own fieldwork, include such things as only collecting water once a day, deciding to bathe children less frequently, leaving utensils to be washed until the following day, cooking only one meal, and preparing less time-consuming types of relish. These adaptations may not sound particularly dramatic, but they can seriously affect the standard of care in the household.

Disease and Malnutrition

In view of the connection that the original IRDP (SMC) data had suggested between women's work, weaning, and children's nutritional status, our 1988 interviews asked women various questions about weaning, the preparation of weaning
foods, and the health of children immediately before and after weaning. The same
women were also asked questions about malnutrition and disease, and especially
about the relationship between diarrhea, child nutrition, and child health.

All twenty women interviewed distinguished between two different conditions associated with loss of weight, loss of appetite, declining health, and other
specific bodily changes. The first of these was referred to as chifimba (swelling) or
nsala (hunger), and seems to correspond in its severe form with kwashiorkor. This
condition was said to be caused by lack of food, and all the women interviewed
claimed that it was new and had not existed in the past. The second locally
recognized condition was variously referred to as ulunse, ulondo, ukondoloka
(thinness), and amankowesha (caused by sexual contamination). This condition
corresponds well, in its developed form with marasmus. This is said to be caused
by a lack of food, but more specifically by early or rapid weaning made necessary
by the mother becoming pregnant again while the child is still breast-feeding.
Richards recorded the existence of this condition in the 1930s, and noted that it was
considered very dangerous for the mother to become pregnant again before the
previous baby had been successfully weaned (Richards 1939:67). The symptoms of
the condition in the 1980s were said to be coughing, wasting, and diarrhea, just as
they were in the 1930s. Several women claimed that particular kinds of diarrhea
were caused by "dirty breast," that is, by the contamination consequent on sexual
relations between husband and wife being passed to the child via the breast milk.

Women in this area, as in many other parts of Africa, have a very clear perception that the health of a young nursing child is threatened by a mother's second
pregnancy because this will result in premature weaning. Careful spacing of children (at least two years apart) is thought to be essential, and many parents practice
coitus interruptus while the mother breast feeds in order to ensure that she will not
become pregnant. Weaning is seen as a critical stage for children, and mothers introduce children to solid foods in the form of thin porridge very early, so as to
"make the body strong." Women clearly recognize that wasting can be caused by
lack of food, just as they know that diarrhea is related to bad water and contaminated food. However, when the decline in a child's health takes a particular course,
it is seen as symptomatic of a crisis in social relations, rather than simply as a result
of dietary insufficiencies. Thus, weaning has long been a focus of concern for
women, and they realize that "family planning" and birth spacing are also critical
to child health. Information gathered from fifty women revealed very regular spacing of births, with gaps of about two years, except where migration, illness, still
births, or infant death had disrupted the pattern.[21] Thus, with regard to weaning
and birth spacing, some of the messages of modern medicine accord very well
with local common sense.

All of the twenty women interviewed in the nutrition survey said that diarrhea (ukupolonya/shiki) was caused by bad food or bad water—several respondents also said that it was related to weaning and the appearance of the first teeth.
A number of specific kinds of diarrhea were identified, including those named il-
onda and punku. There was general recognition among informants that food left

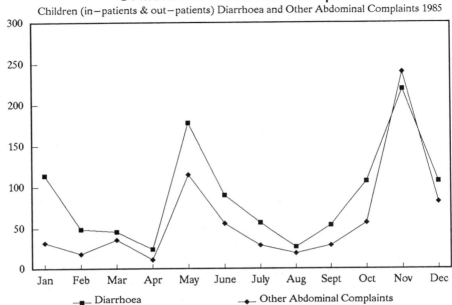

Figure 7.1

standing, lack of water for washing, and lack of personal hygiene all contributed to the incidence of diarrhea. It appears that most women see a general connection between the problem of child diarrhea, general child welfare, and the time available to prepare food and carry out domestic tasks.

A large number of studies from around the world have identified a link between diarrhea and malnutrition in young children. We attempted to investigate the links between women's seasonal workload, malnutrition, and diarrhea by examining the records for inpatient and outpatient treatment, and the village children's clinic records for Chilubula mission hospital over a three-year-period from 1985–1987 (Figures 7.1–7.5). The results show a peak in admissions and outpatient treatments for malnutrition, diarrhea, and a number of other complaints in November and in May for children under 5 years. When the records from the village clinics are analyzed, differences emerge between children in different age groups. The percentage of children losing weight by month increases significantly in November for all age groups. There is also a smaller peak for all age groups in July. Children 12–23 months of age and 24–59 months of age also show an increase in the numbers losing weight in May. The differences between age groups are even more marked when the percentages below the lower line of weight for age are examined. Children 0–11 months of age and 24–59 months of age show a small increase in the percentages in May and June, whereas children 12–23 months of age show a small increase in May and a dramatic rise in November. Taken together, these graphs bear out women's perception that the children who are being weaned

Chilubula Mission Hospital

Children (in−patients and out−patients). Upper Respiratory Tract Infection (1985)

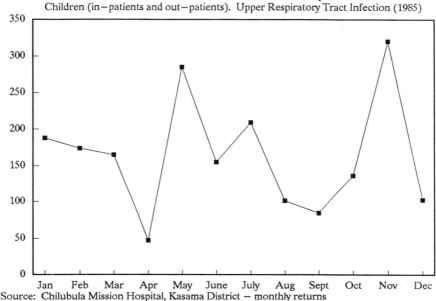

Source: Chilubula Mission Hospital, Kasama District − monthly returns

Figure 7.2

are the most vulnerable (12–23 months of age); children who are still being breast-fed (0–11 months of age) seem to be least affected. This is also in accordance with the fact that children who are being breastfed are not so susceptible to disease from contaminated food. Overall, these findings may be explained in terms of a set of interacting factors, including seasonal workload, food availability, and disease patterns. It is certainly the case that this is the way they are interpreted by local people and local medical personnel. From August to November, women are extremely busy in field preparation. Citemene preparation is heaviest in August and September, while land preparation on ibala (permanent) fields starts in October and is at its height in November. Maize planting also begins in November. Women are often unable to take weaned children to the distant fields with them, leaving them at home with boiled sweet potatoes and other foods to eat during the day. These foods are susceptible to contamination, and the young children who are eating without adult supervision may not only eat contaminated food, but may simply not eat enough food. The beginning of the rains in November increases the likelihood of food and water contamination, and children who lose their appetite through diarrhea, malaria, or other sickness may be unable to ingest enough food, particularly if they are not eating under adult supervision. All of these factors interact to cause a decline in child welfare that begins to show markedly by the month of November. Similarly, the smaller peaks in malnutrition and diarrhea in May, and the increasing percentage of children losing weight by month and falling below the

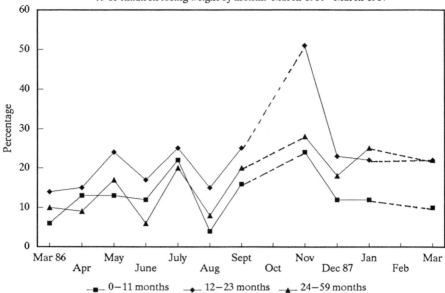

Village Children's Clinic, Chilubula
% of childlren losing weight by month: March 1986−March 1987

Figure 7.3

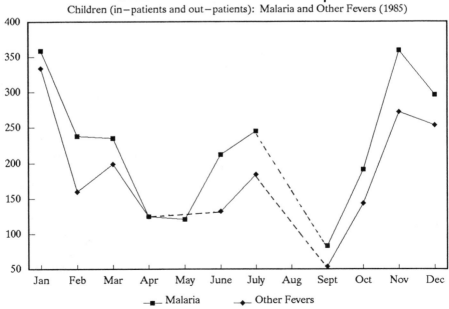

Chilubula Mission Hospital
Children (in−patients and out−patients): Malaria and Other Fevers (1985)

Figure 7.4

Village Children's Clinic, Chilubula
% of children falling below lower line on weight for age chart, March 1986–March 1987

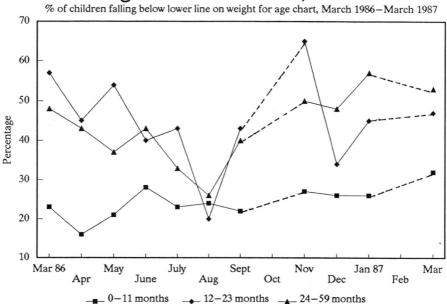

Figure 7.5

lower line of weight for age during the months of May through July may be the result of increasing workloads for women who are harvesting millet, maize, and then beans during this period. These findings suggest, therefore, that there may be a series of factors, including women's workload, contamination of food and water, and disease patterns, which in combination bring about a general decline in child health, especially for children who are being weaned. This is borne out in part by the fact that the children admitted to Chilubula hospital over the three-year period who were suffering from malnutrition and/or diarrhea appeared to come from all types of households, and from all socioeconomic levels. However, local medical staff pointed out that rates of morbidity are higher for poorer households. There is no data available to demonstrate this, although it would be in accordance with findings from elsewhere. Better-off households are able to mobilize more resources to deal with illness and the special needs of sick children, and they can substitute for the labor of particular individuals more easily, by buying labor if necessary. Thus, malnutrition in children under five years of age cannot be said to be simply caused by women's workload, although this should be seen as one of several key factors.

In the next chapter, we consider how the maize boom has affected both the sexual division of labor in households and the resources and decision-making powers of women and men. The following chapter thus provides the wider context in which the changes discussed in this chapter should be set.

8

From Millet to Maize: Gender and Household Labor in the 1980s

During the period 1975–1988, the production of maize in the Northern Province increased by about 850%. The reasons for this dramatic increase have been discussed elsewhere (Sano 1988; 1989), but from an historical perspective, recent government policy seems rather familiar. Between 1973 and 1985, a significant contraction in industrial sector employment and a decline in urban real incomes encouraged the government to instigate a "back to the land" policy, not all that different from the policy pursued in the 1950s. In essence, people were to be attracted back to the land by a revived agriculture, spurred on by a call to develop Zambia and make the country self-sufficient. In fact, there is little evidence that this policy has met any general success although, as we argued in Chapter 6, there are some indications of a recognition by the people of this area that migration to urban areas is no longer a viable option for most.

In order to persuade people to return to or stay "on the land," and to revive agriculture in the Northern Province, it was necessary, as it had been necessary in the past, to develop a viable cash crop that might raise the level of rural incomes and improve the purchasing power of rural producers. To this end, the Government of Zambia instituted a policy of promoting hybrid maize cash-cropping in the province. This policy, supported by marketing controls and consumer and transport subsidies, has been extensively criticized, not least for encouraging the production of maize to the virtual exclusion of everything else and for subsidizing its production in areas where it is comparatively disadvantaged in terms of input and transport costs (Dodge 1977; Kydd 1988).[1] With the removal of subsidies in the 1990s as part of the government's economic recovery program, the question now facing rural producers in the Northern Province is whether hybrid maize production is any more reliable a household strategy than migration to an urban area.

The increased production of maize in the province in the 1980s was largely brought about through targeting and encouraging small-scale farmers. These

TABLE 8.1.   Maize Sales to the Northern Cooperative
Union by District (90 kg bags)

	1975/76	1985/86
Isoka	32,034	126,751
Kasama	66,076	97,828
Mbala	40,992	244,073
Chinsali	15,921	97,793
Mpika	14,606	79,793
Luwingu	10,275	9,415
Mporokoso	5,806	1,348
Kaputa	654	1,232

(*Source:* Adapted from Bolt and Holdsworth 1987: Section 2.)

farmers continued to find hybrid maize production attractive throughout a series of years when inputs were delayed in arriving, transport deteriorated, bags of maize were left uncollected, producer payments were chronically late, and the deterioration of the retail distribution system meant that there was little to buy in the province even for farmers with money in their pockets. Looked at from the point of view of the producers, rather than that of the economists or government policy makers, the tremendous success of maize cash cropping is rather hard to credit. However, it would appear that this latest effort to develop the Northern Province, to get people to remain on the land and invest in farming, and thus create a settled agriculture and a productive population has, in some ways, been much more effective than it was in previous attempts. In particular, the rise of maize cashcropping has, in some areas, fundamentally altered the relationship between permanent field cultivation and citemene cultivation, and thus, the gender division of labor.

As always, however, we should guard against provincial-level generalizations. Levels of maize production are very variable across the province, and are much affected by microecological factors, infrastructural provision, marketing possibilities, land-use patterns, and the socioeconomic structure of the farming household (Table 8.1). Even at the present time, the agricultural production systems of the Northern Province continue to defy easy generalization and researchers working within the province are quick to stress that findings specific to particular areas cannot be generalized to the whole province. The existence of this almost unmanageable degree of variability is a continuing testament to the versatility of the agricultural production systems of the province.

A substantial amount of quantitative data was collected by ARPT (Kasama) and IRDP (SMC) in the 1970s and 1980s on the social and economic parameters of the farming systems of the province. Data exists on extra- and intrahousehold labor inputs, land use and cropping patterns, produce disposal, agricultural investment, and off-farm income and expenditure. This invaluable data provides the background for the qualitative material presented in this chapter. This material is based on thirty semistructured interviews with male household heads, thirty interviews with returned male migrants, and thirty semistructured interviews with women farmers (of whom seven were female heads of households), as well as on participant-observation, case histories, and event histories collected over the period 1986–1990.[2] Through this material we have investigated the changes that increasing commercialization and maize production have brought for gender relations and strategies of household production. In an area in which male absenteeism has so often been seen as a source of social and economic malaise, we now need to ask what have been the effects on households of the growth of a male-dominated commercialized agriculture. In order to begin our investigation of the changing dynamics of gender and household relations, it is necessary to start with the politics of resource allocation and acquisition.

The Politics of Land Use and Land Supply

In most areas of the province, land is readily available and is not a factor limiting production.[3] Pockets of particularly fertile land, especially in the northeast of the province, are very densely settled and land acquisition in these areas is difficult (Pottier 1988; Sichone 1991). This raises problems both for the expansion of agriculture and for the devolution of land to following generations. In most of the rest of the province, there is little difficulty in acquiring land (Milimo 1983). However, dambos, riverine areas, and zones with specialized soil types can be subject to contestation. In addition, local people recognize different land types, as well as variations in fertility and soil type; thus, they can experience local shortages of specific kinds of land (see Chapter 2).

The overall settlement pattern of the province now shows significant clustering of the population along the roads and the line of rail, a major change from the earlier periods surveyed in this book. The siting of major roads on watersheds has led to the denudation of tree cover in areas close to the road. These areas then sometimes become subject to overutilization and erosion, while those away from roads have retained some of their forest cover and are underutilized. The increasing importance of transport, as well as marketing and supply infrastructure, means that land close to roads is much sought after, and that settlements there are more stable because people do not want to move away from transport and supply routes.[4] In this way, cash cropping has increased both settlement stability and population concentration, and has also increased competition for land close to settlements and roads. However, if maize cash cropping is to be further developed and extended in the province, this will almost certainly entail a massive program of infrastructural development in order to open up areas that are currently underutilized.

Land use patterns in the province have been changing in response to a variety of factors throughout the last century, but increasing commercialization and the introduction of hybrid maize have brought about very specific changes within the different agricultural systems of the province. The central change has been the increasing area of land under ibala (permanent field) cultivation, both for maize and, in the case of some of the more commercialized farmers, for so-called "traditional" food crops.[5] However, this new focus on permanent, as opposed to citemene, cultivation raises fresh questions about how the agricultural systems of the province are to be represented and understood.

As we argued in Chapter 2, citemene systems in the province have always incorporated permanent or village gardens (ibala or mputa). These permanent gardens were used for growing a wide variety of crops, including maize, millet, cassava, pumpkins, and relish crops. In many areas, they were predominantly cultivated by women, although men often helped with the ridging. The importance of the permanent garden in the citemene system was well recognized by agronomists, but largely ignored by government officials and policy makers whose understanding of agriculture in the province was very much dominated by the practice and symbolism of "cutting and burning." The size of ibala gardens grew in some areas of the province as cassava production increased (Vedeld 1981; Trapnell 1953); although in other areas they had always been large, and a number of researchers noted from an early date that in the "Northern Grassland" and "Lake Basin" systems, it was difficult to distinguish the ibala gardens from the main gardens. In a survey conducted in Mwinesano village northeast of Kasama in 1981, Vedeld noted that it was not always possible to make a clear distinction between a "village garden" and a "permanent field." He also suggested that researchers often had different opinions about exactly how these two types of field were to be categorized (Vedeld 1981:44).

Vedeld does not discuss how this problem has arisen, but he implies that it is merely a minor technical difficulty. However, what is at issue here may be something more than a straightforward confusion of categories. We should begin by asking ourselves why this confusion has come about. Part of the answer to this question concerns the manner in which the agricultural production systems of the province have been constructed as objects of knowledge. The relative neglect of the role of permanent fields or gardens in earlier accounts of these systems has produced a situation in which permanent cultivation is represented, in the 1970s and 1980s, as a largely new development in this area and one associated with the introduction of hybrid maize. There is no doubt at all that there has been an increase in ibala-cropped area in recent years. IRDP (SMC) estimated that the average ibala area grew by 20% between 1980–86 (Bolt and Holdsworth 1987: Section 7). However, the issue is not simply whether the area under permanent cultivation has increased, but how these areas should be classified, and whether there are any meaningful distinctions to be made between permanent fields and village gardens.

Part of the problem appears to lie in the analytical association made between permanent fields and maize cultivation. This leads to an overemphasis of the importance of maize in the farming systems, and, as a number of researchers working in the province (e.g., Evans and Young 1988; Bolt and Holdsworth 1987; Bolt and Silavwe 1988) have already pointed out, has operated to discourage attempts to develop improved varieties of so-called traditional crops.[6] In reality, permanent cultivation is not straightforwardly synonymous with maize cultivation, and many other crops—including millet, cassava, pumpkins, beans, local maize, and relish crops—are grown on ibala (permanent) fields. Essentially there is little difference between these fields and the permanent village gardens of the past, except that the modern ibala gardens tend to be larger and to have a greater percentage of their total area devoted to staples, often at the expense of relish crops.

However, it is clearly not appropriate for all permanent or semipermanent fields simply to be placed in a single category. Whereas women controlled the produce of village gardens in the past, they may have little control over crops grown on the permanent fields around settlements at the present time, partly because these fields, which have been reclassified as ibala gardens for the production of staples, have grown at the expense of the older type of village gardens. There is increasing competition over land around settlements, and poorer households, especially female-headed households, can experience difficulties when trying to gain control over plots close to settlements, partly because they do not have the necessary labor. Within the agricultural systems of the province, household reproduction has always been dependent on inputs drawn from different and differentiated sources. In the context of rapid socioeconomic change, the categorization of different kinds of fields and sources of food and income becomes a highly politicized issue. As the value of these different resources changes unequally, individual men and women seek to redefine their ownership and control of them through the relabeling and recategorization of fields, gardens, and other resources.[7] This is another reason why researchers may find it difficult to distinguish one category of field from another. For the individual household members concerned, as we shall see below, the specific designation or labeling of a particular field or garden is never simply a technical matter of classification. The fundamental import of this becomes clear if we now turn to the question of land classification in the larger context of land acquisition and control.

Acquiring Land and Controlling Labor

Rural households in the Northern Province are not best treated as bounded entities that pool resources and act as a joint fund. Conjugal partners have a complex and shifting set of rights in each other that include rights to labor, land, and resources as a corollary of their relationship with each other. Conflicts over land and labor bring into play both traditional and nontraditional forms of authority as the individual partners seek to impose their own interpretations on a changing situation. There are a number of ways in which individuals can assert the legitimacy of their claims to land at the village and household level (through residence, descent, first clearance, labor investment, and conjugal rights), but they can also call on the authority of headmen, local chiefs, Village Productivity Committees, local councils, and local courts.[8] However, none of these claims, in whatever arena, are ever unproblematic or undisputed. Rights regarding land are not fixed, but rather multiplex and historically determined, often being bound up with the biography of particular individuals. Processes of contestation and countercontestation are thus ongoing.

Land registration in the Northern Province is the exception rather than the rule, although this is beginning to change in some areas (see footnote 3). Returning migrants have little difficulty in acquiring land and they are most likely to begin this process by approaching the local chief or the Village Productivity Committee.[9] Powers to allocate land in the province still rest with chiefs and headmen, even though the government theoretically stripped chiefs of their powers after independence (Van Donge 1982:94). By whatever route land is allocated, most individuals then exercise their claim to the land by clearing it and cultivating it. However, some follow this first stage by applying to be granted title deeds to the land. This course of action tends to be taken only by those who are educated and wealthy enough to be aware of the advantages that can accrue from registration, or by those who wish to be able to bequeath their land to particular individuals, an issue that we saw was raised in the context of peasant farming in the 1950s. Because inheritance is still governed largely by customary law—the existence of a will has not always proved to be the most secure way of guaranteeing the rights of an individual in the face of opposition from the kin group—many men who wish to make sure that their sons inherit, as opposed to their matrikin, take this course of action.

The fluidity of the legal situation in Zambia with regard to land thus parallels and participates in the multiplex claims established through customary precedent rather than serving to impose a fixed structure on a previously unclarified situation.[10] The fact that elites are able to manipulate customary tenure arrangements for their own benefit is common in many African states (Bassett and Crummey 1992; Downs and Reyna 1988), but it is an increasing feature of land acquisition in the Northern Province, particularly because farming has become an investment opportunity for retiring migrants who recognize that the declining standards of urban living effectively preclude them from subsisting on their pension in town. There is, however, nothing necessarily illegal about what in some areas amounts to "land-grabbing." Discussions take place between prospective landholders and local chiefs or Village Productivity Committees, who are legally empowered to conduct such negotiations (Pottier 1988:117–124). Powerful and educated individuals have little difficulty persuading chiefs or village committees of the legitimacy of their claims, particularly if they are backed up by arguments of birth, residence, or attachment to the area of the individual concerned or his family. There are stories of money changing hands or "help" being provided in other ways, but the main point is that this method is an easy way for those with money or influence to acquire land in the rural areas. The people who already live in that area are often unaware of the consequences of granting land—sometimes large amounts of land—to an individual who will then begin the process of registration. Most local people do not register their land, and they assume that others farming around them have exactly the same kinds of rights to land as they do. This does not mean that local people are unaware of the existence of land registration or of the fact that some people are acquiring control over large areas of land, but they are not always fully informed of the potential risks of allowing a dual system to emerge. These risks are well recognized, however, by returning migrants who have to go through the process of acquiring or reactivating rights to land. Of thirty returned male migrants who were interviewed, nine had no particular worries about the land situation in the area, seven thought that the government's current policy on land registration and farm ownership was positive, two felt that current government policy was an improvement on colonial policy, and twelve were extremely anxious that land registration was leading to a situation in which local people were losing land to outsiders. As one man said of the present situation in the area: "The right to land by birth is being overridden by the right to land by Government lease."

Despite these recent changes, the continuing role of a chief and/or a Village Productivity Committee in landholding and acquisition means that this process is still very much bound up with the identities and personal influence of particular individuals. In any discussion about land, various interested parties will push claims and interpretations. The ability to make these claims or interpretations stick is often a function of local structures of power, influence, and personality (see Pottier 1988: Chs. 5 and 6). It is not surprising that under these circumstances poorer members of the community and women—even in the matrilineal areas—are often disadvantaged in this process unless they are championed by others more power-

TABLE 8.2. Land Acquisition: Methods Used

Methods used/source	
Headman via VPC	4
VPC alone	3
Chief via VPC	1
Chief via headman	2
Father via VPC	2
Father via headman	2
Brother via chief	1
Brother	1
Aunt	1
Uncle-in-law via VPC	1
Government via back-to-land Scheme for pensioners	6
Government via back-to-land Scheme for unemployed	6

VPC: Village Productivity Committee.
(*Source:* Interviews with Male Heads of Households.)

ful than themselves. This is particularly important when it comes to their ability to claim land close to settlements for use as village gardens, because without such land it can be difficult to grow sufficient relish crops for household consumption and/or for sale (discussed later and in Ch. 7). Local histories of dispute and dispute resolution take over, so that particular claims and interpretations become powerful simply because they accord well with decisions taken in the past. It is always true that the fluidity of claims relating to land are such that any event, particularly a death, can reopen apparently closed issues, and new decisions can overturn old ones.[11] Lines of power, authority, and precedent are difficult for certain groups of people to manipulate.

Although land was nominally nationalized after Independence, and the government has attempted to legislate on property rights and to develop machinery for land allocation, there has been little solidification of claims procedures. Officials, and those who can attach themselves to or manipulate state structures, have been able to use customary arrangements to gain preferential access to land and to the resources that follow from landholding. At the local level, political connections of this kind are often personalized, and are structured according to local lines of power and authority. Success in claiming rights to land or rights to produce often involves access to a network of social relations that incorporates kin, neighbors, local officials, bureaucrats, and politicians. Access to this network is not necessary in all cases, nor does it always have to be direct. In this situation, there are some who operate at a disadvantage, among them many women whose networks tend to be more localized than those of men and whose access to wider political and economic resources is more limited, indirect, and insecure.

Thirty male heads of households were interviewed in the villages around the Chilubula mission (Kasama District) in 1989. They were asked how they had acquired the land they were currently farming. The results are given in Table 8.2.

Interviews were conducted close to a government resettlement scheme, which accounts for the relatively large number who had acquired land directly from the state.[12] Leaving this group aside for the moment, let us investigate the responses to this question by the rest of the sample who had acquired their land through the regular local mechanisms. These farmers often couched their answers to the question of land acquisition by stressing specific interpretations of past decisions and negotiations. For example, one individual was keen to stress that he had acquired his land via Chief Mwamba, and that this had been approved subsequently by the Village Productivity Committee. It was clear that this farmer wished to assert his customary rights to land by virtue of allegiance to his chief (Richards 1939: 244–248), and to emphasize the legitimate role of chiefly authority as a way, in part, of modifying the power of those running the Village Productivity Committee. Another man, however, seemed to prefer to distance himself from traditional authority structures. His account stressed that although land had been allocated to him via the headman and the Village Productivity Committee, it had really been inherited from his father. In this case, this individual wanted to emphasize his rights to a specific piece of land in this matrilineal area through a focus on the legitimacy of father-son inheritance. Thus, the issue here is not so much one of specifying the regulations governing land acquisition or the different routes by which individuals can acquire land, but rather one of investigating the various and variable interpretations that individuals seek to impose when describing a particular transaction or set of transactions. In many instances, a man who chooses to respond by saying that this land came to him from the chief via the headman may very well, under other circumstances, have been able to describe his land as coming to him from his father. It is, therefore, not possible to collect data on land acquisition without being aware that the same transactions can sometimes be described and interpreted in several ways.

None of the farmers interviewed reported having paid anything for their land, although three said they had paid between K50–K450 for land registration. Out of the sample of thirty, twelve farmers said that they had title deeds to their land. Of these twelve, eight were resident on a government scheme. When asked who would inherit the land they were currently working, nine said it would be inherited by one of their sons, five said their children, two said it would be a nephew, one said a brother, one said a member of the extended family, and twelve said that they didn't know. However, of the twelve who did not know who would inherit their land, five of them explicitly stated that this was linked to the fact that the land was not registered in their name. It is impossible to say anything definitive based on such a small sample, but informant responses do confirm more general impressions gathered during fieldwork that land and land registration are becoming important issues in the area, and that some men are looking for ways to exclude matrilineal kin from inheriting valuable property after their death. This implies, in turn, that the value of land, particularly cleared land, is changing and that a market in land may start to emerge.[13]

Of the thirty male household heads interviewed, twenty-three had been migrant workers, and two had worked for the local mission. Again, these figures confirm a wider impression formed during fieldwork that many of the more commercialized farmers in this area are returnees. This is partly because wage employment is still the most secure way of acquiring the capital needed to engage in hybrid maize production, and partly because farming has recently become an attractive option both for pensioners and for those seeking a more secure income than that provided by a declining urban wage. The average length of time that farmers in the sample had been engaged in full-time farming was 6.9 years, but 50% of the sample had been farming for five years or less. However, eight individuals in the sample had farmed elsewhere at some point in their lives and a further five held land elsewhere, although only three of them claimed to be actually farming it. Of the eight who had farmed elsewhere, five had done so in the villages of their in-laws, while two had farmed in their natal villages. One man who had farmed elsewhere had moved to the Chilubula area so that he could acquire a farm registered in his name that would provide security for his wife and children in the future; and another man had moved to the area because he had inherited a farm from his brother. One farmer who admitted to holding land elsewhere explained that he had recently set out to acquire land in his natal village, but that he was not planning to farm it in the near future. What these stories demonstrate is that residence options are still fluid, and remain so partly because land is not yet a constraining factor and can be acquired through a variety of mechanisms.

Residence choice is a complex matter, and individual men may move several times in their life in response to changing circumstances.[14] As we discussed in Chapter 6, women who were temporarily unmarried may also have moved from place to place, activating a range of kinship relations and offering their labor in return for support. However, their situation with regard to landholding was, and is, complex and far from secure. Even in the past, women from both matrilineal and patrilineal groups in the province effectively acquired land through marriage. Young wives in matrilineal communities remained part of their parents' household until they and their husband had proved that they had the knowledge and the maturity to manage the responsibilities and uncertainties of production and consumption (see Chapter 3 this text; Richards 1939:124–127) A woman's access to land, once she was no longer part of her natal household and had ceased to eat food grown on gardens cleared by her father, depended greatly on the labor of her husband and other male kin. There were certainly other ways of getting land cleared, as discussed throughout this book, but this does not alter the fact that rights to land were strongly associated with first clearance. Richards discusses the fact that a millet garden was commonly spoken of as belonging to the man because he had originally cleared it. The rights to the millet crop itself, however, were made up of a series of complex claims, and were not entirely determined by who controlled the land on which it was grown. A wife might have the use of the crop because of the labor she had contributed to its production and because of her responsibilities for feeding the household. Relatives of both the husband and wife were able under certain circumstances to lay claim to part of a crop (Richards 1939:188–192). Ultimately, the question of rights over crops was to be seen, as Richards pointed out, in the context of a situation in which notions of authority and power had always been expressed in terms of the right to levy tribute (ukutula) on the labor and food supplies of others. Thus, even in the matrilineal groups of the Northern Province, senior men controlled land allocation, and although inheritance was nominally in the female line, this tended to mean that men could inherit rights from the matrilineal kin group; it did not mean that women passed on their land or other property to their children.[15]

Of central importance to women were the village gardens. These, as we have seen, were distinct from citemene fields, and were, according to Richards, frequently thought of as the property of the woman because they were created largely through her labor. Therefore, the woman had a greater degree of control over the foodstuffs produced on them than she had over the produce of the citemene fields, although if a husband had contributed any labor to the village garden, he too could make claims (Richards 1939:185). Generally, it seems that in the past, as in the present, men made very little effort to try and control crops like cassava, sweet potatoes, and pumpkins, and were much more interested in rights over the main staple, finger millet. Nonetheless, Richards noted in the 1930s that the complexity and shifting nature of rights to crops gave rise to considerable domestic tension, and that this was exacerbated by the fact that millet and/or relish crops could be sold for money. She also pointed out that men tended to consider any money acquired through the sale of produce as theirs alone (Richards 1939:192–193).

Watson reported for the patrilineal Mambwe in the 1950s that women acquired rights to land through marriage: "There are no 'estates of holding' for women; they work land whose rights are held by men" (Watson 1958:99). The man provided the garden and the woman provided beer and food for the household. At the time of Watson's research, the crops for which there was a local market were largely grown in the kitchen or village gardens (ivizule), which women controlled, and women used the money from these sales to buy clothes and personal goods (Watson 1958:110). By the 1970s, however, this had changed. Pottier noted that cash crops were no longer grown on kitchen gardens, and that women's control over crop sales was increasingly disputed, giving rise to domestic tension, particularly at harvest time (Pottier 1988:115–118). He also reported that the undervaluing of village or kitchen gardens by the local Village Productivity Committee officials, as compared to the communal gardens and the main gardens, gave rise to a situation where inputs and technological assistance were withheld from the former (Pottier 1988:117–118). Although women still controlled these gardens, they were disadvantaged in their attempts to improve productivity. Women's invisibility in the eyes of local state officials—whether Village Productivity Committees or agricultural extension officers—and their decreasing ability to influence decision making, further undermined their access to land. This situation is serious for women in the province because although they are in some circumstances, as we shall see later, increasing their control over citemene production, they have lost control over the village gardens and the crops grown on them, which were the basis for their strategies of agricultural adaptation and household reproduction in the 1940s and 1950s.

Women's and men's rights to resources are changing in the context of a degree of commercialization of agriculture that is without precedent in the Northern Province. Rights to land have always been bound up with kinship obligations, labor investment, and conjugal negotiation, and this was so for both matrilineal and patrilineal communities. The degree to which individuals could impose particular interpretations on a situation varied considerably and was often fairly congruent with existing lines of power and authority both in the local community and in the community-at-large. In the 1930s and 1950s, increasing commoditization was already affecting negotiations between conjugal partners and kins people. The penetration of the cash economy both heightened domestic tension and demanded a clearer delineation of rights and obligations, particularly as they related to the complex nexus of land, labor, and food. This process continues to the present day, and takes on new aspects as the structures of the state begin to intervene more thoroughly.

As we have shown, the farming systems of the Northern Province have undergone any number of changes over the last 100 years in response to changes in labor availability, the development of a market, and the constraints and possibilities inherent in the natural resource base. The introduction of hybrid maize must then be seen as just one of these changes, but this is not to deny its specific effects. These appear to have been most marked in terms of gender relations, and particularly in regard to the extent to which women can exercise control over their labor and over the crops that their labor helps to produce.

Data collected by ARPT indicates some interesting shifts in the control over crops, especially when viewed with some historical perspective. This data, collected from three hybrid maize-producing sample points in the province, showed that male household heads claimed to control 52% of all the cultivated land used by a household; wives claimed 24%, and a further 24% was said to be jointly controlled. In terms of control over crops, household heads said they controlled 88% of the hybrid maize area, 54% of the local maize area, 20% of the permanent finger millet area, 30% of the citemene finger millet area, 17% of the cassava area, and 56% of the beans area. Wives had the greatest control over citemene finger millet (38% of the area) and cassava (47% of the area) (Bolt and Silavwe 1988). The degree to which husbands and wives control the cultivation of land and or the disposal of particular crops obviously varies from one region of the province to another. There is also some variation within areas, most particularly between households of different generations and those of differing socioeconomic status. At the present time, some crops, like cassava and pumpkins, are thought of as women's crops, whereas others, notably maize, are classified as men's crops. These classifications are part of a changing set of rhetorical structures that form the basis for household decision-making and resource allocation (Moore 1993a). However, what is clear is that men have much greater control over cash crops, especially hybrid maize, and that as cleared land becomes more critical because of its potential value in cash crop production, it becomes increasingly necessary to try to identify who controls specific pieces of land and particular crops.

Central to any understanding of the complex changes taking place within maize cash-cropping households is an analysis of the labor demands of this crop and the specific gender division of labor that has come to be associated with its production. Hybrid maize production is very labor intensive, and in the present situation, male household heads appear to be seeking to extend their control over the labor of their wives and other household members in order to secure the additional labor required. This makes women increasingly vulnerable to demands on their labor time and decreases their ability to make claims on strategic resources based on the amount of labor they have invested in production.[16]

Gender and Household Labor

Labor has long been recognized as one of the major constraints to increased agricultural production in the Northern Province, and most farmers depend on family

TABLE 8.3. Mean Labor Input On-Farm For Wives and Husbands (hours) 1982-83

Farmer Category	A	B/C	D	E	F
Lukulu Sample					
Wives	513	462	—	744	625
Husbands	189	500	—	713	397
Chunga Sample					
Wives	468	202*	430	457	513
Husbands	179	152	438	351	355

A : Selling no maize or selling for no more than 1 year out of 6; B: Selling 1–10 bags for less than four years out of six; C: Selling 11–30 bags for less than four years out of six; D: Selling 1–10 bags for four or more years out of six; E: Selling 11–30 bags for four or more years out of six; F: Selling over 30 bags for four or more years out of six.

*Low figure due to missing data for October and November.

(*Source:* Adapted from Evans and Young 1988: Section 6.)

labor for cultivating their fields. The vast majority of farmers in the province are hoe-cultivators, and the number and size of household fields are limited by the available labor and the level of technology. In a few areas, notably in the northeast of the province, ox-technology is used for ploughing, ridging, and planting. This has enabled expansion in the size of cultivated areas, but has not solved the chronic labor shortage partly because this technology is available to only a few, and partly because weeding and harvesting still have to be done by hand (IRDP 1985; Francis 1988; Geisler et al. 1985; Kerven and Sikana 1987:Section 5d). At all levels of technological input, hybrid maize production appears to require increases in both women's and men's on-farm labor as illustrated in Table 8.3.

The above figures show that women spend more time in on-farm activities than do their husbands, and when we combine this with the hours women spend on domestic labor (see Table 7.9), it is clear that women's overall labor input increases greatly with commercialization and hybrid maize production. In general, women spend the largest amount of on-farm labor time in planting, weeding, and harvesting; while men's input is greatest in land preparation. These findings accord well with what we know about the "traditional" sexual division of labor, which emphasizes the importance of male labor in cutting trees and clearing land. However, data from IRDP (SMC) analyzed by Evans and Young shows that husbands in farmer categories E and F make a significant contribution to both weeding and harvesting, which suggests that a degree of flexibility in the sexual division of labor has been introduced or has become necessary with hybrid maize production (Evans and Young 1988: Section 6).[17] This view is consistent with that expressed by many women informants who said that they preferred the old system in which women's and men's jobs were clearly defined. A number of them complained that since the previous division of labor had begun to break down, men could ask women to do anything and they found it very difficult to refuse to perform these new tasks.

There are, in fact, a number of new tasks that women have to perform, including fertilizer application, weeding, shelling, and bagging. Some tasks, like weeding, are not really new, but the nature and extent of weeding required for cultivation under permanent or semipermanent conditions, and for the successful

TABLE 8.4.   Mean Monthly Labor Inputs (Hours) by Farmer Category

Farmer Category	1		2		3	
Field Type	I	C	I	C	I	C
Month						
October	144	12	130	0	227	0
November	197	0	139	0	426	0
December	142	4	197	12	573	9
January	139	0	217	48	338	17
February	131	10	149	49	226	0
March	111	0	161	0	185	0
April	55	122	121	128	294	11
May	63	382	60	365	229	276
June	91	252	117	403	396	128
July	46	315	108	363	367	212
August	30	285	57	203	153	149
September	10	206	72	105	34	117
Area (hectares)						
Cultivated	1.46	—	2.19	—	4.92	—
Means Hours/Hectares	794	—	698	—	701	—

Category 1: selling no maize to the official market; 2: selling 1–30 bags; 3: selling 30v bags; I: Ibala; C: Citemene.

(*Source:* IRDP (SMC) data, Bolt and Holdsworth 1987: Section 5.)

production of hybrid maize, is quite different from that required in citemene cultivation. There are also other tasks, notably that of crop transportation, which have become very much more burdensome because of the poor technology available to most farmers. However, interviews with women informants suggested that it was not simply the introduction of new tasks that concerned them but that changes in cropping patterns and field use, combined with the overwhelming importance of gaining access to cash income, had produced a situation in which men were seeking to control women's labor much more directly. This labor control was of two types. First, there was a drive to control the total amount of labor that wives devoted to certain crops; and second, there were attempts to control the timing of women's labor input. These two forms of control are clearly related, but individual farmers, women and men, were emphatic that they were not the same thing. In order to examine what this distinction is about, it is necessary to turn to some of the data on the seasonality of labor input.

It is clear that under present cropping patterns, as with earlier agricultural regimes, there is a strong seasonal dimension to labor requirements (Table 8.4). November to January is a peak labor period for all farmers. Land preparation on ibala fields begins in October and is at its height in November. Maize planting begins in November and continues through the following month, and finger millet is planted in December and January. The first bean crop of the year is planted in November and groundnuts are planted in December. Fertilizer application begins in mid-December and continues until February. Maize weeding starts in mid-December. The end of January until March is a slightly less hectic time when cassava and a second crop of beans might be planted. However, labor requirements rise again dramatically during the period from April through July because of harvesting. The finger millet harvest finishes before the maize harvest, while the second crop of beans is ready some time in May or June. From June until September, the work of cutting trees and stacking branches continues on the citemene plots, and the cycle begins again.

Table 8.4 shows that the introduction of hybrid maize and the increased production on permanent fields has produced a situation in which household labor is more or less continuously utilized throughout the year. To a certain extent, the demands of citemene and permanent cultivation are complementary. This is most apparent when looking at the figures for Category 1 farmers. However, with increasing commercialization, competition for labor emerges. The continuous nature of the labor required on the permanent fields clashes with the peak labor demands for the citemene fields (Bolt and Holdsworth 1987: Section 5). This is made more serious by the fact that the worst clash occurs at harvest time when female labor is needed simultaneously for finger millet and for maize. The greatest pressure, then, is on women's labor, a fact that, as we indicated in Chapter 7, may have consequences for the welfare of some of these households. The more commercialized (Category 3) farmers seem to deal with this problem, in part, by reducing their labor hours on citemene fields. This does not necessarily mean that they are producing less finger millet, because the crop is also grown on ibala fields, but the very fact that the crop is produced on permanent fields, and may even be fertilized, apparently reduces women's control over it. As we have already indicated, women have the largest and most secure claims over finger millet when it is produced on citemene fields. Any reduction in their control over this crop would automatically affect their ability to manage food stocks, hire labor, and raise income through brewing beer. Diminished labor time on citemene fields may also reduce women's control over beans and groundnuts, which are part of the citemene cycle.

In addition to the total increase in on-farm labor and the resulting conflicts between citemene and ibala cultivation, increasing hybrid maize production also generates incompatibilities between crops within the ibala regime. This is partly because hybrid maize yields are highly dependent on the correct timing of inputs and requires a rigidity of regime that is simply not necessary with a crop like cassava.[18] This means, of course, that the timing of labor inputs and the control of that timing is crucial to successful production of the crop. Women interviewed complained that the pressure from hybrid maize cultivation during the period of December through March interfered with their work on finger millet and beans. The additional labor demanded for weeding and fertilizer application caused a crisis in the timing of women's labor input (see also Bolt and Holdsworth 1987: Section 5). Evans and Young (1988: Section 6) reported women expressing anxieties on these points, and their findings accord well with the results of our own fieldwork. Geisler et al. also recorded that women complained of an overlap in January and February when the maize requires weeding, and millet, groundnuts and beans have to be sown in the citemene fields. Their informants reported leaving the maize field unweeded because they were away in the citemene fields, which were at a great distance from the house and the permanent fields (Geisler et al. 1985:13). Many of the women we interviewed indicated that they often had to leave work undone, particularly weeding, or that they grew smaller amounts of beans, groundnuts, cassava, and relish crops than they would have wished. There was a strong feeling that maize was competing with other priorities, although these views were not expressed by successful female heads of households and the wives of successful farmers to the same degree. Contributing to women's concerns over labor allocation and timing was the increasing distance that many had to walk to their citemene fields (see Chapter 7), a problem that could no longer be solved by living in mitanda because their attention was also needed on the permanent fields close to their homes.

That women's labor, and increasing amounts of it, is vital to successful commercial farming is well-recognized by male household heads. One man was adamant that his wife should only cultivate on ibala fields cleared by him, arguing that if she became involved in any own-account farming activities, this would seriously interfere with the amount of labor available for the maize crop. Two other men interviewed in 1989 about their maize production explained independently that they had had a very bad year because their respective wives had been absent during key periods in the agricultural season. Women are crucial both as labor providers in their own right and as supervisors of the labor of others (both household and extrahousehold labor), and men recognize this. To what extent men are able to control the amount and timing of the labor of their wives, and of other household members, is a question that raises larger issues of residence and remuneration.

The Ties That Bind: Residence, Remuneration and Reward

Previous research has demonstrated that adequate agricultural production is positively correlated with the developmental cycle of the household, and with its overall size and composition (Stolen 1983b:336–338). For the Northern Province, Evans and Young (1988: Section 5), for example, found that there was a positive correlation between the total number of potential workers in a household and total cultivated area. However, when comparing mean household size across farmer categories, they could discern no clear relationship between household size and different levels of commercial maize production. Such results are always difficult to assess, of course, because they depend on the definition of the household and on the definition of the worker. Evans and Young employed a rule of thumb that to be counted as household members, individuals had to be resident for more than six months of the year, and they defined a working adult as someone over eighteen years of age (although the IRDP labor data records the labor input of all household members over the age of seven years). On this basis, they calculated that the mean household size for Lukulu was 5.5 and for Chunga 6.2, the mean number of adult workers per household across farmer category was 1.95 and 2.65, and the total ratio of adult (18v years) males to females was 0.75 and 0.98 respectively. A further analysis showed that in Lukulu, the number of working males was positively correlated with the total cropped area, while the number of working females seemed to be inversely related, although the relationship was not significant. In the case of Chunga, the reverse was true, with the number of working females being positively and significantly associated with the total area under cultivation, but the number of working males was only weakly associated.

The relationship between the sexual composition of the household and the total cropped area as a proxy for agricultural production is obviously of interest because of the long-held contention that male absenteeism is at the root of food insecurity in the province. No firm conclusions from Evans' and Young's work can be drawn on this point: the differences in relationships in the two sample points could be the consequence of weaknesses in the data, of the definition of working adult, of relative differences in levels of citemene dependence, or of different degrees of dependence on plough technology.

However, their data does provide an interesting contrast with some of our own findings. The average household size among the 24 male-headed households involved in cashcropping (out of a total sample of 30) was 8.7, and the average number of working adults (defined as those older than fifteen years) was 4.87 per household. These figures are much higher than the provincial average and those of Evans and Young. This is in part because of differences in methodology, because we seem to have been much more inclusive in our definition of household membership, and partly to the fact that we collected household composition data at one point in time.[19] This means that it is relatively easy to deal with the regular comings and goings of school children, and to make a decision regarding their residential status, but it is not so easy to decide how to categorize sick children who have been sent to live with their grandparents, those elderly relatives resident in the household who will be moved on to another relative if they become too burdensome, divorced sisters who regularly appear with their children, and nephews who have come to stay until they can find some way of supporting themselves. Like Evans and Young, we followed the six months per year residence rule, but we found it hard to apply in practice. What does one make, for example, of the case of a man who asserted that his mother was not a member of the household, even though she had been living there for the last year? As Evans and Young conclude (1988:Section 5), the very notion of a static household labor force is a very questionable one under such circumstances. This extraordinary degree of household fluidity and residential mobility has, as we have indicated, a long history in this area. In order to try and overcome some of the difficulties of household definition and composition, we asked more general questions about how many members of the household the individual male farmer thought had made significant contributions to the 1988/89 season's crops, and how many people he felt he was obliged to support and feed with this one particular harvest. The aim of these questions was to get some idea of the amounts of intrahousehold labor the farmer felt he had at his disposal, and the total number of relationships of dependency he felt involved in. In this way, many individuals were mentioned who were clearly not members of the household in any straightforward sense, but this wider questioning did help to clarify the context in which farmers felt they were operating when making decisions about labor and resource allocation.

As a result of these questions, some very interesting data emerged on household composition that cannot be easily explained in terms of the dominant discourse of social breakdown that has re-emerged in the writings of contemporary agriculturalists and development experts in the Northern Province. It has been asserted by a number of writers (Sharpe 1987; Evans and Young 1988:Section 5) that the high preponderance of two-generational households in this area is a result of male labor migration and more generally of commercialization of the economy. We argued in the previous chapter that there is little direct evidence to support this view, at least for the Bemba-speaking people, and that their settlements probably

contained a significant number of two-generational households from the 1930s onwards, and perhaps earlier. This is partly because two-generational households are
a feature of specific stages of the developmental cycle and of village fission, particularly in this highly mobile society. The kind of complex, multigenerational
household that researchers imagine was a feature of past social organization would
probably have been recorded as several interacting households if they had been
subjected to modern analytical methodologies. This is because the complex households based on joint housekeeping described by Richards, and evidenced in the
work of Ann Tweedie (see Chapter 3, this text), are probably best understood as
a set of overlapping activities and interactions that could appear in very different
forms depending on whether they were approached from the point of view of consumption, production, distribution, or accumulation.[20] Strictly speaking, in any
event, they were matrilocal family groups (bakalamba) comprised of a number of
interconnected units, not properly defined as separate households, and yet not ju-
diciously conceived of as a single unit. This accounts for Richards' repeated use of
the English term "house" to describe these interacting groups, a term drawn from
British history through which she tried to capture the intermediate status of an entity that is neither a household nor a descent group, and yet contains something of
both. A close reading of Richards' text and an examination of her kinship diagram
for the village of Kasaka as well as Tweedie's for the village of Kanyanta show quite
conclusively that these "houses" were made up of a number of constituent units
that were rarely more than two generations deep; although some were made up of
grandparents and grandchildren living together (Richards 1939: Chs. 7 and 10; see
also Chapter 3 this text).[21]

    Of the twenty-four male-headed households in our sample engaged in cash
cropping, fifteen might possibly be defined as complex/extended or multigenerational, and others could easily have been defined in that way had they been sampled in another year or at another moment in their developmental cycle. The
composition of these fifteen households reflected two things: first, the household
heads' need for labor, and second, the desire of certain individuals to attach themselves to kinsmen who are successful in the hope that they will be able to benefit
from this success. How these benefits are defined varies greatly: a divorced sister
may need to feed her children; a young man may hope to get his school fees paid;
a son may hope to inherit land; a nephew may need to raise enough capital to begin
farming himself; and an elderly parent may just need somewhere to live. There
may come a time when a successful entrepreneur would prefer to slough off his
relatives, but the risks of cash cropping under hoe agriculture means that he has
little hope of being successful unless he can bind labor to him. This process of creating a following and with it a settlement has a long history in societies of the
Northern Province (Richards 1951:173). The Bemba system has always been a
wealth-in-people system, and if the chiefs of the nineteenth century needed to
bind followers to them, the peasant farmers of the 1940s and 1950s had equally to
establish a group of kin on their farm in order to make a success of their enterprises. The returning migrants in the 1950s and 1960s who wished to be headman
had to persuade others to remain with them in their villages or risk losing them to
another who felt he had more to gain through the process of fission (Richards 1939;
1958; Harries-Jones and Chiwale 1963; Kapferer 1967; Kay 1967). The situation is
very little different for successful cash cropping farmers in the 1980s and 1990s,

many of whom have established separate farms outside villages. They need, just like their counterparts in the 1940s and 1950s, to bind labor to them and thus they must invest in kin and in social networks.

In this society, as in many other African societies, the ability to mobilize labor is connected to the political ability to command a following and this, in turn, is related to personal status. In the past in Bemba society, dissatisfied village and household members simply left and placed themselves under the care and control of kin or patrons elsewhere (Richards 1939:143–144). In the present circumstances of maize cash-cropping, both household heads and their kin have reasons for coming together, but on terms that are open to constant renegotiation. This process, as we have already suggested, is not one that is easily captured by a discourse that asserts that kinship is breaking down, nor is it explained by a thesis that contends that matrilineal societies are particularly vulnerable to individuation and nuclearization with the emergence of capitalism and commercialization.[22] The picture that emerges from our research findings demands a different approach.

For example, one man interviewed in 1989 had a household made up of himself, his wife, three children, both his parents, his parents-in-law (this was a cross-cousin marriage), his wife's brother (15 years of age) and a boy of twelve years. This farmer still needed to employ workers for ploughing, harvesting, shelling, and bagging. Another individual had his wife's sister and her four children, plus his own brother living in his household, as well as his own wife and children. A third man lived with his wife and five children, plus two of his sister's daughters, as well as a small child belonging to one of them. In general, although we have no exact figures for the area, we found that cross-cousin marriage is still common and that the ties between brothers and sisters are a crucial part of any set of social networks.[23]

One man who produced 200 × 90Kg bags of maize in 1989 had a household of thirteen people: himself, his wife, six children, his mother, three male laborers (one of whom, Ben, was the son of his sister's daughter), and Ben's wife. This man paid his laborers in cash and in kind: K150 per month, plus two tins of maize. This household had a ratio of eight producers to thirteen consumers, but the farmer still needed to employ piece workers during key points in the agricultural cycle, and still complained that he was labor short. This farmer was not the only one to employ permanent laborers who lived as household members, there were two others in our sample. But, the distinction between laborer and dependent kin is not always an easy distinction to make as Richards noted for an earlier period (Richards 1939:142). Labor acts as an arena in which relationships of status and dependency are mutually reinforced. Under these circumstances, some seek to cut themselves off from their kin, some try to turn relations of obligation into contractual relations, and yet others work to ensure that what were once formal terms of engagement become socially imbricated. What is most interesting in this context is that many individuals prefer to leave social relationships as open as possible, try to avoid defining them in one way as opposed to another, and seek to exploit such ambiguities as exist in order to make these relationships work for them. This means, of course, that the terms of these relationships are constantly open to renegotiation and contestation. At the same time, such processes are always unequal ones and are frequently congruent with the existing lines of power and authority within the community.

For example, one man reported that his household consisted of himself, his wife, and their four children, but that he was assisted in his farm work by his parents and his mother-in-law who lived in the same village, and that he paid these latter relatives at least one bag of maize each. Here was a relationship in which remuneration was formalized under mutually agreed terms even between close kin. This example might be contrasted with that of the man who supported his mother-in-law, younger brother, and sister's son with food even though they lived in another village at some considerable distance from his own and did no work on his farm. There were three men in the sample who provided food for close kin who lived at some distance from them. All of the twenty-four men who were engaged in cash cropping were clear that they provided no remuneration to household members for their work on the farm, although three mentioned that they provided some "tokens" as a sign of appreciation for their work, and a further two explicitly stated that household members were dependents and were rewarded with clothes and sustenance. This situation found interesting parallels in the case of a number of farmers who had difficulties acquiring and retaining labor. The competition for local labor could be stiff, and one man spoke of having to give his workers lunch in order to encourage them to stay, while a second, who employed eight permanent workers, said that it was crucial to make sure that laborers were treated fairly so that they did not get lured away by other farmers. In both of these cases, the farmers were aware that workers could change their allegiance, but what would bind them to the farmer was not higher wages, but certain degrees of obligation and respect, combined with particular benefits or the promise of future benefits.

The sharing and distribution of food has always been a feature of political and social relations in Bemba society (as we saw in Chapter 3), and the obligations of men to women, and of men of status to their dependents were in the past expressed in terms of maintenance and sustenance. But food has never been the only component of sustenance, cloth and clothes have often assumed equal importance. Cloth (originally bark cloth) was an important part of the marriage transaction, and a woman who was not clothed by her husband would have grounds for divorce (Richards 1939:113, 219; 1940). This accounts, in part, for the extraordinary emphasis on clothes in Bemba society, and for the way in which they were often presented as the rationale for early labor migration (see Chapter 6 this text; Richards 1939:216–218; Wilson 1940), and for the fact that the ability to clothe the family is presented by present day farmers as one of the great benefits of maize cash cropping. Among the twenty-four farmers engaged in cash cropping in our sample, clothes still constituted one of the largest items of annual expenditure, sometimes greater than the amount spent on agricultural inputs, just as they did in the late 1950s (see Chapter 3). When some of these farmers spoke of giving their household members "tokens" in appreciation for their work on the farm, they were usually referring to clothes or cloth (fitenge). The provision of clothing, as well as food, is one of the major ways in which the obligations of maintenance and sustenance are made tangible and concrete. It is no exaggeration to say that food and clothing symbolize, as they did in the past, social relations.

The present-day system of labor recruitment and household formation then, displays many continuities with the wealth-in-people system that sustained the political structures of this area in the past, but it is perhaps now a more difficult system to play and sustain. It is still a system based on a complex nexus of obligations of maintenance and sustenance, dependency, and respect. Kin may be counted as laborers, and workers can be treated like family members, but in many cases, the exact nature of that relationship will be left open. It is also the case that relationships change and transmute. Farmers with failing fortunes may need to rid themselves of dependents, but they can also find themselves deserted.

We should not exaggerate the degree of continuity here, however. A number of social and economic changes that have taken place in the Northern Province during the last sixty years have limited the extent to which individuals can mobilize labor by drawing on social relationships based on dependency. In the past, of course, the ultimate relationship of this kind was slavery, but slavery is long gone and the increasing mobility of the society, coupled with social and economic differentiation, has transformed the context in which individuals are operating. Chiefs could once demand tribute (mulasa) labor from their subjects, and the authority of senior kin was something to be reckoned with. But, by the 1930s, individuals were bemoaning the passing of this Golden Age (Richards 1939:144–145, 256–260). What has most evidently grown over time has been the need for cash in order to secure household reproduction. With this has come involvement in wage employment, and an increase in the number of nonagricultural activities in which people are engaged. The sheer diversity and range of social networks in which individuals are now involved, straddling both town and countryside, the impact of education and class formation, and the appalling unpredictability of a life-world dominated by the vagaries of a declining market economy are all factors contributing to a situation in which it is now extremely difficult to control the labor of dependents. But in the context of an underdeveloped regional labor market and the evident risks of cash cropping, farmers have little option. These dependents, like the farmers themselves, are trying to make ends meet, diversifying their opportunities, and seeking openings wherever they can. The result is that Bemba households, like others in the Northern Province, are not homogeneous, single production and consumption entities, but a nexus of overlapping interests and activities whose (sometimes very temporary) coherence is itself an achievement and not something pregiven.[23] The cooperation implied in conjugality and evidenced in the sexual division of labor is something to which individuals aspire, but it is often a feature of a particular stage in an individual's life and in the developmental cycle of the household rather than a fact of life or of social organization.

We would argue that this was also the case in the past, where marriage provided the opportunity for men to link themselves to groups of cooperating women and to develop their own network of supporters based on the conjugal unit and its descendants (see above and Chapter 3). Thus, conjugality and the two-generational household are not something entirely new in the Northern Province, but rather they have been part of a repertoire of strategies that have come and gone within a number of time dimensions. Dimensions of time that are simultaneously biographical, structural, and historical. In the past, as in the present, marriage established women's and men's rights to each other's labor, but these rights were not absolute. At the basis of Bemba society, as Richards noted, were groups of cooperating women, and it was through the channels established by these women that the system of redistribution operated (see Chapter 3). The cooperative groups established by men (linked together by marriage) for productive purposes tended to be much more fragile, in part as a consequence of greater male mobility (Richards 1939:130–

131). In addition, a man's rights over his children were fluid and open to interpretation. Rights over the labor of spouses and children obviously varied greatly from group to group in the province, and the situation among the Mambwe, for example, was quite different from that among the Bemba. With labor migration, the ability to control the labor of sons and sons-in-law apparently declined still further. Thus, while there was much to be gained by cementing networks based on the conjugal unit and its descendants, the Bemba household as it played out in practice was never a single cooperative production and consumption unit under the control of a male household. Membership of the household shifted constantly, and obligations were owed to kin in distant villages, both matrilineal and bilateral. However, the difficulty for present-day farmers, as for the peasant farmers of the 1940s and 1950s, is that the potential success of their enterprise depends on the degree to which they can weld their household into something that resembles a joint utility function. It is for this reason that networks based on the conjugal unit and its descendants were as important in the past, as they are in the present. Attempts to weld households into joint utility functions sometimes appear almost doomed to failure—except perhaps in the case of very successful commercial farmers—not only because the male farmers themselves must invest in diverse social networks to survive, but because their own household members must do the same. The argument we make here is not that the changes of the last sixty years have brought about the dissolution of the Bemba household, but rather that they have exacerbated tendencies that already existed and have made it effectively impossible for a single production and consumption unit modeled on the nuclear family to emerge, except in the case of very successful farming (and urban) households. We can see why this should be so if we look at the current relationship between on-farm and off-farm activities, and between household and nonhousehold labor.

Making Ends Meet: Agricultural and Nonagricultural Pursuits

Recent studies in the province have stressed that access to extrahousehold labor is the key factor in being able to expand cash-crop production. The more commercial households certainly employ more nonhousehold labor than do their less successful counterparts.[25] However, labor availability is a problem for all, and a number of farmers (women and men) in our sample complained that they needed more extrahousehold labor, but simply could not afford to pay for it. Maize is the crop that many farmers see as being expensive in terms of labor. As one man said: "The high expensive inputs in maize production reduces the crop to a less profitable enterprise. One of the costs is labor which is required a lot more compared to other crops almost throughout the growing period up to harvesting and packing time."

The vast majority of nonhousehold labor employed on farms is casual and is often employed on a piecework basis. Out of twenty-four male householdheads in our sample engaged in cash cropping, five employed permanent laborers and all used extra casual labor. None of the thirty women farmers sampled employed permanent laborers, although eight households were engaged in cash cropping. However, 17 of these households employed casual labor, and this figure does not include those women who brewed beer for direct consumption by the work-party or those who made use of church, Women's League or cooperative working

TABLE 8.5. Extrahousehold Labor Use by Farmer Category

Farmer Category	1	2	3
Total hours farm labor	1199	1615	3199
Hours extrahousehold labor	88	130	629
Hours household labor	1111	1485	2570
Size of household	6	6	7
Number of persons Aged 19v	2	2	3
% Female-headed household	26	18	16

Category 1: Selling no maize to the official market; Category 2: Selling 1–30 bags of maize; Category 3: Selling 30+ bags of maize.

(*Source:* Bolt and Holdsworth 1987: Section 5.)

parties.[26] In all, twenty women used casual labor for their citemene fields, five used it for cassava gardens, five for maize, two for beans, one for soya beans, one for potatoes, and one for vegetables. Remuneration levels varied greatly among the male and female farmers sampled, but kapenta (dried fish), salt, sugar, clothes, school requisites, and maize were popular means of payment. Some paid money, especially for ridging, and rates of between K2 for 3m of ridge and K5 for 25m of ridge were quoted. A great deal depended on the exact type of work to be done, and several farmers said that they paid more for ridges made from new ground than for those made on previously tilled land, while others claimed to pay more for ridges for cassava roots because they need a greater depth of soil. Thus, it was not possible to make any meaningful comparison between wage rates.

Evidence from elsewhere in the province suggests that most extrahousehold labor is used for maize, and this would appear to be the case for those in both our samples who were engaged in cash cropping, with large amounts of casual labor being employed for cultivation, weeding, harvesting, shelling, and bagging. A few farmers were able to afford to hire tractors for ploughing, which relieved their labor problems greatly. However, many farmers also reported using substantial amounts of labor for vegetable production, and for planting and ridging of cassava and sweet potatoes. Many of the women farmers, both cash-croppers and non-cash-croppers, used a great deal of casual labor for citemene cultivation. This may partly reflect the stage in the development cycle of these households because some women's husbands were elderly or sick, and few had resident children under the age of 18 years (see footnote 26). Out of the thirty women farmers interviewed, there were seven female-headed households in the sample altogether. However, it is probably also a reflection of the fact that much of the labor done on citemene fields, after the initial cutting of the trees, is done by women and organized by women through their own exchange networks. It is, therefore, possible that women farmers report higher levels of casual labor use on citemene because they are more closely involved in this form of cultivation. These findings correspond well with other information gathered during the research, which shows that women are actively engaged in a series of exchanges that incorporate agricultural labor, income generation, commodity acquisition, and social networking (see Chapter 7).

When the thirty women farmers in the sample were asked to describe the relationship between their on-farm and off-farm activities, they all made a direct link between various sorts of income generation, different types of labor investment, and household reproduction. For example, twelve women spoke of brewing beer as a way of gaining access to agricultural labor, especially for citemene fields, but only four of them in 1988/89 had given their laborers beer to consume directly at the end of the working day. The rest brewed beer in order to get access to cash or commodities either for household consumption or to remunerate laborers. The millet used to brew the beer came from household stocks, except in the case of one woman who had worked for another woman to get the millet she needed, and a second individual who bought K50 worth of millet in order to get her beer brewing enterprise started. A further seven women said that they engaged in petty trading either to raise money for labor payments or to purchase commodities for labor remuneration and/or for home consumption. In addition, five women said that they worked for other women for food and commodities (including clothes, sugar, salt, and fish) sometimes for household consumption, and sometimes to engage in further exchanges and convert the commodities they had acquired into labor, cash, or other commodities (see Chapter 7). Women cooperated in these enterprises; relations of this kind were recorded between women and their sisters, mothers-in-law, daughters, and wives of their brothers-in-law. The ethic of cooperation and reciprocity is strong even outside close kin, and ideally if you work for others you must at some point offer them work. One elderly woman put it rather well when she said "I do not work for them since I have nothing to pay them." This is precisely the distinction most women would make between the redistributive economy they know so well and rely on, and the stigma of begging (see Chapters 3 and 7). It is also interesting in this regard that three women from wealthier households pointed out that they only worked for other people when they had to take part in church or Women's League group activities. These women have no need to work for other women, but they do need to involve themselves in social networking in order to gain access to information about local activities, trading opportunities, and supplies, as well as to avail themselves of group labor on their farms.

There are a number of things to be said about women's heavy involvement in the economy of redistribution. First, it means that there are many demands on their labor time. Second, if agriculture is to be successful, women must invest in off-farm activities in order to provide labor and other agricultural inputs. Third, it is clear that, apart from a very few extremely successful commercial households, the income from agriculture is either insufficient for household consumption needs or that women do not have access to this income. This must be the case because women spend so much time working for food and commodities. Most women reported that their husbands controlled all the money from maize sales and from other cash crops, including vegetables if these crops were deemed to be part of the cash cropping enterprise. The twenty-four male household heads involved in cash cropping also claimed to control all the money from sales, but they acknowledged that their wives did get some money from beer, vegetable, and wild food sales. Only three women (taking both samples together) were engaged in own-account cash cropping on ibala fields. Two were producing maize and one was producing beans, and in each case their husbands referred to these enterprises as separate from their own farming activities.[27] The degree to which women's income is separate from that of their husbands and the extent to which the parties cooperate and take on joint responsibility for expenditure varies from one household to another. The women farmers interviewed were clear that they cooperated with their husbands in farming enterprises, but that there was much negotiation over how labor and other resources were to be allocated. The question of beer brewing is a good example, with women brewing both for labor on the citemene and the ibala fields. Men do not have complete control either over the millet necessary for brewing or the timing of women's activities. The result is a great deal of give and take over each stage in the procedure over such issues as where the millet will come from, when will the beer be brewed, and for what purpose it will be used. If a woman agrees immediately to her husband's request for beer for labor on the cash crops, then she expects reciprocity at a later stage. None of this may be very surprising, but it does emphasize the extent to which both women's and men's farming activities are dependent on off-farm activities.

A number of studies throughout the province have emphasized the importance of off-farm income for people's livelihoods in general, and ARPT (Kasama) and IRDP (SMC) have stressed that the level and timing of this income has a critical influence on the opportunity cost that individuals place on agriculture (Bolt and Holdsworth 1987; Bolt and Silavwe 1988). What is most noticeable about this income is its relative regularity throughout the year when compared with agricultural income. There is considerable variability in the levels of off-farm income both between areas and between farmer categories, but ARPT (Kasama) calculated off-farm income as 43% of total income for all sample points. They also reported that male household heads claimed to earn 67% on average of annual off-farm income, wives earned 18%, and 5% was earned jointly. It was noted that the main off-farm income source for women was beer brewing, which contributed 59% of their total annual off-farm income, and which provided women with the means of making small purchases during the year, especially food. When income was matched with expenditure, it was found that clothing was the highest expenditure item (29% of total expenditure) followed by food and milling (19%), and groceries (11%). When expenditure was broken down by gender, husbands spent most of their money (32%) of clothes, followed by groceries and consumer durables. Female household heads spent most of their money on clothes, while wives spent it on food (38%). Off-farm expenditure also showed a massive rise in August apparently because of the anticipation of monies from forthcoming crop sales. A detailed analysis of income and expenditure, both agricultural and nonagricultural, showed that although agricultural income boosted total income, much of the profits were reinvested in the next crop, leaving off-farm income, in cash and kind, as the major source of household purchasing power during the year (Bolt and Silavwe 1988: Section 7).

This last point accords well with our own research findings that also show that very few male household heads took out official loans of any kind in 1988/89, although six of them received free fertilizer and seeds under a government scheme.[28] The individual farmers interviewed were adamant that it was crop receipts that dictated agricultural investment from year to year. The same point is stressed by ARPT (Kasama) (Bolt and Silavwe 1988: Section 7). Many of the women farmers interviewed complained that they could not engage in cash cropping because they could not pay for the labor or for fertilizer and seed. One woman even claimed that ". . . most farmers have dropped their crop enterprises on their farms because of expensive agricultural in-puts." Some of them were members of cooperatives, but they felt that these offered them little assistance, and while some stated that they could not apply for loans because they knew they would be considered a poor risk, others showed very little interest in loans in general. Many informants expressed a great deal of hostility toward cooperatives and loan schemes, and out of thirty returned migrants interviewed, only four of them had anything positive to say, and none of them had taken out a loan. One man said ". . . getting loans is risky because fertilizer and seeds have become very expensive to the point of a farmer giving up completely on cash-cropping." The general conclusions to be drawn from this are that agricultural income is invested largely in agriculture, and because it comes in a lump sum in September, it is also used for large nonagricultural expenditure, like clothes. Male household heads in our sample reported using a significant percentage of their annual crop income on clothes. However, agricultural income does not provide for household needs during the year, and these expenses must be covered by nonagricultural income. This means that off-farm activities are an essential part of everyone's livelihood. The fact that the levels of off-farm income increase with commercialization (Bolt and Silavwe 1988: Section 7) shows that successful agriculture is part of a general and diversified strategy of successful entrepreneurship, and that thus there is no benefit to household heads in trying to prevent their household members from engaging in diversification. Given that diversification is essential for household reproduction, it is inevitable that household members will struggle both to generate income and to retain control over it. This must be the case, because they have no guaranteed claims over the resources of any other household member. Lines of power and authority are important here, as are conventional expectations, the power of personalities, and the gender, age, and social status of individuals. Women's responsibility for feeding the household, and especially their children, is a genuine burden under such circumstances. They, therefore, struggle both to benefit from their marriages and from joint cooperation with their husbands, while simultaneously guarding their access to independent sources of income and to alternative resource networks.

As the need for diversification within the household has increased, women have become particularly vulnerable because, relative to men, their control over the labor and resources of others has been eroded much more quickly. This is perhaps most evident with regard to their loss of control over the labor of their children. A mother's control over the labor of her children was never a guaranteed right, but it was always something that she could expect to draw on, especially in the case of female children and domestic labor. This was after all, as Richards noted, the cornerstone of the joint housekeeping enterprise.

One clear example of the problems women now face is provided by looking at the effect that schooling has had on the labor resources of households. Young men may attach themselves to a household in order to try and get their school fees paid, and it is not unusual for young men to be paid for by their maternal uncles. However, school fees are not always forthcoming, nor are funds for the compulsory uniforms that pupils must wear. A number of cash-cropping farmers in our sample reported that one secure source of casual labor were schoolgirls and schoolboys who worked either for paper and pencils or for the cash to purchase these items. It is obvious that school-going children who are working for others in order to get the means to go to school are not working for their own households. School attendance often deprives households of the labor of children and young adults, but in a situation of resource scarcity it does so doubly. This affects the overall supply of household labor, but it affects women in particular because they had a degree of control of children's labor in the past that is now denied them. In particular, they were dependent on the domestic labor of girls and young women so that they could be released for other activities, both on- and off-farm. Evans and Young noted that the economic contribution of children is variable, but that it is an important factor in a household's capacity to generate income and expand production. They observed that the contribution of children to domestic work is particularly crucial, and that in the categories of farming household in which the contribution of young persons older than eighteen years of age to domestic work is greatest then the wives' input in this sphere is lowest (Evans and Young 1988:Section 6). Bolt and Holdsworth, also using IRDP (SMC) data, calculated that the contribution of children older than seven years to total on-farm work was 34%, 22%, and 25% for farmer Categories 1,2, and 3. They concluded, therefore, that labor contributions from this age group decrease with commercialization in both relative and absolute terms. They also produced some fascinating evidence to show that in Category 1 farming households, teenage sons (13–18 years of age) spend the greatest amount of time away on citemene, followed by wives. But, in category 3 households, wives have the largest responsibility for citemene production (Bolt and Holdsworth 1987:Section 5). This provides evidence for the effect of schooling on teenage labor input, but it also suggests that it is women's labor that makes up the shortfall in teenage and child labor on citemene. It is perhaps best in the long run if we cease to imagine households as joint utility functions or havens from a heartless world, and come to realize that for women and for men, they are part of the strategy of marriage, which with all its emotional and material rewards and disadvantages, is itself a form of survival through diversification.

Conclusion

Our discussion of agriculture and nutrition in the province has directed our attention successively to history, culture, and politics—both the politics of gender within households and the community, and the politics of development and national agricultural policy. It is in the intimate context of lives lived in farming households that larger processes and policies have their effects, and indeed, to a certain extent, their origins. In the first decade of this century, for example, colonial policies concerning settlement and residence gave rise to rules governing whether and when husbands and wives could sleep together in mitanda. The regulation of citemene cultivation appeared to imply the regulation of intimate life. The lives of the people of Northern Province have always been bound up in larger processes and policies, whether initiated by chiefs, colonialists, or development consultants. In recognition of this fact we have tried not only to demonstrate the methodological and theoretical validity of a constantly shifting analytical focus, a perspective that continually moves between the level of the household and the level of the national economy, but we have also attempted to explore the nature of the links between these different levels and how they might have been experienced by ordinary people.

The links between these different levels have warranted a revolving engagement with history, culture, and politics. In most anthropological re-studies, an earlier situation is contrasted with a later one, and history emerges as those changes that have taken place over the period of time separating the two. The result is that two snap-shots are placed side-by-side. An often oversystematized version of "the past" is then contrasted with a version of "the present." Elements of change and continuity are identified: some features of the society may look very familiar from the account of the past, others may have disappeared altogether. What tends to be ruled out in this approach is an appreciation of the continuous and often indeterminant nature of the making of history, and the constantly present tension between what people do, what they say they do, and what they "ought" to do, which makes any society what it is at any given moment in time. In the case of northern Zambia, the dominant metaphor used to describe change has been that of social breakdown, and it has been particularly applied to the relations of kinship. We have attempted to move away from and beyond this metaphor, arguing that it obscures rather than illuminates the changing nature of social relations in this area.

We can never know exactly what kinship relations looked like in the past, but almost certainly they were more fluid than the jural model used by anthropologists would suggest. We know that kinship relations, as they affect strategies of diversification, household composition, labor recruitment, and systems of cooperation and sharing, remain important and necessary to the people of this area in the context of a modern, if failing, economy and a postcolonial state. To conceptualize these current relations of kinship in terms of breakdown, would obscure the degree to which they have been continuously and creatively reworked in the context of labor migration, cash cropping, and the wage economy, as well as the extent to which they remain central to all these processes of change. The question is not really whether kinship was broken down or not, whether continuity triumphs over change or vice versa; the more important point is to look at the way in which certain problems and solutions (albeit often partial ones) emerge and re-emerge over long periods of time. Here some analysts might argue for the existence of certain cultural givens that endure and that provide an underlying set of structural features. We wish to emphasize, however, the recursive nature of certain problems—such as labor recruitment, the politics of residence, and household reproduction—and the way in which they keep surfacing and resurfacing in changing times. These problems, which seem like repetitions of old problems, are not exactly the same, but as they emerge they are formulated in terms of existing discourses, and they take shape in the light of previous histories; as such, they are grafted onto a version of the past to be remade in the present. This process of grafting is only possible because problems and solutions are mutually interdependent, as are past and present. This does not mean that the present is an intentional emulation of the past or that the future is necessarily determined by certain structural features of the past. The issue here is one about the nature of repetition and the practical and discursive need for recognizability. However "new" a situation may be, it will have to be appropriated to a certain extent in terms of a set of practices and discourses that are already known.

Our study has raised these issues in terms of the construction of history and of historical writing. We have delineated the dominant discourses that have, over 100 years of history, served to represent this area and its people, and we have paid special attention to the representations produced within the disciplines of anthropology and ecological science. We have not, however, seen these discourses and representations as having determined the history of this area, or as having ruled out any other versions of that history. We have tried to create a detailed account of social and economic change in the area that takes into consideration the influence that these representations have had on, for example, the construction of agricultural policy. At the same time, however, we have recognized that history has not been entirely captured by these representations and discourses, but has also continuously evaded them, just as the citemene cultivators of this area were both captured by the demands of the colonial state, yet also found new areas of uncultivated space in which they could continue to carve out their own world. We have also tried to indicate that it is not only the powerful who construct representations of the Other: the less powerful also do so. The people of the Northern Province have constructed their own representations of their world, produced accounts of the interventions made in this world by external others, and made their own history.

We have been led to an appreciation of the depth of historical understanding of Zambian rural people themselves who have seen come and go a variety of interventions designed to control, reform, and stimulate their agricultural practices. The politics of settlement, residence, and agricultural production are as pressing for the present-day government of Zambia as they have ever been, yet so many of the solutions paraded as "new" in the 1980s and 1990s have been tried before. This is not to deny the enormous changes that have taken place in this area, nor to deny the profound effects of the rise of the maize cash-cropping industry on social and economic organization. But maize has not solved the problems of household reproduction for the majority of the people of the province, and thus a number of policy discussions recur that are familiar to us from an earlier period. The need for a back-to-the-land policy, the desire for a settled peasantry, the impossible cost of agricultural development, and the difficulty of finding a crop or crops that would raise income and guarantee livelihoods are all features of debates in the 1990s, as they were in the 1950s. As President Frederick Chiluba's government faces its first real test of popularity, the political demands made by farmers who want to know where the benefits of development are, seem all too familiar. For farmers in the Northern Province, they echo with the voices of their fathers or perhaps of their younger selves. The "progressive farmers" of the 1990s are very similar to those of the 1950s. They have been labor migrants and they have capital or pensions to draw on. Successful farming still depends largely on having access to an off-farm income. Some of the former migrants draw on skills they learned "in town," while others continue to engage in wage labor once they have officially retired to be farmers. It is true that the young men are no longer migrating in such numbers and that many of them in the province are turning to farming in the hope that it will provide a living. It is not yet clear how these young people will fare, but reports emerging from Zambia in 1992 suggest that the government is acutely aware of their predicament, and once again there is a tremendous urgency to set up farming schemes. These schemes look for all the world like the peasant farming schemes of the 1950s.

Some of the continuities between the 1950s and the 1990s may be more apparent than real. However, what is important for people themselves is that those continuities are apparent. It should not surprise us then that when people respond to new development schemes and new policies, they bring their history with them. This point is nowhere more apparent than when we come to look at citemene itself. Citemene, against all odds, is alive and well in the 1990s. Despite all the exhortations of the experts, most farmers, with the exception of a few well-off individuals, continue to incorporate it into their agricultural strategies. Given the uncertainties of input delivery, pricing policy, the national economy, and the climate, this is not surprising. In a recent article, Barrie Sharpe points out that although the proportion of farmers with citemene plots in the IRDP (SMC) sample fell from 65% in 1982–83 to a low of 34% in 1984–85, it rose again to 46% in 1985–86. The IRDP (SMC) sample points are, of course, in areas where hybrid maize cash-cropping is particularly well-established: this is why they were chosen as sample points (see Chapter 7). A considerable degree of underreporting with regard to the actual number and extent of citemene plots is likely because farmers remain wary, as they were in earlier times, of disclosing their citemene activities. Despite this, and despite the high commitment shown by IRDP (SMC) farmers to maize cultivation, just under half of them were still engaged in citemene production in 1985–86. Sharpe argues that this figure reflects the riskiness of hybrid maize (1990:592–595). We would agree with Sharpe in that we see the continuance of citemene as part of a strategy of flexibility and diversification. The need for such a strategy to ensure household reproduction is as great now as it was in the past for the majority of farmers in the province, although the nature of citemene is two-fold. As we have argued in earlier chapters, citemene is not only part of a practical strategy of diversification, it is also a powerful metaphor that encapsulates the possibility of such a strategy. It is in this sense that we have argued that the practice of citemene has been an integral part of the representations and self-representations of the people of the Northern Province over at least the last 100 years.

However, citemene itself has changed. The ritual meanings associated with it have altered, the sexual division of labor that characterized it has changed, and its role in the agricultural system is still undergoing modification. Nonetheless, citemene practices have continued despite changes in Government agricultural and residence policies. As a system of agricultural production, it is compatible with a variety of residence arrangements, but in order for this to work it has been necessary for people to make changes in cropping patterns, in the sexual division of labor, and in the mix of field types. Semipermanent cultivation of ibala fields is increasing in extent in the province, and often at the expense of the village gardens formerly cultivated by women. This makes women vulnerable because it undermines their control of resources and cuts down on crop diversity. One major change, however, is that as men become more heavily involved in hybrid maize production, and manage to control the income it produces, women become much more strongly tied to citemene in order to secure resources for themselves and for household reproduction. In this sense, the gender connotations of citemene are altering as it becomes less associated with an exclusive male identity. However, what history has taught the farmers of the Northern Province is that it is possible to adapt to changes, and that for the time being, there is every reason to continue cutting down trees.

NOTES

CHAPTER 1

The Colonial Construction of Knowledge:
History and Anthropology

1. See especially Werbner (1967); Epstein (1975); and Gluckman (1954); see also Richards' own articles on this subject. Richards (1940a; 1951; 1960; 1961; 1968; 1971).

2. Succession to Bemba chieftainships was, in fact, rarely as orderly as this account implied. Other writers have drawn attention to a degree of competition and civil strife which succession frequently brought to the surface: Werbner (1967); Roberts (1973).

3. Roberts discusses at length the role played by the development of long-distance trade in the formation of the Bemba political system; Roberts (1973: Ch. 6).

4. For a demonstration of this compulsion on the part of early BSAC officials, see Gouldsbury and Sheane (1911) and Gouldsbury (1915; 1916).

5. For a general account of this process in colonial Central Africa, see Chanock (1985).

6. Through a detailed analysis of the hymns and symbolism of both the Lumpa Church and the Ba Emilio movement, Hinfelaar, himself a Catholic priest with long experience in the Northern Province, makes a convincing argument for the survival of non-Benan'gandu ritual beliefs and their assimilation into forms of Christianity (Hinfelaar 1989). See also earlier work by Garvey on the relationship between Bemba chiefs and Catholic missionaries (Garvey 1977).

7. We have combed through a range of sources in search of evidence for the practice of citemene cultivation by Bemba (as opposed to other ethnic groups) in the precolonial period. The absence of such evidence does not necessarily indicate the absence of citemene cultivation. It is more likely to be indicative of the fluidity and indeterminancy of ethnic identity in parts of the region in the late eighteenth and nineteenth century when the Bemba were coming into being, as well as indicative of the "slave route" bias inherent in travelers' accounts. From Lacerda (cited in Burton 1873) onwards, most travelers passed through on the margins of Bemba territory, remarking less on cultivation than on the problems of provisioning their parties caused by Bemba raiding, depopulation, and famine. Lacerda described the mound cultivation of millet by Bisa between the Lùangwa valley and the Chambeshi river (Burton 1873:92), and cassava production near Lake Bangweulu (Burton 1873:100). In 1831–32, Gamitto, traveling the same route westward from the Luangwa valley, noted Bisa villages and gardens that had been raided by the Bemba (Gamitto 1960; Vol. 1:166), and the destruction of cassava fields near Lake Bangweulu (Gamitto 1960, Vol. 1:188). By the 1860s, when Livingstone made the journey west from the Luangwa, Bisa agriculture had been further disrupted by Ngoni raids, but he does describe Bisa cultivating millet in ash-manured "small round patches" (Livingstone 1874, Vol 1:166). More gen-

erally, however, he echoes Gamitto in finding the area one in which it is almost impossible to buy food, and in which the people appear to subsist largely on wild foods. When he reaches the stockade of the Bemba chief Chitapankwa at the end of January, however, the chief provides his party with maize and groundnuts, and persuades them to stay to sample the green maize which is just ripening (Livingstone 1874, Vol. 1:189–90).

In 1883, French naval officer Vistor Giraud took the northern route into Bemba country. He found the same problems of provisioning his party as did previous travellers, but on approaching the large stockaded capital of the Bemba paramount, he noted that fields of maize and sorghum were becoming more common, and he described their mode of cultivation (Giraud 1890:266). But according to Giraud, millet (ulezi) formed the basis of the Bemba diet, and he provides us with a brief description of a form of citemene cultivation. A small area was cleared of trees and was fenced. Millet seed was broadcast in these clearings "sans plus se preoccuper de la pluie ou de beau temps" (Giraud 1890:266). He makes no mention of any preparatory burning of branches.

8. For a full analysis of this important period, see Roberts (1973:Ch. 8) and Henry Meebelo's study of early colonial encounters in the area (Meebelo 1971).

9. It is rather difficult to know exactly who these refugees were since by this stage 'Bemba' had come to be used as a generic term for all people inhabiting what was nominally Benan'andu-controlled territory. Marcia Wright makes this point in her study of life histories, reminding us that in a situation in which slavery was common, ethnic identities could be extremely fluid (Wright, 1993).

10. "The Bisa cultivate on the ground, while their neighbors, the Bemba, cultivate in the air."

11. ". . . climb in the trees like monkeys and die of hunger," while the Bisa "cultivated the soil so as to have lots to eat."

12. "Food is something sacred, wasting it infuriates the people: for them, in the past, the destruction of the grain-bins was the signal of war."

13. ". . . our Bemba so keen on adventures and travelling."

14. "today a messenger comes to recruit men for government work, tomorrow another arrives to enrol men for work on the mines, a third arrives to collect millet and millet flour."

15. "Bwana, we understood nothing, on the one hand we're told to cultivate and take food to the Boma, and on the other, our young men, who are the only ones capable of cultivation, are taken from us, what do you want us to do?"

16. For the fascinating story of Bishop Dupont, see Pineau (1937); Garvey (1977); Hinfelaar (1989). Many thanks to Father Dan Sherry for providing us with a a copy of Pineau's book.

CHAPTER 2
The Colonial Construction of Knowledge: Ecology and Agriculture

1. See Beinart (1984) and Chapter 5, this text, for further discussion of this issue.

2. Both Schultz (1976:64) and Allan (1965) have also noted the importance of historical and economic factors, as opposed to environmental conditions, in determining the specific nature of local systems.

3. The Bemba have a number of indigenous terms for different sizes of first year ash gardens. The following list given by Stromgaard (1985a:45) accords well with the terms collected by the authors during field research in 1988.

> Ubukula—Big size, first year ash garden.
> Ichikumba bukula—Medium size, first year ash garden.
> Chikumba—Small, first year ash garden.
> Akakumba—Smaller, first year ash garden.
> Chikuka—Very small, first year ash garden.

4. It is important to note here that Stromgaard has reported in a recent article that households do not open new citemene fields every year, but rather every other year (Stromgaard 1985a:57). He suggests that this may be because of diminishing forest cover. We
discuss this point further in Chapter 3.
5. Stromgaard has reported the results of experiments to determine the effects of the so-
called Bemba practice of lopping instead of felling trees on woodland recovery rates
(Stromgaard 1988b:371–372). He concluded that lopped trees regenerate faster and produce a greater biomass of vegetation over a 16-year period than do felled trees.
6. Richards lists the different soil types recognized by the Bemba, but points out that despite
a certain amount of rudimentary knowledge, many cultivators procede on the basis of
trial and error (Richards 1939:280–287). Kerven and Sikana have conducted a recent investigation into indigenous soil and land classifications in Northern Province (Kerven
and Sikana, 1988). They emphasize the contextually derived and nonhierarchical nature
of local classifications, and they stress, as Richards did, that different farmers have very
different levels of knowledge and expertise. See Reid et al. (1986:4.5.2) for a recent discussion of soil types and their classification in the Northern Province.
7. This is because the dominant species of the main Julbernadia-Brachystegia genera that
compose the Miombo woodland are fire-tender and newly established woody growth and
sucker shoots are particularly vulnerable (Trapnell 1953:102). Thus, both established and
regenerating woodland are destroyed by repeated late burning (Peters 1950:26–27). Allan
has argued that under normal citemene practices the woodland was not repeatedly late
burned. Only the areas around new cuttings were protected from fire until late in the
season, elsewhere burning started earlier. Thus, any one particular piece of woodland
would only be subjected to late burning occasionally (Allan 1965:75).
8. The first census in Zambia was in 1963, and it mainly used chief's areas as the basis for
enumeration. A sample census was undertaken in 1950, but prior to that, estimates of the
African population were made using lists of taxable males, and, at an earlier period, hut
counts. Colonial officials took local censuses while on tour, but the quality of these were
variable, and some officials clearly did little more than guess. Problems with population
figures are discussed further in Chapter 6. For further discussion, see Kuczynski (1949)
and Ohadikike (1969).
9. The debate about the relationship between population increase and agricultural intensification is an old one. What is clear is that no single causal mechanism can be preferred
over others except in determinate historical circumstances. See Boserup (1970); Hill
(1977).
10. Sonkwe is *Sorghum Caffrorum* var. *Breviaristatum*.
11. Trapnell notes that in the 1930s, small, separate gardens were made for mwangwe, depending on the number of children, in areas close to the Paramount or other important
chiefs in the Kasama area (Trapnell 1953:47). The connection between political authority
and labor availability is not made, but is clearly relevant.
12. In 1976, tests were undertaken by the Intercropping Research Programme at Mazabuka.
Some of the results are discussed in Haug 1981:93–94.
13. It has also been suggested that the system was introduced into the region as a result of
intrusions of populations from the north. See Willis (1966); Vesey-Fitzgerald (1963); Trapnell (1953:56).

CHAPTER 3

Relishing Porridge: The Gender Politics of Food

1. The question of labor migration and levels of remuneration are discussed in Chapter 4.
However, Otter, District Commissioner of Kasama, noted in 1932 that local wages for
unskilled labor were 3d per day, and that this meant that a man would have to work

six to seven weeks to earn his tax as opposed to three weeks or less on the mines
(ZA 7/4/28:Awemba TR Aug., 1932).

2. "D'autre part de nombreuse jeunes gens vont aux mines, dont la femme, enfants et beau-
   parents restent ici, sans être soutenir. D'autres malins s'ils reviennent au pays ont soin de
   n'arriver au village qu'après le kutema de sorte qu'ils viennent vivre en parasites sur les
   biens d'autri" (WFD: Chilubula 30.1.1931).
3. This data is contained in Richards' papers, which are housed in the library of the London
   School of Economics. In this chapter, we make extensive use of unpublished and unan-
   alyzed data collected by Richards and Ann Tweedie. Wherever possible, this data has
   been presented in "raw" form in tables in the chapter in order to provide researchers
   with access to the original information.
4. Increasing attention has been given to intrahousehold food allocation and age and sex
   inequalities in recent research, Harriss (1990).
5. Both sets of data are unreliable. The calculations are based on respondents' recall of the
   height of grain in the granaries after harvest. After measuring the size of each granary,
   the cu.ft. of grain was calculated and this was later transformed into weight of grain in
   lbs. What is clear, however, is that Richards and Gore-Browne did not use the same equa-
   tion when transforming cu.ft. into lbs. Furthermore, Richards' data do not take into account variations in household size, but we can assume that grain supplies did not
   increase exponentially with household size. Gore-Browne calculated her figures on the
   basis of what she called a ten-month food supply. What she meant by this is very unclear,
   but it seems to have been a way of excluding from her calculations early maturing grain
   varieties that would have been eaten in the period just prior to the new harvest. In any
   event, her figures, which are based on both observation and recall, appear to relate to a
   single annual harvest. Taken together these two data sets are problematic, but they do
   provide an impressionistic sense of variations in agricultural yield for the main staple in
   some areas in the 1930s.
6. See Chapter 7, Tables 7.4, and Table 7.5 for comparisons of the diet from the 1930s to
   the 1980s.
7. The only figures for sales in the 1930s that exist are those contained in the one-off comments of missionaries and administrators in their records and reports. There are no consistent data sets.
8. "Relish" is the term commonly used for a sauce of meat, fish, or vegetables eaten with the
   staple porridge.
9. In her original notes Richards actually describes the village as having "ample man-power"
   with a ratio of 26 men to 40 women (AR: V35). This is quite inconsistent with what she
   says in her published work, but the inconsistency is perfectly explicable in terms of the
   struggle to deal with a very variable and complex situation in the context of the overwhelming pervasiveness of the discourse on male absenteeism (see Richards and Widdowson 1936:195). Richards seems to have found herself quite frequently in the situation of
   tacking back and forth between her data, which did not conclusively show that food production levels were determined by male absenteeism and the received view of the time.
10. Richards also kept dietary records for five households in Kasaka from September 13 to
    October 3, for four households in Matipa from November 18 to 27, 1933, and for 5 house-
    holds in Kasama Village (adjacent to the township) from February 14 to February 22, 1934
    (AR:V1–73).
11. It is instructive to compare Richards's data with a seasonal food calendar supposedly
    compiled by local residents in Mpika District at the request of the District Commissioner
    in 1937. The District Commissioner was responding to a survey of food conditions in the
    Northern Province undertaken by the Diet Committee (SEC 1/1042:1937).

January: We eat millet if there is any left and cassava. The relishes are mushrooms and cassava leaves. We also eat forest roots and fruits.

February: We eat cassava meal with white beans and bean leaves as relish and also pumpkins and its leaves, and native cucumbers.

March: We eat mealie meal and cassava meal with beans and luwanga (wild spinach). This is the month for mealie meal. In the valley people are now eating lupunga (a wild millet grass) with fish and meat.

April: We eat millet meal which is plentiful and relish of groundnuts, peas and sweet potatoes.

May: We eat millet meal with fish and dried mushrooms. These have been smoked dried and are very lasting. We are also eating kaffir-corn (sorghum) meal in this month. Finkamba and Mponso beans.

June: We eat meal of all sorts as they are all ripe. Fish and beans are the relish.

July: We eat mealie meal and millet meal. The relish is fish and chikanda (a tuberous edible root). Rice is now ripe and it is eaten in the valley.

August: We eat the meal from mealies, kaffir-corn and millet. The relish is fish, meat, peas and beans.

September: We eat the same food as in August. Fowls are plentiful now and they are often used for relish. Meat and fish in this month too.

October: We eat the same food as September.

November: We eat whatever kind of meal is left as well as forest fruits and honey, fowls and game meat as relish. Salt has now been made and this is added to other relishes.

December: We eat wild figs and other forest fruits because meal is scarce. Cassava meal usually lasts longer and we eat it in this month. Salt is still plentiful.

12. Richards noted that wage earners received a money allowance for food beside their wages, but that it was inadequate for the purpose during the hunger season (Richards and Widdowson 1936:190). Prices tended to rise steeply during the hungry months, but there is insufficient data to say whether or not the food allowance was adequate at other times of year (Richards and Widdowson 1936:181).

13. Out of 167 households questioned during the survey, 37 had no millet stores. Of these 37, 19 received gifts from relatives and friends, 1 received gifts and bought grain, 3 obtained their supplies by barter, 1 by working, 1 both worked and bartered, 6 received gifts and also worked, 2 received gifts and used them for brewing beer for working parties, and 4 received no supplies of millet (Preston Thomson, 1954:35).

14. Preston Thomson does not say so directly, but from data given elsewhere in her report it seems as though information on food transactions and exchanges were collected for all three villages over a period of 12 months, but only 6 days of each month were spent in each village. Thus, her data should be taken as indicative (Preston Thomson 1954:24; 46).

15. The importance of recording food sharing at the point of consumption and its relevance for understanding interhousehold exchanges in matrilineal communities has been emphasized also by Megan Vaughan in her study in southern Malawi (Vaughan 1983:277–278).

16. Ann Tweedie was commissioned to do a study into Bemba economics in February 1959. She began work in April 1959 and left for England in July 1960. She collected household consumption and other economic data for three villages, and did some comparative work on a further five. In the three main villages, she employed research assistants who kept weekly income and expenditure forms for all individuals for a period of one year. Individual village diaries were also kept and food consumption was recorded. All three villages were in the Kasama district. Her work was originally planned by Henry Fosbrooke, Director of the Rhodes-Livingstone Institute, as a rural study along the lines of the urban budget surveys that had been undertaken in Central Africa by David Bettison.

17. The tremendous increase in marketing and trading is obvious from the fact that in the Kasama rural district in 1959, 161 general dealers licenses, 71 hawkers licenses, 31 tea room licenses, and 5 butchers licenses were issued as compared with none in 1937 (ATW: Private papers; Some Aspects of Bemba Economics). Also in 1959, the Development Commissioner approved 48 loans to individuals with a total value of £11,700. Of these, 34 loans were toward the cost of trade goods. (ATW: Private papers; List of loans, Development Commissioner's Office).

18. Data on food consumption were actually collected over a longer period of time, but this particular time period was chosen to facilitate comparison with Richards' and Gore-Browne's data, which were also collected at this time of year. The name of the village and all personal names have been altered to protect identities.

19. A similar situation was noted by Stromgaard when he conducted research in the early 1980s, recording the use of guns, spears, nets, and traps. The most common animals caught were duiker and bush pig. He argued that, even in the 1980s, hunting was a lucrative source of income, albeit only for the young and the strong (Stromgaard 1985a:56). We did not encounter any individuals who claimed that hunting provided them with a regular income, but given the legal controls on hunting, this is perhaps not surprising. What is certain is that research in Northern Province continues to seriously underestimate the amount of routine hunting and gathering that local people are involved in and thus tends to downplay the role of hunted and gathered foods in the diet.

20. Many households had granaries next to their mitanda if the fields were very far from home, and they then carried the grain from the mitanda granary (nkoloso) to the village when required (Richards 1939:86). This system of dual storage was still in use in the 1950s and in the 1980s, but with decreasing frequency because of the fear of theft. However, it should be realized that in the absence of ox-carts and other forms of transport, the labor required to get crops in from the field is considerable (see Reynolds 1991). Tweedie notes that ukupula is of particular importance for the livelihood of unattached women in the village, and suggests that it should be seen as a way of sharing out the labor of the available men among all the women.

21. For further theoretical elaboration of the links between the sexual division of labor and the necessity for a system of redistribution, see Moore (1993a).

CHAPTER 4

Cultivators and Colonial Officers: Food Supply and the Politics of Marketing

1. "Independent settlements" are discussed in more detail in Chapter 5. Their development was related to the institution of a new "parish" system of administration in the late 1940s.

2. It is important to note here that the period of World War II is only poorly covered by the written archives, yet the war itself undoubtedly had a major influence on food policy in Northern Rhodesia, as well as on the process of differentiation taking place within the Northern Province. See Gann (1964) for an account of the impact of World War II on Northern Rhodesia.

3. Richards also made a brief study of Bisa villages in the Bangweulu swamps. See Richards (1939) and Richards and Widdowson (1936).

4. See Chapter 2 for the definition of these different ecological zones.

5. What is noticeable here is that villagers were growing varieties of sorghum that ripen at different times. As noted in Chapter 2, this is a way of reducing the gap between one main harvest and another, and increasing the amount of staple available during the hungry months. This kind of mixed cropping was an essential part of a developed citemene system, but it has almost entirely disappeared at the present time.

6. Government prices for 1946 were set as 2/-per tin for cassava meal, 2/6 for other meal, 2/s for grains, and 4/6 for nuts/beans. These were prices for town, in the country the corresponding prices per tin were 1/6 for cassava meal, 2/- for other meal, 1/6 for grains and 4/- for nuts and beans (See 2/227 Vol. I: 11–13/4/46).
7. Sara Berry has developed a more elaborate and Africa-wide version of this argument (Berry 1993).

CHAPTER 5
Developing Men: The Creation of the Progressive Farmer

1. When villages did move, they often relocated within a few kilometers, sometimes less, of their original location, see Harries-Jones and Chiwale (1963), and Tweedie (1966) on this point. Villages that "disappeared" over time mostly did so for political rather than economic reasons. This is borne out by the fact that chiefs' villages, notably the capitals of Chitimukulu and Mwamba, remained in the same areas over generations, although minor changes in location could result from the death of a chief or from another misfortune.
2. For the "development" intiatives of other colonial states in the postwar period, and African responses to these initiatives see for Tanganyika Iliffe (1979:Ch. 14) and Feierman (1990, Chs. 6 and 7); for Kenya, see Berman (1990:Chs 6 and 7) and Berman and Lonsdale (1992 (Book 2), Ch. 12); for Nyasaland, see Mandala (1990, Chs. 6 and 7).
3. For the most detailed account of agricultural policy in the late colonial period in Northern Rhodesia see Makings (1966); also Hellen (1968); Dodge (1977); and Baldwin (1966). For a thorough examination of the impact of these policies on the process of rural differentiation in the Southern Province, see Chipungu (1988).
4. For a detailed study of the effects of the Parish System in this area, see Tweedie (1966).
5. This was part of the larger formulation of a National Development Plan: see Makings (1966:205).
6. In the Southern Province, where there was already a significant African maize cash-cropping industry, a rather more ambitious program of African Improved Farmers was instituted in 1946. This was extended to the Central Province in 1952. See Chipungu (1988:95) and Makings (1966:216).
7. These issues were also raised in relation to the Serenje scheme, which lay outside the boundaries of the Northern Province, but was a comparable area in many ways. Anthropologist Norman Long produced a detailed study of the effects of this scheme in his book, *Social Change and the Individual* (Long 1968). For peasant farming schemes in general see also Dodge (1977), Makings (1966), and Hellen (1968).
8. Makings made a similar point about peasant farming schemes in general, arguing that most participants "over-reached" themselves (Makings (1966:222).
9. In 1961 a survey of 1000 peasant farms all over Northern Rhodesia also found low incomes, implying that this was not only a problem in the Northern Province. (Dodge 1977:27 Makings 1966:221).
10. This problem also arose on the Serenje scheme and was described in detail by Long (1968:Ch. 3).
11. Only the records of the Abercorn discussions appear to have survived.
12. In the more productive Southern Province, the lure of development and of government subsidy was rather more powerful than it ever was in the Northern Province. See Chipungu (1988:Ch. 5).
13. On the rise of nationalism see Rotberg (1965). There is a large literature on the politics of labor on the copper-belt, see, for example, Berger (1974); Epstein (1959); Parpart (1983); Perrings (1979); and Harries-Jones (1975).
14. The politics of colonial development schemes are closely tied-up with the whole question of colonial conservation policy and the resistances that this engendered. This issue has been examined for Southern Africa as a whole by William Beinart (Beinart, 1984). It is explored by Mandala (1990); Feierman (1990); and Berman and Lonsdale (1992).

The specific question of the relationship between colonial agricultural improvement, settler schemes, and the rise of nationalism in Zambia has been addressed in a debate that revolves largely around the history of the Southern Province and the effects of the ANC/UNIP split in that area: Dixon-Fyle (1977), Chipungu (1988). For discussions of the peasantry and nationalist politics in Northern Province see Bratton (1980:Ch. 7). For the Luapula Province, see Bates (1976:Chs. 4 and 5).

15. For a full account of this loan and an economic analysis of the Mungwi scheme, see Johnson (1964).
16. For this theory see Scott (1978) and Chipungu (1988).
17. In 1969, Simon Kapwepwe, a Bemba-speaker and spokesman for Northern interests, resigned his position as vice-president of UNIP. He and others formed the United Progressive Party in 1971 (Bratton, 1980:212–213).
18. See Bratton (1980:269). For a comparable study of the disappointments of rural people in the Kabwe district see Muntemba (1978). Pottier provides a detailed study of government development intiatives in the Mbala district in the 1970s, and a range of local responses to these, see Pottier (1988).

CHAPTER 6

Migration and Marriage

1. For example, see Richards (1940); Wilson (1941); Mitchell (1961, 1969, 1987); Epstein (1958; 1981); Powdermaker (1962); Gluckman (1961). The functionalist and structural-functionalist paradigms that inform much of this work account for the emphasis given to such notions as stabilization and equilibrium. In general, westernization and modernization were presented, even in the most sympathetic accounts, as potentially deleterious; for a critique see Magubane (1971). However, it must be remembered that many anthropologists allied themselves with the liberal critics of colonial policy and wanted to insist on the recognition of urban dwellers as urban dwellers rather than as primitive tribesmen; see Ferguson (1990b:616–619).
2. For a study that comes to conclusions somewhat similar to our own, although it is framed in very different theoretical terms, see Colin Murray's book on Lesotho: Murray (1981).
3. Much has been written on labor migration in Zambia, and particularly on the Copper-belt. For examples of this work and for authors who rely on phase models to explain changes in patterns of labor migration, see Ohadike (1969); Berger (1974); Heisler (1974); Bates (1976); Perrings (1979); Parpart (1983); Chilivumbo (1985).
4. These figures were given to us by returned migrants. However, it is also clear that in many cases the journey by foot took much longer. It is worth noting here that Wilson (1941:49) recorded the journey time from Kasama to Broken Hill (Kabwe), which is much further south than the Copper-belt, as seven weeks on foot, whereas the journey by lorry took one week and cost 35s. In 1937, Mr. Jobling opened up a recruitment agency at Kasama for Tanganyika Sisal Plantations Ltd. Workers were transported free to the plantation and brought back to Kasama at the end of their contract. However, only 572 men were recruited in the year, despite the convenience of the offer of free transport. It appears that people stated to officials that they preferred to go independently, and that they did not like the system of remitted pay (SEC 2/1300:1937).
5. This point is discussed further in Ferguson (1990a:402–405), in which relevant figures are given.
6. The following three paragraphs are direct translations of life-histories collected near Chilubula mission.
7. We do not have reliable data for polygyny during this period, so we do not know how many men were straddling the rural-urban divide with one wife in town and another in the country. Also, we have no way of cross-referencing these figures with information on province of origin for workers, so we do not know if there were significant variations by

province in the number of married workers. However, life histories and archival sources suggest a very high rate of accompaniment for workers for the Northern Province from the 1920s onward. Chauncey (1981:137, 142) indicates that 25% of miners were accompanied by dependents at Bwana Mkubwa in 1926 and 20% were married at Roan Antelope in 1927. Parpart (1986:143) gives a figure of 30% married in all mines in 1931. These percentages show that a significant proportion of workers were accompanied from an early date, but the number of married workers in the mine labor force continued to grow as time progressed.

8. Wilson (1941:20) states that fifteen and a half years old is the average age of first employment for boys.

9. Taxable males were those considered able to work. Boys under the age of about 15 years old and elderly men were excluded, as were lepers, the insane, and anyone suffering from an infirmity. Decisions about age were obviously somewhat ad hoc and were made by touring officers as they went about collecting taxes and registering the population.

10. A comparison of figures for absent males taken from tour reports for Chief Nkolemfumu's area in the Kasama District shows a fairly steady progression in the proportion of men away working, but these percentages include men working in the province: 1934: 52%; 1936:66%; 1937:46.6%; 1939:41%; 1948:57%; 1949:56.7%; 1949:66.03%; 1950:71.56%; 1951(W):65%; 1951(E):83%; 1952(W):80.4%; 1952(E):86.6% (SEC 2/793 Kasama: TR Sept./ Oct. 1952.

11. However, we do not have time series data by chief's area or by village to demonstrate this finding, largely because the data does not exist. Even where figures are available for a series of years (sequential or nonsequential) in the three-decade period, changes in chief's and district boundaries, plus variations in the recording methods used by touring officers, makes most of the data useless. The net result is that the data is very patchy and even where it exists it is noncomparable over time. The estimation of 10–35% is made using the crude method of calculating the figures available on proportions of men employed in the province by area whenever and wherever those figures are available in the archival sources we consulted over the three-decade period. Opportunities for formal employment in the province during the period 1960–1990 would seem to have decreased despite the expansion in the size of urban centers because of stagnation and recession in the Zambian economy.

12. Pottier (1983:9–10) has shown the importance of the International Red Locust Control in providing local employment around Mbala during the colonial period.

13. As stated earlier, the figures are so patchy and unreliable that it is not worth trying to create a time-series using projection and other methods.

14. Many tour reports simply give figures per chief's area and those with data on individual villages are the exception rather than the norm. The introduction of the parish system in 1947 made the problem of accurate recording and subsequent reanalysis of the figures even more difficult. As we have seen, aggregate figures are particularly misleading because of the high degree of local variation, and the parish system exacerbated this problem. However, parishes were not formally drawn up in some areas for several years after the legislation was passed, and many census books continued to work on a village basis. This resulted in some tour reports, including that for Mpepo's area, continuing to provide figures based on village-level enumeration.

15. There are only two instances in the Mpika Tour Reports in which an indication of the number of absent adult females is given. In the first, 27.6% of the registered adult women were recorded as absent from villages in Mpepo's area in 1953; in the second, 33.7% of the adult women were recorded as absent from Chikwanda's area in 1958 (SEC 2/844 Mpika: TR April 1953; SEC 2/849 Mpika: TR Nov./Dec. 1958).

16. Generally speaking, the increase in food production was more noticeable in the Mpika District, which had closer contact with the line of rail. However, the parts of Kasama,

Luwingu, and Chinsali Districts close to roads, urban centers, and special resources (e.g., fish) were also involved in producing foodstuffs for sale. Strangely enough, it was the more marginal areas of the province that were most productive once opened up by infrastructure (see Chapter 4). However, it is important to remember that surplus production and/or the amount offered for sale in the province continued to be erratic. One year sales might be large and the next they were not. For example, in 1957, 248 bags of all types of produce were purchased locally, as compared to 3,618 in 1955 and 984 in 1956 (SEC 2/94:1957). Colonial officials tended to extrapolate from two or three figures of this sort to statements about long-term trends. They were almost inevitably proved wrong. Food had to be imported into Kasama District in 1956 and 1957 from Tanganyika to meet requirements for the government employees, schools, etc., but this was in large part due to a refusal on the part of local people to sell to the government as part of their political protest against colonial rule.

17. The Annual Report for Native Affairs, Kasama District for 1956 (SEC 2/111) gives a figure of 37% of the total population absent from the district, and this includes 64% of taxable males, 27% of adult women, and 32% of children.

18. Cassava is a time-consuming crop to process and thus increases the amount of time women have to spend in food preparation (see Chapter 7). However, much of the cassava sold to the government and other buyers was unprocessed or only partially processed, although some was sold as cassava flour. It is worth noting here that Watson argued that migrant labor had not adversely affected food production in the Mambwe area because of the fact that a lack of gender specialization meant that women could substitute for men (1958:225). However, he took the view that if the sex ratio increased beyond 2 women for every man then both the economy and the social life would be disrupted (Wilson, 1958:34). He had no grounds for asserting this figure, but his argument is nonetheless interesting when applied to the Bemba case and looking at our information on sex ratios and on labor substitutability (see Chapters 3 and 8 and text above).

19. Richards found that 49% of 144 marriages collected at random from six villages in 1934 were cross-cousin marriages; see also footnote 23, Chapter 8. She also recorded two other types of preferential marriage among the Bemba: the wife's brother's daughter (mpokeleshi) marriage and the grand-daughter (umweshikulu) marriage where a man has a right to the daughter of his son, real or classificatory. This latter type of marriage is probably best explained in terms of the importance of consolidating political alleigances through marriage—Richards hints at this (Richards 1940b:44–46). For further discussion of Bemba marriage, see Gouldsbury (1915,1916); Richards (1939); Richards and Tardits (1974); Tardits (1974); Labrecque (1931).

20. See Tardits (1974:22) for a very different view of what Bemba marriage payments do, or rather do not, legalize. Richards's overlegislation of Bemba marriage had much to do both with the influence of structural-functionalism, and with a failure to recognize that the colonial state was not codifying traditional custom, but rather creating it; see footnote 10, Chapter 8.

21. Richards took the straightforward anthropological view of the time that matrilineal societies were more vulnerable than patrilineal societies because they were not based on ties of property or land between males. She also thought that "European influence" had increased the tensions in the matrilineal system between the father and the mother's brother, thus producing further strains (Richards 1939:116; 1940:8–11).

22. However, see the essays in Bloch and Parry (1989) for a critique of this position.

23. Richards recorded that money was used as a payment for ritual services connected with the removal of taboos, and for ceremonies surrounding the founding of the chief's village (Richards, 1939:220).

24. The colonial government tried hard to encourage the use of money as a generalized medium of exchange, and although traders were willing to cooperate, they sometimes found

that people preferred barter. An officer touring Chitoshi's and Munkonge's areas in 1932, reported a conversation with Jobling, the government contractor, who had expressed a willingness to pay cash in exchange for food, but found that people would not sell except for salt and trade goods (ZA 7/4/28 Awemba:TR Aug. 1932).

25. Richards found that in an analysis of forty-six marriages in the Mpika and Kasama Districts, an average of 3.4 years of brideservice had been done, in addition to ten marriages in which the husband said he had worked "continually." From a list of eighty-two marriages, 64% of the contracts were fulfilled by service and 36% were fulfilled by a cash payment (Richards 1940b:52).

26. An order was issued by Chief Chitimukulu in 1936 that "no woman shall remove to another part of the Territory or out of the Territory without first obtaining the consent of her headman . . . No person shall help or encourage any woman to leave the District for purposes of prostitution" (Bemba Native Authority Ordinance, 9/1936 in SEC 2/1297.

27. Mpango is not paid in the case of a man marrying a woman who has been married before. In the case of a man inheriting a wife, neither nsalamu nor mpango are paid. This is still the situation today.

28. The following communication from the Provincial Commissioner of Kasama gives a clear indication of the interdependence of marriage and divorce. PC Secretary 14/12/36, Re:Native Marriage Registration states: "Native Authorities have not shown any great enthusiasm for the scheme, though it is thought that it may be of assistance in checking the number of unmarried women who go to the Railway line . . . the idea . . . comes a little strange to tribes where marriages are matrilocal. Where this is so, the family of the wife regard the husband with a certain amount of suspicion until he has proved himself capable of supporting the woman, and the right to take back the woman is reserved until this is proved. A certain amount of fear is felt lest this right should be lost to a husband who holds a certificate of marriage. Once, however, it is seen that a divorce granted in the Chief's court, on valid grounds, rescinds the certificate this fear will be allayed. . ." (SEC 2/406;Vol. I).

29. Richards includes some discussion of the impact of European law on divorce and comments that European legal concepts were often applied to situations where they were not really applicable. "Fixed residence with one parent or the other, as implied by a European divorce award, means little or nothing in Bemba society. As Candamale, an ex-chief in the Chinsali area said, "Yes, the bwanas always gave the children to the father if the mother had committed adultery. But it did not matter. The children went from one to the other as our custom is" (Richards 1940b:107).

30. There are virtually no figures for divorce rates based on court hearings after 1930 because the institution of Native Courts at that point meant that thereafter marriage cases were heard by chiefs in their courts.

31. The data is very sporadic, but the Native Commissioner of Kasama writing to the magistrate of Kasama 7/6/26 noted that of 59 divorce cases adjudicated by the chief, 39 were on grounds of mutual consent, 9 were for desertion, 4 were for adultery, 2 were for ill-treatment, and 5 were for incompatibility. The Native Commissioner of Chinsali, writing to the same magistrate on 11/6/29, noted that of 119 divorces granted by Chiefs in the period April 1925–March 1926, 14 were on grounds of desertion, 3 were for adultery, 4 were for cruelty, 1 was for leprosy, and 97 were on the basis of mutual consent (ZA 1/9/2/3). Colonial officials were generally outraged that so many people divorced on grounds of mutual consent, but chiefs upheld this possibility because it accorded well with local understandings about the processual nature of marriage and its initial instability. Divorce by mutual consent tended to mean by mutual consent of the kin groups involved and not just by mutual consent of the partners.

32. Wilson (1941; 1942) noted the difficulty of defining marriage in an urban situation. "The distinctions we draw . . . between postitutes and concubines, and between them and peripatetic, long-term and life-long wives are necessary for understanding, but they are often difficult to draw in particular cases. No woman, except the last, confine themselves to one category alone, and there is a constant tendency for one type of union to pass gradually into another. The factors of stability and of instability operate continuously, pulling against one another, in every case" (Wilson 1942:66).

33. The Church was very much against polygyny, and the White Father's Mission in particular strongly disapproved of widow inheritance and the sexual rites required to lift the death from a bereaved spouse (ukutamfye or ukubule mfwa) (Richards 1940b: 98–99).

34. Polygyny was not permitted by the Catholic Church and the area where we worked was and is a Catholic stronghold. However, the Status books for the Chilubula parish do record cases of disputes arising as a result of Christian converts taking a second wife, very often because they had to inherit her.

35. Watson (1958:40) recorded for the Mambwe that the introduction of cash into the region led to brideservice being replaced by bridewealth. This did not happen for the Bemba, but what is of interest here is that Watson did not argue that this led to a breakdown of Mambwe society, even though he noted that the elders were not able to control this transition. Watson (1958:226) actually argued that high marriage payments and virilocal marriage led to stability in family life, especially when compared with the Bemba. We would argue that the Bemba system has turned out to be no more unstable than the Mambwe system over time, and that Watson had, in fact, only as much evidence for Mambwe stability in the 1950s, as Richards did for Bemba instability in the 1930s. Much rests on interpretation and the construction of objects of knowledge, as well as on the misleading effects of systematizing process and negotiation.

CHAPTER 7

Working for Salt: Nutrition of the 1980s

1. In fact, many studies were conducted all over Zambia in this period. The National Food and Nutrition Commission was established by an Act of Parliament in 1967. A National Food Consumption survey and a UNDP/FAO Nutrition Status Survey were conducted in 1969–72 (see F.A.O./UNDP, 1974a; F.A.O./UNDP 1974b; F.A.O. 1977). It is not possible to use these surveys as baselines for the more detailed work done in the Northern Province in the 1980s because the data is unreliable in some instances and is too highly aggregated to be useful for our purposes. The National Nutrition Surveillance Programme began in 1979 and was designed to provide an ongoing data collection system which would allow health officials to target areas of special need (see Freund 1985; Kauppinen 1985a, 1985b; Kauppinen and Mweemba 1985). Relevant studies from other areas of Zambia include IFPRI (1985); Perez (1984); Gobezie (1984a); Herthelius (1984); Freund and Kalumba (1984); ARPT (1985); and Kwofie (1979).

2. See Popkin (1981) for evidence that increased workload for women can affect nutrition.

3. A large number of studies in Africa note that the well-being of women and children under conditions of increasing commercialization and developing small-holder production depend on access to household income, and some measure of control over household spending and consumption (Henn, 1983; Bruce and Dwyer, eds 1988) In such situations, conflicts between wives and husbands may become acute and women's control over economic decision making may decline noticeably (Jones 1986; Whitehead 1981;). A number of studies have noted that rural malnutrition is closely associated with such processes (Chambers and Singer 1981; Longhurst 1984; Schofield 1979).

4. In Zambia as a whole, female-headed households make up approximately 33% of all rural households (Government of Zambia, 1980 Census Results), therefore, the figures for the Northern Province are slightly higher than the national average. However, this has to be seen in the context of a massive decline in the mining industry in the 1980s and in levels of male labor migration from the province in general (see Chapter 6).

5. In recent years, it has become popular to claim that there is a link between cassava grown as a staple and male absenteeism because of the crop's lower labor requirements. It is worth pointing out that this is far from straightforwardly the case, and that in many ways it is simply another example of the "myth of male absenteeism" discussed in Chapters 2 and 3. Evans and Young note that in the patrilineal, Nyamwanga IRDP (SMC) sample point, Chunga, where mean household size is greater and the number of adult males per household is significantly higher than in other sample points, households grow more cassava (Evans and Young, 1988: Section 4).

6. It would be hazardous to use these figures to say anything definite about the adequacy or inadequacy of the Northern Province diet at any period. Any such exercise would have to incorporate the changing views of nutritionists on the question of nutritional requirements. In general, estimates of the number of calories required to maintain health have declined dramatically since Richards made her assessment of the Bemba diet (see Dasgupta and Ray 1990). In addition, physiological research on individual adaptation has produced general uncertainty on the issue of what constitutes a nutritional "requirement" (see Pacey and Payne, 1985:Chapter 3; Dasgupta and Ray, 1990). However, what is clear is that nutritional standards in the Northern Province have not improved over time, and most importantly they have not improved with the introduction of hybrid maize. In 1969, 29.4% of children under the age of 5 years in the Northern Province were less than 80% weight-for-age, and in 1985, 29.2% were less than 80% weight-for-age. (Sharpe, 1990:588–589).

7. Sharpe (1990:593) estimates, on the basis of IRDP (SMC) data, that hectarages of relish crops might have declined by as much as 20% between 1980 and 1986.

8. The method used was that of a semistructured interview. The research assistants who worked on this survey were Mr. J. Nkumbula and Mr. C. A. Sichilya; their professional skills and hard work are gratefully acknowledged. The data from the survey was supplemented by participant-observation, case histories, and oral histories obtained during the period of research.

9. Mr. C. A. Sichilya, September 1988.

10. Hunting in Zambia is illegal without a license. Many informants bemoaned state-imposed restrictions on hunting, although these restrictions were often disregarded.

11. Women much prefered to be "paid" in kind rather than in cash because they found it difficult to limit the demands that their male relatives could make on available cash, and also because the absolute scarcity of commodities and the high cost of transport meant that it was much easier to exchange beans and other goods for a blanket rather than to be paid in cash and then have to travel long distances to town to buy a blanket. The complex exchange systems in beans and fish existing in the Mambwe area are described by Pottier (Pottier 1988: Ch. 7).

12. We were unable to get reliable figures for the actual amounts of income from each source. Many of the activities involved, however, are relativity small-scale and contingent, which makes informant recall an unreliable basis on which to base figures for total income. There is also some evidence to suggest that there is a significant underestimation in the number of households involved in selling gathered foods, vegetables, and fish.

13. Exchange labor groups organized around the membership of churches are common in some areas, but they work on a rotation basis and do not involve remuneration in beer or cooked food (see Long 1968). In some areas, the more commercial farmers are banding together to organize labor exchange groups to combat labor shortage. These groups also

work on a rotational basis and do not involve remuneration. The main point to note here is that communal work arrangements vary enormously from one area of the province to another, and even within the same area and/or village. This point was made by Stolen in her study of villages in Mbala and Kasama Districts. In Mulenga village (Mambwe) in Mbala District during the 1980/81 season, 50 beer-for-work parties were arranged comprising 708 man-days and 69 oxen-days distributed among 21 households. The number of work-for-beer parties per household varied from 1–6, with a village average of 2.4. The number of participants in each party varied from 9–26 when manual work was done, and between 2 men and 2 pairs of oxen and 7 men and 7 pairs of oxen when ploughing was done (Stolen 1983a:99). In the Bemba villages in the Kasama District, the number of beer parties was very much lower (average 218 man-days) and no draught animals were used (Stolen 1983a:101). Stolen accounts for these differences by arguing that the patrilineal Mambwe system is more conducive to male cooperation and to individual expansion and accumulation (Stolen 1983a:110).

The types of labor cooperation observed during our research are discussed in Chapter 8.

14. Two types of beer are made: chipumu or bwalua—a thick beer made entirely from millet, and katata—a lighter beer made either from maize with a millet yeast or from fermented millet.

15. Pottier makes exactly the same point (1988:128), but see also Hedlund and Lundahl 1984:64; and Colson and Scudder 1988.

16. Evans and Young (1988:Section 6) calculated that farmers paying five cups of salt for four to six hours work in 1987, were effectively paying an hourly wage of Ko.4–0.5 as opposed to the going cash wage of K1.20 per hour. They also provide an example of a farmer who hired three local women to weed maize for one dress each (K40 each) and five local women to harvest maize (10 baskets) for one dress each, and two more women from a neighboring village to shell maize for a citenge (cloth) each (K35 each). See Chapter 8 for remuneration levels collected during our field research season in 1989.

17. A full analysis of what is happening to kinship in the province would have to include information on topics such as clan organization, kinship networks, succession, inheritance, rights in persons, theories of shared substance, and notions of the person.

18. This is a familiar argument for Africa, where it is often suggested that small-scale entrepreneurs of all kinds have to try and avoid the claims of kin that, if allowed to continue, will cause their enterprises to fail. The same sort of argument was made by Long (1968) in his analysis of Jehovah's Witnesses in Serenje District, all of whom severed links with their kin (see also Poewe 1978; 1979). However, we criticize this argument from a number of points of view in Chapter 8.

19. The IFPRI (1985) survey in the Eastern Province found that female-headed households, although having lower incomes and a lower duration of food supply, nevertheless have better child nutrition at each income level. Also, their income seems to be more directly transferred into food consumption than is the income of male-headed households. This is despite the fact that women in female-headed households often have higher workloads than do women in male-headed households.

20. These compromises are practically invisible as far as most of the standard methods of collecting labour data are concerned. See Reynolds 1991 for a discussion of the strengths and weaknesses of methods of labor data collection.

21. Ideally this small sample would be supplemented by an analysis of the parish records collected by the White Fathers. At Chilubula Mission, the "status books" are extremely detailed and show family formation from the 1890s to the present day. We have not been able to analyze this very large potential sample of births, but a superficial examination confirmed the impression of great regularity in birth spacing. There are, no doubt, many problems with this material because parishioners may not have declared "illegitimate"

births to their parish priest. But, in an area in which demographic data of any kind are hard to come by, these records must be regarded as having great potential value.

.

CHAPTER 8

From Millet to Maize: Gender and Household Labor in the 1980s

1. For the country as a whole the cost to the Government of subsidies on maize and fertilizer price differentials and of subsidies for maize and fertilizer handling (distributed via the cooperatives and Namboard) totalled K485.5 million in 1986, as compared to K103.7 million in 1981 and K54.1 million in 1976 (Sano, 1988).
2. The research assistants who conducted the interviews were Mr. Boniface Sambo, Mr. Henry Musonda, Mr. Francis Mubanga, and Mr. Abraham Chakota Sichilya. We gratefully acknowledge their professionalism, hard work, and dedication.
3. All land in Zambia is vested in the president. The colonial land holding system was changed in 1975, but its dual structure was retained, so that land is divided into State land and Reserve and Trust lands. The State land corresponds to the former Crown land and comprises some of the best land in Zambia. The Reserve and Trust lands correspond to the former Reserves and Native Trust lands. These lands are mainly controlled through customary law, although they are officially invested in the president (Mvunga 1980). These lands can now be registered under title to individual owners, and the titles may be bought and sold. Even so the land market in the Northern Province is little developed.
4. This was most apparent during village regrouping in the 1970s, when those who were already well-situated with regard to services and markets refused to participate (Bratton 1980:Ch. 6; Bwalya 1979).
5. IRDP (SMC) data shows that the increase in maize sales in the province is the result of expansion of cropped areas rather than of increased yields. In fact, there is evidence to show that yields for the more commercial farmers have fallen over the period 1980–1986. This fall has occurred despite the rise in the level of fertilizer application. The reasons for this fall require further investigation, but would appear to be due to a greater incidence of maize monocropping without proper rotation and to late delivery of fertilizer (Bolt and Holdsworth 1987: Section 7).
6. At the end of the 1980s, the ARPTs of a number of provinces began to press for increased resources for research into traditional food crops. Work of this kind is now being carried out and reports from Zambia in 1992 indicate that the results have been promising.
7. For an excellent case study of a similar process in West Africa, see Carney and Watts 1990; Carney 1988.
8. For a discussion of land rights and land tenure among the Bemba see Richards 1939:266–274 and Gouldsbury 1915; for the Mambwe see Watson 1958 and Pottier 1988; for the Ushi see Kay 1964 and Gatter 1990; for the Lala see Long 1968 and Peters 1950.
9. The Village Productivity Committee is supposed to be an elected body that makes day-to-day decisions about village affairs. VPC registration started in 1972 under the Registration and Development of Villages Act (1971), see Chapter 5. Village headmen are members of the VPC because of their traditional status, all other members are elected. One or two women are usually included. Chiefs generally welcomed the setting up of VPC because it enhanced their role as mediators and effectively worked to reinstate their powers—particularly with regard to land—even though they were supposed to have been dipossessed of these powers after independence (Bratton 1980:75–79). It is a dramatic example of the Zambian State's ambivalent relationships to chiefs; like its precursor the colonial state, it found itself both wishing to relieve traditional authorities of their power and needing them to carry out various functions. In the late 1980s, not all villages had functioning VPC.

10. Strictly speaking, of course, there is no such thing as customary law or customary land tenure in the sense of a traditional law of land tenure. These customary arrangements were created under colonial rule, as administrators sought to fix and define what previously had been fluid and open to interpretation; see Chanock 1985; 1991; Colson 1971; Moore, 1986.

11. For a discussion of these issues in the context of a brilliant case study which has had a profound influence on the way anthropologists approach customary law and land rights, see Moore 1986.

12. The government (see footnote 3 above for a discussion of landholding) has recently set up a number of schemes, not unlike the Peasant Farming Schemes of the 1950s, designed to try and develop agriculture and get people back onto the land. As in the 1950s, participants must abide by rules set down for progressive agriculture and in return they receive some extension help and occasional inputs, including credit provision. However, these schemes are usually undersubscribed, as they were in the past, and it is not difficult to get a place on a scheme. These schemes are few and far between in the Northern Province, but as we conducted research close to one—partly as a way of ensuring that we would be able to interview a reasonable number of farmers involved in cash-cropping—the land acquisition method of some of our sample is far from typical, because in general only a very small number of farmers live on land ceded to them under government schemes.

13. In the past, inheritance was not an important consideration, at least among the Bemba, because there were few forms of inheritable wealth. According to Richards, a man inherited a bow on the death of his maternal uncle, and in the case of a woman, her girdle (mushingo) was handed on. We recorded no cases of these objects being inherited in the recent past. However, Richards noted even in the 1930s that "nowadays money is often divided between a man's own children rather than his nephews" (Richards 1951:174). This was taken by Richards herself to indicate a breakdown in the principles of matrilineal descent even at that early date, but it must be remembered that Bemba kinship was essentially bilateral, and that rank and/or status frequently cut across the principles of matrilineal kinship, as Richards herself always stressed. What should be noted is that cleared land in a prime location is now a form of inheritable wealth of some significance and that individuals do try to exclude certain persons or categories of kin from inheriting it, at least in some cases. There are still, however, cases of nephews inheriting land. But, we should be cautious of interpreting father-son inheritance at the present time as an indication of the breakdown of matrilineality, it is perhaps better understood as an increased emphasis on the bilateral aspects of Bemba kinship that have always existed.

14. Pottier made a similar argument for the Mambwe, although many of the men whose complex labor histories he described did not in fact ever go as far as the major labor centers (see Pottier 1983; 1988:43–46).

15. In other matrilineal societies, women inherit land directly from female kin and/or from the matrilineal kin group, but this was never really the case among matrilineal groups in the Northern Province because residence patterns seem always to have been very fluid and land was not a scarce resource. Additionally, more men were in charge of land allocation, because rights to land came via the chief, and after that were accorded by the action of headman on the basis of residence and clearance (Richards 1939: 244–248).

16. There are a large number of studies from Africa that document the fact that under increased cash-cropping male smallholders demand that their wives do more work as family labor; for examples from Zambia see Muntemba 1982; Colson and Scudder 1988. Under these conditions, resource allocation within the household and negotiations between husbands and wives can become very fraught; see Whitehead 1981; Dwyer and Bruce 1987; Jones 1986. As sales of maize, beans, and other crops increased in the Northern

Province throughout the 1970s and 1980s, there was an increasing necessity to delineate rights to the sale of crops from particular fields cultivated by the conjugal household (see Pottier, 1988:114–116; Geisler et al. 1985:15). The same process is described for the Southern Province in Colson and Scudder 1988.

17. However, Evans and Young (1988: Section 6) note that there is some variation between samples with regard to these findings, and that the results from the Chunga sample, in particular, showed a low level of substitutability between husbands and wives, and thus a more rigid sexual division of labor.

18. ARPT (Kasama) has collected data on the relationship between yields and the timing of inputs and labor operations. Figures collected during the 1986/87 season showed that yields of MM752 fell 29% with a six week delay in planting. Part of this delay was due to late delivery of seed. Data also shows that few farmers apply fertilizer when planting, and many do not fertilize within the first few weeks of growth, which is when maize takes up its most important nutrients. Again, an important reason for this appears to be late fertilizer delivery (Kerven 1988). Agronomic tests indicate that no weeding reduces yields by 25% or more, and data from IRDP (SMC) and ARPT (Kasama) show that weeding in all their sample points was generally late and inadequate (Bolt and Holdsworth 1987; Bolt and Silavwe 1989). Thus, whatever problems farmers may have with the supply and the timing of labor, their farming enterprises are rendered a great deal more risky by late delivery of inputs; something all analysts recognize as a chronic problem in the Northern Province.

19. The average household size in the province was 4.5 persons in 1969 and 4.8 persons in 1980, although it is worth noting that 24.7% of the households in the province in 1980 had 7 or more members (1980 census). The IRDP (SMC) data has been collected over time, and is thus more accurate than our own, because changes in household size and composition can be recorded and potentially correlated with variations in agricultural yields and cropping patterns. For further figures, see Table 8.5.

20. There is extensive literature on the complexities of analyzing and defining African households; see for example, Guyer 1981; Guyer 1988; Guyer and Peters 1987; Whitehead 1981; Moore 1993a. However, what is notable is that it is much easier to recognize these difficulties from a theoretical point of view than it is to do anything about them in practice. More often than not, researchers find it necessary to work with the notion of a bounded household unit when collecting data in order to demonstrate that this unit is not in fact bounded and has multifocii. The only methodological alternative is, of course, a network approach (including such methods as social field and organizational analysis) or an interactionist one.

Economic anthropology is still beset with the polarity between the individual and the collectivity (most usually the household) as analytic units. The difficulties are well demonstrated by Gatter's (1990:Ch. 3) sensitive approach to the Ushi concept of ulupwa (strictly speaking a bilateral kin group), which he ends up equating with household, but which he points out can only be understood in relation to a specific ego (see also Poewe 1978). The nearest Bemba term to household is probably also ulupwa (bilateral kin group), and this is because it is marriage that operationalizes the bilateral relations that reproduce the ulupwa through time. There is thus an established relation between the notion of the ulupwa and the concrete conjugal unit with its offspring. However, as Gatter points out, it is necessary to distinguish between the ulupwa as a cognitive category (which has a potentially broad membership) and the actual group of kin with whom individuals interact as a result, primarily, of co-residence. This was essentially the position taken by Richards (1940; 1951:175–176). Modern commentators—especially those concerned with development problems—consistently make the mistake of not distinguishing between kinship and descent, in the Northern Province, and thus they fail to realize that multiple models exist that are operationalized in different contexts.

21. "The father is the head of the household, but this term does not imply a large kraal or set of huts which we find among the Southern or some of the Eastern Bantu. A man lives in his hut with his wife and small baby, children over three being sent to their maternal grandmother to be brought up under her charge and later building huts of their own" (Richards 1951:175).

22. Both Long (1968) and Poewe (1978; 1979; 1981) have argued that there is an affinity between Protestant ideologies and capitalism, and an incompatibility between matrilineal kinship and the latter. This incompatibility gives rise ultimately to the rejection of matrilineal ties and attempts to retain resources within the nuclear family. The result is a thesis that stresses that matrilineal systems are vulnerable to certain forms of economic change. In essence, this argument is a version of earlier anxieties about whether matrilineality was doomed in Africa (Richards 1940; Watson 1958; Douglas 1969). However, see Gatter 1990 and Holy 1986 for critiques of Poewe's thesis, albeit from different perspectives. The recent stress in the development literature on two generational households is probably an accurate reflection of the number of conjugal couples living in separate huts or houses, but as argued above, this is not a new situation, and the constant references to two generational households in the literature is simply a way of pointing to what researchers perceive as the breakdown of kinship.

23. The strength of brother-sister ties was stressed by Richards (1939:115–117), as well as being noted by Tweedie and Harries-Jones and Chiwale (1962). Richards calculated that out of fourteen young couples resident in Kasaka village, eight were cross-cousin marriages (Richards 1939:160). Out of a sample of 144 marriages investigated in 1934, the rate of cross-cousin marriage was 49% (Richards 1940:44). We have no area figures for polygyny, but out of the thirty men sampled, three had more than one wife. Richards gave a figure of 32% in 1938 among pagans and 14% in 1934 among pagan and Christian families mixed (Richards 1940:59). The 1980 census records that 14.8% of men in the rural areas of the Northern Province have more than one wife, but we know that significant variations exist from one area of the province to another.

24. In this sense, we agree with some aspects of the argument that suggest that a structural tension between productive individualism and wider consumer sharing is characteristic of units of this kind (Poewe 1978; 1981), but we would be cautious about asserting it a as a specific feature of matrilineal societies, and we would not see it as having given way under capitalism to a conjugal/nuclear family. The only exception we would make is for very successful commercial farmers, who seem to have succeeded in transforming their household into something resembling a joint utility function, but we would suggest that this is as much a consequence of their success as its cause. In our view, there comes a point when a household head can effectively persuade other household members to work with him and to invest in an overarching household enterprise headed by him, simply because his activities have been so successful and the rewards are so great. However, the problem is to establish exactly what this point is and how it comes about!

25. Bolt and Holdsworth (1987: Section 5) show that although maize absorbs the greatest (and increasing) proportion of labor with commercialization, its labor input per hectare is relatively low compared to other crops. Also, total labor input per hectare for all crops on ibala fields falls with commercialization. This is probably due, despite greater labor availability, to increased competition between crops with commercialization (see discussion above). However, data collected also show a fall in cultivation to harvesting ratios with commercialization (Category 3 farmers spend only 40% of labor time on cultivation as opposed to 65% of total labor time for Category 1 farmers), which implies higher returns to labor with increasing commercialization.

26. The differences in the level of engagement in cash cropping are a feature of sampling decisions. The male household heads were chosen on the basis of their involvement, at different levels, in cash cropping; the women farmers were chosen on the basis of village

residence and of an involvement in both farming and migration. The result is that the women farmers come from a much older age cohort (av. 60.2 years compared with av. 40 years for the male household heads). This inevitably means that a smaller number are involved in cash cropping because at the present time, successful cash cropping is very much associated with the developmental cycle of households and with their ability to command labor.

27. Women's own-account farming of this kind seems to be increasing in the province, and it has been reported also from Luapula and the central provinces. This may be a result of women's improved access to credit provision. However, it was not common in the area in which we worked except among the better-off, female-headed households.

28. Goldman and Holdsworth (1990:568) argue on the basis of the IRDP (SMC) data that cash cropping among small-scale producers has not been stimulated by credit availability, because only about 25% of farmers report using credit. The rest rely on off-farm income and agricultural receipts. Overall, they argue that farmers have responded positively to government pricing policy on maize, but they also point out that over the period 1974–1986 real (deflated) maize prices remained relatively constant, as did gross margins per hectare for maize production (1990: 558–563).

REFERENCES

Alder, J. R. 1958. *A Report on an Investigation into Chitemene Control in the Abercorn District.* Lusaka: Department of Agriculture.

Allan, W. 1949. *Studies in African Land Usage in Northern Rhodesia.* Rhodes-Livingstone Papers, No. 15.

Allan, W. 1965. *The African Husbandman.* Edinburgh:Oliver and Boyd.

ARPT, 1984 *Causes and Characteristics of Female Farming in Mansa and Nchetenge Districts, Luapula Province.* Adaptive Research Planning Team, Lusaka: Ministry of Agriculture and Water Development.

Baldwin, R. 1966. *Economic Development and Export Growth: A Study of Northern Rhodesia, 1920– 1960.* Berkeley, CA: University of California Press.

Bassett, T., and D. Crummey eds. 1992. *Land in African Agrarian Systems,* Madison, WI: University of Wisconsin Press.

Bates, R. H. 1976. *Rural Responses to Industrialization: A Study of Village Zambia,* New Haven, CT: Yale University Press.

Beinart, W. 1984. Soil Erosion, Conservationism and Ideas about Development: A Southern African Exploration, 1900–1960. *Journal of Southern African Studies* 11:52–83.

Berger, E. 1974. *Labour, Race and Colonial Rule.* Oxford: Clarendon Press.

Berman, B. 1990. *Control and Crisis in Colonial Kenya: The Dialect of Domination.* London: James Currey.

Berman, B. and J. Lonsdale. 1992. *Unhappy Valley: Conflict in Kenya and Africa (Book 2): Violence and Ethnicity.* London: James Currey.

Berry, S. 1993. *No Condition is Permanent: The Social Dynamics of Agrarian Change in Sub-Saharan Africa.* Madison, WI: University of Wisconsin Press.

Bloch, M. and J. Parry, eds. 1989. *Money and the Morality of Exchange.* Cambridge: Cambridge University Press.

Bolt, R. and Holdsworth, I. 1987. *Farming Systems Economy and Agricultural Commercialisation in the South Eastern Plateau of Northern Province, Zambia.* Adaptive Research Planning Team, Kasama: Economic Studies, Vol. 1.

Bolt, R. et al. 1989. *Food Availability and Consumption Patterns in Northern Province ARPT Trial Areas: First Report on ARPT 1988 Nutrition Data Collection.* Kasama: Adaptive Research Planning Team, Northern Province.

Bolt, R. and Silavwe, M. 1988. *Farming Systems and Household Economy in Northern Province Plateau Region ARPT Trial Areas.* Kasama, Adaptive Research Planning Team, Northern Province.

255

Bolt, R. and M. Silavwe 1989. *Maize Production in Northern Province: A Review of Issues from Farm Level Production to Provincial Policies*. Kasama: Adaptive Research Planning Team, Crop Brief No. 1.

Boserup, E. 1970. *Women's Role in Economic Development*. New York: St. Martin's Press.

Brammer, H. 1976. *Soils of Zambia*. Lusaka: Department of Agriculture.

Bratton, M. 1980. *The Local Politics of Rural Development: Peasant and Party-State in Zambia*. Hanover, NH: University Press of New England.

Brelsford, W. V. 1944. *Aspects of Bemba Chieftainship*. Rhodes-Livingstone Communication No. 2.

Brelsford, W. V. 1945. Making an Outlet from Lake Bangweulu. *The Geographical Journal*. 106 (1–2):50–58.

Brelsford, W. V. 1946. *Fishermen of the Bangweulu Swamps. A Study of the Fishing Activities of the Unga Tribe*. Livingstone: Rhodes-Livingstone Institute.

Bruce, J. and D. Dwyer, eds. 1988. *A Home Divided: Women and Income in the Third World*. Stanford, CA: Stanford University Press.

Burton, R. F. 1873. *The Lands of King Cazembe: Lacerda's Journey to Cazembe in 1798*. London: John Murray.

Bwalya, M. C. 1979. Problems of Village Re-Grouping: The Case of Serenje District. In: D. Honeybone and A. Marter eds. *Poverty and Wealth in Rural Zambia*. Lusuka: Institute for African Studies, University of Zambia.

Carney, J. 1988. Struggles over Crop Rights and Labor Within Contract Farming Households in a Gambian Irrigated Rice Project. *Journal of Peasant Studies* 15 (3):334–349.

Carney, J. and M. Watts. 1990. Manufacturing Dissent: Work, Gender and the Politics of Meaning in a Peasant Society. *Africa* 60(2):207–241.

Chambers, R. and H. Singer. 1981. Poverty, Malnutrition and Food in Zambia. Country Case-Study for *World Development Report*, 1981.

Chanock, M. 1985. *Law, Custom and Social Order: The Colonial Experience in Zambia and Malawi*. Cambridge: Cambridge University Press.

Chanock, M. 1991. Paradigms, Policies and Poverty: A Review of the Customary Law of Land Tenure. In: Mann and Roberts eds. *Law in Colonial Africa*. Portsmouth, NH: Heinemann.

Chauncey, G. 1981. The Locus of Reproduction: Women's Labour in the Zambian Copperbelt 1927–1953. *Journal of Southern African Studies*. 7(2):135–164.

Chilivumbo, A. 1985. *Migration and Uneven Development in Africa: The Case of Zambia*. Lanham Md: University Press of America.

Chipungu, S. N. 1988. *The State, Technology and Peasant Differentiation in Zambia: A Case Study of the Southern Province, 1930–86*. Lusaka: National Education Company of Zambia.

Chipungu, S. N. ed. 1992. *Guardians in Their Time: Experiences of Zambians Under Colonial Rule, 1890–1964*. London: Macmillan.

Clifford, J. 1988. *The Predicament of Culture: Twentieth-Century Ethnography, Literature and Art*, Cambridge, MA: Harvard University Press.

Clifford, J. and Marcus, G. E. eds. 1986. *Writing Culture: The Poetics and Politics of Ethnography*. Berkeley, CA: University of California Press.

Colson, E. 1971. The impact of the colonial period on the definition of land rights. In: V. Turner ed. *Profiles of Change: African Society and Colonial Rule*. Cambridge: Cambridge University Press.

Colson, E. and S. Thayer. 1988. *For Prayer and Profit*. Stanford, CT: Stanford University Press.

Comaroff, J. and Comaroff, J. 1991. *Of Revelation and Revolution: Christianity, Colonialism and Consciousness in South Africa*. Chicago: Chicago University Press.

Comaroff, J. and Comaroff, J. 1992. *Ethnography and the Historical Imagination*. Boulder, CO.: Westview Press.

Dasgupta, P. and D. Ray. 1990. Adapting to Undernourishment: The Biological Evidence and its Implications. In: J. Drèze and A. Sen eds. *The Political Economy of Hunger, Vol. I*. Oxford: Clarendon Press.

Dixon-Fyle, M. 1977. Agricultural Improvement and Political Protest on the Tonga Plateau, Northern Rhodesia. *Journal of African History* 18:579–596.

Dodge, D. 1977. *Agricultural Policy and Performance in Zambia: History, Prospects and Proposals for Change.* Berkeley, CA: University of California Press.

Douglas, M. 1969. Is Matriliny Doomed in Africa? In: M. Douglas and P. Kaberry eds. *Man in Africa.* London: Tavistock: 121–136.

Downs, R. and S. Reyna, eds. 1988. *Land and Society in Contemporary Africa.* Durham, NH: University Press of New England.

Epstein, A. L. 1958. *Politics in an Urban African Community.* Manchester, U.K.: Manchester University Press.

Epstein, A. L. 1975. Military Organisation and the Pre-Colonial Polity of the Bemba of Zambia. *Man* 10:199–217.

Epstein, A. L. 1981. *Urbanisation and Kinship: The Domestic Domain on the Copperbelt of Zambia 1950–1956.* London: Academic Press.

Evans, A. and K. Young. 1988. *Gender Issues in Household Labour Allocation: The Transformation of a Farming System in Northern Province, Zambia.* London: Report to ODA's Economic and Social Research Committee.

F.A.O. 1977. *National Food and Nutrition Programme of Zambia.* Rome: F.A.O.

F.A.O./UNDP. 1974a. *National Food and Nutrition Programme of Zambia: Food Consumption Survey.* Rome: F.A.O.

F.A.O./UNDP. 1974b. *National Food and Nutrition Programme of Zambia: Nutrition Status Survey.* Rome: F.A.O.

Feierman, S. 1990. *Peasant Intellectuals: Anthropology and History in Tanzania.* Madison, WI: University of Wisconsin Press.

Ferguson, J. 1990a. Mobile Workers, Modernist Narratives: A Critique of the Historiography of Transition on the Zambian Copperbelt. Part I. *Journal of Southern African Studies* 16 (3):385–412.

Ferguson, J. 1990b. Mobile Workers, Modernist Narratives: A Critique of the Historigraphy of Transition on the Zambian Copperbelt. Part II. *Journal of Southern African Studies* 16 (4):603–621.

Fields, K. E. 1985. *Revival and Rebellion in Colonial Central Africa.* Princeton: Princeton University Press.

Francis, P. 1988. Ox Draught Power and Agricultural Transformation in Northern Zambia. *Agricultural Systems* 27:35–49.

Freund, P. J. 1985. *Evaluation Report on the National Nutrition Surveillance Programme.* Lusaka: Institute of African Studies, Univerisity of Zambia.

Freund, P. J. and K. Kalumba. 1984. *UNICEF/GRZ Monitoring and Evaluation Study of Child Health and Nutrition in Western and Northern Provinces, Zambia.* Lusaka: Institute of African Studies, University of Zambia.

Gamitto, A. C. P. 1960. *King Cazembe and the Marave, Cheva, Bisa, Bemba, Lunda and other Peoples of Southern Africa.* Vols. 1 and 2. Translated by I. Cunnison. Lisbon: Junta de Investigacoes du Ultramar, Centro de Estudos Politicos e sociais nos 42/43.

Gann, L. H. 1964. *A History of Northern Rhodesia.* London: Chatto & Windus.

Garvey, B. 1977. Bemba Chiefs and Catholic Missions, 1898–1935. *Journal of African History.* 18:411–426.

Gatter, P. 1990. Indigenous and Institutional Thought in the Practice of Rural Development: A Study of an Ushi Chiefdom in Luapula, Zambia. Unpublished PhD Thesis, University of London.

Geisler, G. et al. 1985. *Needs of Rural Women in the Northern Province.* Lusaka: Report Prepared for NCDP/NORAD.

Gertzel, C. 1980. Two Case-Studies in Rural Development. In: W. Tordoff ed. *Administration in Zambia.* Manchester: Manchester University Press, 240–259.

Giraud, V. 1890. *Les Lacs de l'Afrique Equatoriale.* Paris: Hachette.

Gladstone, J. 1985, Audrey Richards, Teacher. *Cambridge Anthropology* 10:(1):10–11.

Gladstone, J. 1986. Significant Sister: Autonomy and Obligation in Audrey Richards' early fieldwork. *American Ethnologist* 13:338–363

Gladstone, J. 1987. Venturing on the Borderline: Audrey Richards' Contribution to the Hungry Thirties Debate in Africa. *Bulletin of the Society of the Social History of Medicine*, 40.

Gladstone, J. 1992. Audrey I. Richards (1899–1984): Africanist and Humanist. In: S. Ardener ed. *Persons and Powers of Women in Diverse Cultures*, Oxford: Berg Press, 13–28.

Gluckman, M. (1954). Succession and Civil War Among the Bemba. *Rhodes: Livingstone Journal* 16:6–25.

Gluckman, M. 1961. Anthropological Problems Arising from the African Industrial Revolution. In: A. Southall ed. *Social Change in Modern Africa*. Oxford:

Gobezie, A. 1984a. *Conclusion and Recommendation of the Nutrition Studies of the Two Communities in Luapula Province*. Mansa: Adaptive Research Planning Team, Luapula Province.

Gobezie, A. 1984b. *Mukunta Nutrition Survey Wet/Hungry Season*. Mansa: Adaptive Research Planning Team, Luapula Province.

Goldman, I. and Holdsworth I. 1990. Agricultural Policies and the Small-Scale Producer. In: A. Wood et al. (eds) *The Dynamics of Agricultural Policy and Reform in Zambia*. Ames, Iowa: Iowa State University Press: 555–583.

Gore-Brown, G. 1938. A Rough Record, June 1936 to January 1937, of the Food Supply for Fifteen Adjacent Villages in the north Mpika District (Babemba Tribe). Appendix 7 of *A Report by the Committee appointed to make a Survey and Present a Review of the Present Position of Nutrition in Northern Rhodesia*, Lusaka.

Gouldsbury, C. 1915. Notes on the Customary Law of the Awemba: Part I. *Journal of the African Society* 14:366–385

Gouldsbury, C. 1915. Notes on the Customary Law of the Awemba: Part II. *Journal of the African Society* 15:36–52.

Gouldsbury, C. 1916. Notes on the Customary Law of the Awemba: Part III. *Journal of the African Society* 15:157–84.

Gouldsbury, C. and H. Sheane, 1911. *The Great Plateau of Northern Rhodesia*. London: Edward Arnold.

Government of Zambia, Central Statistical Office. 1990. *Census of Population and Housing*. Lusaka.

Government of Zambia, Central Statistical Office. 1980 *Census of Population and Housing*. Lusaka.

Guyer, J. 1981. Household and Community in African Studies. *African Studies Review* 24: 87-137.

Guyer, J. 1988. A Dynamic Approach to Domestic Budgeting: Cases and Methods from Africa. In: J. Bruce and D. Dwyer eds. *A Home Divided: Women and Income in the Third World*. Stanford: Stanford University Press.

Guyer, J. and P. Peters. 1987. Introduction. *Development and Change*. 18 (2):197–214.

Harries-Jones, P. 1975. *Freedom and Labour: Mobilization and Political Control on the Zambian Copperbelt*. Oxford: Basil Blackwell.

Harries-Jones, P. and J. C. Chiwale. 1963. Kasaka: A Case-Study in Succession and Dynamics of a Bemba Village. *Rhodes-Livingstone Journal* 33:1–67.

Harriss, B. 1990. The Intra-Family Distribution of Hunger in South Asia. In: J. Dreze and A. Sen eds. *The Political Economy of Hunger. Vol. I: Entitlement and Well-Being*. Oxford: Clarendon Press.

Haug, R. 1981. *Agricultural Crops and Cultivation Methods in the Northern Province of Zambia*. Kasama: Zambian SPRP Studies, No 1.

Hedlund, H. and M. Lundahl. 1984. The Economic Role of Beer in Rural Zambia. *Human Organisation* 43(1):61–65.

Heisler, H. 1984. *Urbanization and the Government of Migration: The Inter-Relation of Urban and Rural Life in Zambia*. New York: St. Martins's Press.
Hellen, J. A. 1968. *Rural Economic Development in Zambia, 1890–1964*. Munich: Weltform Verlag.
Henn, J. K. 1983. Feeding the cities and feeding the peasants: What role for Africa's women farmers? *World Development*, 12, 1043–55.
Herthelius, I. J. 1984. *Action Orientated Investigation on Nutrition Status of Children Under Five in Samfya and Nchelenge Districts, Luapula Province*. Mansa: Integrated Rural Development Programme, Luapula Province.
Hill, P. 1977. *Population, prosperity and poverty, rural Kano 1900 and 1970*. Cambridge: Cambridge University Press.
Hinfelaar, H. F. 1989. Religious Change Among Bemba-Speaking Women. Unpublished Ph.D. Thesis. University of London.
Holden, S. 1988. *Farming Systems and Household Economy in New Chambeshi, Old Chambeshi and Yunge villages near Kasama, Northern Province, Zambia: An Agroforestry Baseline Study*. Kasama: Zambian SPPR Studies, No 9.
Hurlich, S. 1986. *Women in Zambia*. Lusaka: Canadian International Development Agency.
IFPRI. 1985. *Maize Policies and Nutrition in Zambia: A Case Study in Eastern Province*. Lusaka: International Food Policy Research Institute, National Food and Nutrition Commission, Rural Development Studies Bureau.
Iliffe, J. 1979. *A Modern History of Tanganyika*. Cambridge: Cambridge University Press.
IRDP. 1984. *Factor Allocation and Technology Adoption in Small Scale Agriculture: A Case Study from Northern Zambia*. Mpika: Integrated Rural Development Programme, Occasional Paper No. 9.
IRDP. 1985a. *Agricultural Commercialisation and the Allocation of Labour Time in Mpika District, Northern Zambia*. Mpika: Integrated Rural Development Programme, Occasional Paper No. 6.
IRDP. 1985b. *The Impact of Ox-Draught Power on Small-Scale Agriculture in Mpika District, Northern Zambia*. Mpika: Integrated Rural Development Programme, Occasional Paper No. 7.
IRDP. 1986. *The Nutritional Impact of Agricultural Change in the IRDP Serenje Mpika and Chinsali Districts*. Mpika: Integrated Rural Development Programme, Occasional Paper No. 14.
Jiggins, J. 1980. *Female-Headed Households: Mpika Sample, Northern Province*. Lusaka: Rural Development Studies Bureau, Occasional Paper No. 1.
Jiggins, J. 1981. Food processing: Finger Millet and Cassava. Appendix 2 for Technical Paper No. 6. In: *Zambia: Basic Needs in an Economy Under Pressure*. Addis Ababa:ILO/JASPA.
Johnson, R. W. M. 1964. Agricultural Development at Mungwi: A Project Analysis. *Agricultural Economics Bulletin for Africa* 5:42–110.
Jones, C. W. 1986. Intra-Household Bargaining in Response to the Introduction of New Crops: A Case Study from Northern Cameroons' In: L. Moock ed. *Understanding Africa's Rural Household and Farming System*. Boulder, CO: Westview Press.
Kalima, L. C. 1983. The Characterisation, Distribution and Extent of the major soils of the high rainfall areas of Zambia. In: Henning Svads (ed.) *Proceedings of the Seminar on Soil Productivity in High Rainfall Areas of Zambia*, Kasama: SPRP Occasional Paper No. 6, 30–63.
Kapferer, B. 1967. *Co-operation, Leadership and Village Structure*. Lusaka: Institute for Social Research, University of Zambia.
Kapferer, B. 1988. The Anthropologist as Hero: Three Exponents of Post-Modernist Anthropology. *Critique of Anthropology*, 8:77–104.
Kauppinen, M. and M. Mweemba. 1985a. *National Nutrition Surveillance Programme, Evaluation Report for the Copperbelt Province*. Lusaka: Ministry of Health.

Kauppinen, M. 1985b. *National Nutrition Surveillance Programme, Evaluation Report for Northern Province, Part I: Health Centres.* Lusaka: Ministry of Health.

Kauppinen, M. 1985c. *National Nutrition Surveillance Programme, Annual Report, 1984.* Lusaka: Ministry of Health.

Kay, G. 1962. Agricultural Change in the Luitikila Basin Development Area Mpika District, Northern Rhodesia. *Journal of the Rhodes-Livingstone Institute* 31:21–50.

Kay, G. 1964. Sources and Uses of Cash in Some Ushi Villages, Fort Rosebery District, Northern Rhodesia. *Rhodes-Livingstone Journal* 35:14–28.

Kay, G. 1967. *Social Aspects of Village Regrouping in Zambia.* Lusaka: Institute for Social Research, University of Zambia.

Kerven, C. 1988. *Input Supply and Demand Survey.* Kasama: Adaptive Research Planning Team Report No. 1.

Kerven, C. and P. Sikana. 1987. *Trial Farmer Sociological Monitoring Nsokolo Area.* Kasama: Adaptive Research Planning Team Community Studies No. 2.

Kerven, C. and P. Sikana. 1988. *Case Studies of Indigenous Soil and Land Classifications in Northern Province.* Kasama: Adaptive Research Planning Team, Northern Province, Zambia.

Kuczynski, R. R. 1949. *Demographic Survey of the British Colonial Empire. Vol. II,* London: Oxford University Press.

Kwofie, K. M. 1979. *Integrating Nutrition Considerations into the Development of Zambia.* Lusaka: NFNC.

Kydd, J. 1988. Zambia. In: C. Harvey ed. *Agricultural Pricing Policy in Africa.* London: Macmillan.

Labrecque, E. 1931. Le Marriage chez les Babemba. *Africa* 4:209–221.

Ladislav, H. 1986. *Strategies and Norms in a Changing Matrilineal Society: Descent, Succession and Inheritance among the Toka of Zambia.* Cambridge: Cambridge University Press.

Lawton, R. M. 1978. A Study of the Dynamic Ecology of Zambian Vegetation. *Journal of Ecology* 66:175–198.

Lipton, M. 1983. *Poverty, Undernutrition and Hunger.* Washington, DC: World Bank Working Paper No. 597.

Livingstone, D. 1874. *The Last Journals of David Livingstone.* H. Waller ed. London: John Murray.

Long, N. 1968. *Social Change and the Individual: A Study of the Social and Religious Responses to Innovation in a Zambian Rural Community.* Manchester: Manchester University Press

Longhurst, R. 1984. *The Energy Trap: Work, Nutrition and Child Malnutrition in Northern Nigeria.* Cornell International Monograph Series, No.13, Cornell University, N.Y.

Luchembe, C. 1992. Ethnic Stereotypes, violence and labour in early colonial Zambia, 1889–1924. In: Samuel N. Chipungu ed. *Guardians in their time: Experiences of Zambians Under Colonial Rule, 1890–1964,* London: Macmillan 30–50.

Magubane, B. 1971. A Critical Look at Indices used in the Study of Social Change. *Current Anthropology.* 12:419–45.

Makings, S. M. 1966. Agricultural Change in Northern Rhodesia, Zambia, 1945–1965. *Food Research Institute Studies* 6(2):195–247.

Malinowska-Wayne, H. 1985. Audrey: Some Re-collections. *Cambridge Anthropology* 10:(1)14–18.

Mandala, E. C. 1990. *Work and Control in a Peasant Economy: A History of the Lower Tchiri Valley in Malawi, 1859–1960* Madison, WI: University of Wisconsin Press.

Mann, K. and R. Roberts eds. 1991. *Law in Colonial Africa.* Portsmouth, NH: Heinemann and James Currey.

Mansfield, J. E. et al. 1973. *Summary of Agronomic Research Findings in Northern Province, Zambia.* Supplementary Report No 7. Land Resources Division, Ministry of Overseas Development, England.

Mansfield, J. E. et al. 1975. *Land Resources of the Northern and Luapula Provinces Zambia—A Reconnaissance Assessment*. Vols. 1–6. Land Resources Division, Ministry of Overseas Development, England.

Marcus, G. E. and Fischer, M. 1986. *Anthropology as Culture Critique: An Experimental Moment in the Human Sciences*. Chicago: Chicago University Press.

Meebelo, H. 1971. *Reaction to Colonialism*. Manchester, UK: Manchester University Press.

Milimo, J. 1983. Socio-Economic Aspects of Small Scale Farmers in Northern Zambia. In: H. Svads ed. *Proceedings of the Seminar on Soil Productivity in the High Rainfall Areas of Zambia*. Kasama SPRP Studies, No.6, 314–329.

Mitchell, J. C. 1957. Aspects of African-Marriage on the Copperbelt in Northern Rhodesia. *Rhodes-Livingstone Journal* 22:1–30.

Mitchell, J. C. 1961. Urbanization, Detribalization and the Stability of African Marriage in Northern Rhodesia. In: A. Southall ed. *Social Change in Modern Africa*. London: Oxford University Press.

Mitchell, J. C. 1969. Urbanization, Detribalization, Stabilization and Urban Commitment in Southern Africa. In: P. Meadows and E. Mizruchi eds. *Urbanism, Urbanization and Change: Comparative Perspectives*, Reading, MA: Addison Wesley Publishing Company.

Mitchell, J. C. 1987. *Cities, Society and Social Perception: A Central African Perspective*. Oxford: Clarendon Press.

Moffat, V. J. 1932. *Native Agriculture in the Abercorn District*. 2nd Annual Bulletin of the Department of Agriculture. Livingstone: Government Printer.

Moore, R. J. B. 1943. *These African Copper Miners: A Study of the Industrial Revolution in Northern Rhodesia, with Principal Reference to the Copper Mining Industry*. London.

Moore, S. F. 1986. *Social Facts and Fabrications: "Customary" Law on Kilimanjaro 1880–1980*. Cambridge: Cambridge University Press.

Moore, H. L. 1993a. Gender and the Modelling of the Economy. In: S. Ortiz and S. Lees eds. *Economy as Process* Lanham, MD: University Press of America.

Moore, H. L. 1993b. Master Narratives: Anthropology and Writing. In: M. Biriotti ed. *What is an Author?* Manchester: Manchester University Press.

Moore, H. L. and Vaughan, M. 1987. Cutting Down Trees: Women, Nutrition and Agricultural Change in the Northern Province of Zambia, 1920–1986. *African Affairs* 86: 523–541.

Muntemba, M. S. 1978. The Underdevelopment of Peasant Agriculture in Zambia: The Case of Kabwe Rural District, 1964–70. *Journal of Southern African Studies* 5:59–85.

Muntemba, M. S. 1982. Women as Food Producers and Suppliers in the Twentieth Century: The Case of Zambia. *Development Dialogue* 1–2:29–50.

Murray, C. 1981. *Families Divided*. Cambridge: Cambridge University Press.

Musambachime, M. C. 1992. Colonialism and the environment in Zambia, 1890–1964. In: Samuel N. Chipanga ed. *Guardians in Their Time: Experiences of Zambians Under Colonial Rule, 1890–1964*. London: Macmillan:8–30.

Mvunga, M. P. 1980. *The Colonial Foundation of Zambia's Land Tenure System*, Lusaka: National Education Company of Zambia.

Northern Rhodesia. 1916. *Annual Report on Native Affairs for 1915*. Lusaka.

Northern Rhodesia. 1918. *Annual Report on Native Affairs for 1917*. Lusaka.

Northern Rhodesia. 1919. *Annual Report on Native Affairs for 1918*. Lusaka.

Northern Rhodesia. 1921. *Annual Report on Native Affairs for 1920*. Lusaka.

Northern Rhodesia, 1931. *Annual Report on Native Affairs for 1930*. Lusaka.

Northern Rhodesia. 1961. *An Account of the Disturbances in Northern Rhodesia, July to October 1961*. Lusaka: Government Printer.

Northern Rhodesia. 1961. *An Account of the Disturbances in North-Eastern Rhodesia, July to October 1961*. Lusaka: Government Printer.

Ohadike, P. O. 1969. *Some Demographic Measurements for Africans in Zambia. Communication of the Institute of Social Research* No. 5, University of Zambia. Lusaka.

Ortner, S. 1984. Theory in Anthropology since the sixties. *Comparative Studies in Society and History* 26:126–166.

Pacey, A. and P. R. Payne 1985. *Agricultural Development and Nutrition*. London: Hutchinson.

Palmer, R. 1983. Land Alienations and Agricultural Conflict in Colonial Zambia. In: R. Rotberg ed. *Imperialism, Colonialism and Hunger: East and Central Africa*. Lexington, Mass: D.C. Heath.

Parpart, J. 1983. *Labour and Capital on the African Copperbelt*. Philadelphia: Temple University Press.

Parpart, J. 1986. Class and Gender on the Copperbelt. In: C. Robertson and I. Berger eds. *Women and Class in Africa*. London: Africana Publishing Company.

Pineau, H. 1937. *Eveque-Roi des Brigands: Monseigneur Dupont*, Montreal.

Perez, L. M. 1984. *An Assessment of the Nutritional Situation in Zambia for Planning Nutritional Interventions*. German Agency for Technical Cooperation. (GTZ).

Perrings, C. 1979. *Black Mineworkers in Central Africa*. London: Heinemann.

Peters, D. U. 1950. *Land Usage in Serenje District: A Survey of Land Usage and the Agricultural System of the Lala of the Serenje Plateau*. Rhodes-Livingstone Papers, No. 19.

Poewe, K. 1978. Religion, Matriliny and Change: Jehovah's Witnesses and Seventh-Day Adventists in Luapula, Zambia. *American Ethnologist* 5(2):303–321.

Poewe, K. 1979. Regional and Village Economic Activities: Prosperity and Stagnation in Luapula, Zambia. *African Studies Review* 22(2):77–93.

Poewe, K. 1981. *Matrilineal Ideology: Male-Female Dynamics in Luapula, Zambia*. London: Academic Press.

Popkin, B. M. 1981. Time-allocation of the mother and child nutrition *Ecology of Food and Nutrition*, 9:1–14.

Pottier, J. 1983. Defunct Labour Reserve? Mambwe Villages in the Post-Migration Economy. *Africa* 53(2)2–23.

Pottier, J. 1988. *Migrants No More: Settlement and Survival in Mambwe Villages, Zambia*. Manchester, UK: Manchester University Press.

Powdermaker, H. 1962. *Copper Town: Changing Africa*. New York: Harper and Row.

Preston, Thomas, B. 1954/1968. *Two Studies in African Nutrition: An Urban and a Rural Community in Northern Rhodesia* Rhodes-Livingstone paper, No.24. Institute for Social Research, University of Zambia. Manchester, UK: Manchester University Press.

Reid, P. et al. 1986. *A Report on the Identification of Zones for Agricultural Research in the Northern Province of Zambia*. Kasama: Adaptive Research Planning Team, Northern Province, Zambia.

Reynolds, P. 1991. *Dance Civet Cat: Child Labour in the Zambezi Valley*. Athens, OH: Ohio University.

Richards, A. I. 1932. *Hunger and Work in a Savage Tribe*. London: Oxford University Press.

Richards, A. I. 1936 The Life of Bwembya, a Native of Northern Rhodesia. In: M. Perham ed. *Ten Africans*. London: Faber & Faber, 17–40.

Richards, A. I. and E. M. Widdowson. 1936. A Dietary Study of North-Eastern Rhodesia. *Africa* 9:166–196.

Richards, A. I. 1939. *Land, Labour and Diet: An Ecnomic Study of the Bemba Tribe*. London: Oxford University Press.

Richards, A. I. 1940a. The Political System of the Bemba Tribe North-Eastern Rhodesia. In: M. Fortes and E. E. Evans-Pritchard eds. *African Political Systems*. London 83–120.

Richards, A. I. 1940b. *Bemba Marriage and Present Economic Conditions*. Livingstone: Rhodes-Livingstone Institute Paper, No. 4.

Richards, A. I. 1951. The Bemba of North Eastern Rhodesia. In: E. Colson and M. Gluckman eds. *Seven Tribes of British Central Africa*. London: Oxford University Press, 164–193.

Richards, A. I. 1958. A Changing Pattern of Agriculture in East Africa: The Bemba of Northern Rhodesia. *Geographical Journal* 124 (3):302–314.

Richards, A. I. 1960. Social Mechanisms for the Transfer of Political Rights in Some African Tribes. *Journal of Royal Anthropological Institute* 90:175–90.

Richards, A. I. 1961. African Kings and their Royal Relatives. *Journal of the Royal Anthropological Institute* 91:135–150.

Richards, A. I. 1968. Keeping the King Divine. Henry Meyers Lecture. *Proceedings of the Royal Anthropological Institute,* 23–35.

Richards, A. I. 1971. The Conciliar System of the Bemba of Northern Zambia. In: A. I. Richards and A. Kuper eds. *Councils in Action.* Cambridge: Cambridge University Press, 100–129.

Richards, A. I. and C. Tardits. 1974. A Propos du marriage Bemba. *L'Homme* 14(3–4):11–118.

Richards, A. I. 1982. *Chisungu.* Manchester, U.K. (first published 1956).

Roberts, A. D. 1973. *A History of the Bemba: Political Growth and Change in North Eastern Zambia before 1900.* London: Longman.

Roberts, A. D. 1981. *Kinship and the Accumulation of Power: The Bemba of N. E. Zambia.* Paper presented to the Conference on the History of the Family in Africa, School of Oriental and African Studies, University of London.

Safilios-Rothschild, C. 1985. *The Implications of the Roles of Women in Agriculture in Zambia.* New York: The Population Council.

Sano, H.-O. 1988. *Agricultural Policy Changes in Zambia During the 1980s.* Working Paper No. 88.4. Copenhagen: Centre for Development Research.

Sano, H.-O. 1989. *From Labour Reserve to Maize Reserve: The Maize Boom in the Northern Province in Zambia.* Working Paper No. 89.3. Copenhagen: Centre for Development Research.

Schofield, S. 1979. *Development and the Problem of Village Nutrition.* London: Croom Helm.

Schultz, J. 1976. *Land Use in Zambia.* Afrika Studies, No. 95. Munchen: Weltforum Verlag.

Scott, I. 1978. Middle-Class Politics in Africa. *African Affairs* 77:321–334.

Scott, J. 1985. *Weapons of the Weak: Everyday Forms of Peasant Resistance* New Haven, CT: Yale University Press.

Sen, A. 1981. *Poverty and Famines: An Essay on Entitlement and Deprivation.* Oxford: Oxford University Press.

Sharpe, B. 1987. *Agricultural Commercialisation Nutrition and Health in Northern Province: A Review of Recent Research and Policy Options.* Mpika: IRDP (SMC)/NFNC.

Sharpe, B. 1990. Nutrition and the Commercialisation of Africulture in Northern Province. In: A. Wood et al. eds. *The Dynamics of Agricultural Policy and Reform in Zambia.* Ames, Iowa: Iowa State University University Press, 583–602.

Sichone, O. B. 1991. Labour migration, peasant farming and rural development in Winamwanga. Unpublished PhD Thesis, University of Cambridge.

Stolen, K. A. 1983a. *Peasants and Agricultural Change in Northern Zambia.* International Development Programme, Agricultural University of Norway, Occasional Paper No. 4.

Stolen, K. A. 1983b. Socio-Economic Constraints on Agricultural Production in the Northern Province of Zambia. In: H. Svads ed. *Proceedings of the Seminar on Soil Productivity in the High Rainfall Areas of Zambia.* Kasama: SPRP Studies No. 6, 330–354.

Stromgaard, P. 1984a. Field Studies of Land Use Under Chitimene Shifting Cultivation, Zambia. *Geografisk Tidsskrift* 84:78–85.

Stromgaard, P. 1984b. Prospects of Improved Farming Systems in a Shifting Cultivation Area in Zambia. *Quarterly Journal of International Agriculture* 23(1):38–50.

Stromgaard, P. 1985a. A Subsistence Society Under Pressure: The Bemba of Northern Zambia. *Africa* 55(1):39–59.

Stromgaard, P. 1985b. The Infield-Outfield System of Shifting Cultivation Among the Bemba of South Central Africa. *Tools and Tillage* 5(2):67–84.

Stromgaard, P. 1988a. The Grassland Mound-System of the Asia Mambwe of Zambia. *Tools and Tillage* 6(1):33–46.
Stromgaard, P. 1988b. Soil and Vegetation Changes Under Shifting Cultivation in the Miombo East Africa. *Geografisk Annaler* 70B(3):363–374.
Stromgaard, P. 1989. Adaptive Strategies in the Breakdown of Shifting Cultivation: The Case of Lamba, and Lala of Northern Zambia. *Human Ecology* 17(4):427–444.
Stromgaard, P. 1990a. Effects of Mound-Cultivation on Concentration of Nutrients in a Zambia Miombo Woodland Soil. *Environment* 32:295–313.
Stromgaard, P. 1990b. Peasant Household Economy in Rural Zambia—The Dilemma of Small-Scale Farmers in Transition. *Fennia* 168(2):201–210.
Tardits, C. 1974. Pris de la femme et marriage entre cousins croisés: le cas des Bemba d'Afrique Centrale. *L'Homme* 14(2):5–30.
Trapnell, C. G. 1953. *The Soils, Vegetation and Agriculture of North-Eastern Rhodesia.* Lusaka: Government Printer.
Trapnell, C. G. and Clothier, J. N. 1957. *The Soils, Vegetation and Agricultural Systems of North-Western Rhodesia.* Lusaka: Government Printer.
Tweedie, A. C. 1966. *Change and Continuity in Bemba Society.* Unpublished Thesis. University of Oxford.
Van Donge, J. K. 1982. Politicians, Bureaucrats and Farmers: A Zambian Case-Study, *Journal of Development Studies* 19;88–107.
Vaughan, M. 1983. Which Family? Problems in the Reconstruction of the History of the Family in Africa. *Journal of African History,* 24:275–83.
Vedeld, P. and R. Oygard, 1982. *Peasant Household Resource Allocation: A Study of Labour Allocation of Peasant Households in Zambia's Northern Province and Market Constraints on Their Increased Agricultural Production.* Kasama: SPRP Occasional Paper No. 3.
Vedeld, T. 1981. *Social-Economic and Ecological Constraints on Increased Productivity Among Large Circle Chitemene Cultivators in Zambia.* Kasama: Zambian SPRP Studies, No. 2.
Vickery, K. 1985. Saving Settlers: Maize Control in Northern Rhodesia. *Journal of Southern African Studies* 11(2):212–235.
Watson, W. 1958. *Tribal Cohesion in a Money Economy: A Study of the Mambwe People of Zambia.* Manchester: Manchester University Press.
Werbner, R. 1967. Federal Administration, Rank and Civil Strife Among Bemba Royals and Nobles. *Africa* 37:22–49.
Whitehead, A. 1981. I'm Hungry Mum: The Politics of Domestic Budgeting In: K. Young et al. eds. *Of Marriage and the Market.* London: CSE Books.
Willis, R. G. 1966. *The Fipa and Related Peoples of South-West Tanzania and North-Eastern Zambia.* London: International Africa Institute.
Wilson, G. 1941. *An Essay on the Economics of Detribalization in Northern Rhodesia, Part 1.* Livingstone: Rhodes-Livingstone Paper No. 5.
Wilson, G. 1942. An Essay on the Economics of Detribalization in Northern Rhodesia, Part 2. Livingstone: Rhodes-Livingstone Paper No. 6.
Worboys, M. 1988. The Discovery of Colonial Malnutrition Between the Wars In: D. Arnold ed. *Imperial Medicine and Indigenous Societies.* Manchester: Manchester University Press, 208–226.
Wright, M. 1993. *Stragies of Slaves and Women: Life Stories from East/Central Africa.* New York: Lilian Barber Press.

Zambian National Archives

District Notebooks

KDH 1/1902 Kasama District Notebooks (Vols. 1–4)
KSD 4/1906 Mpika District Notebooks (Vols. 1–2)

Tour Reports (TRs)

ZA 2/4/1 Awemba Province Tour Reports 1928–29
ZA 7/4/10 Awemba Province Tour Reports 1930
ZA 7/4/19 Awemba Province Tour Reports 1931
ZA 7/4/28 Awemba Province Tour Reports 1932
ZA 7/4/37 Awemba Province Tour Reports 1933

SEC 2/786 Kasama Tour Reports 1933–38
SEC 2/788 Kasama Tour Reports 1940–47
SEC 2/789 Kasama Tour Reports 1948
SEC 2/790 Kasama Tour Reports 1949
SEC 2/792 Kasama Tour Reports 1951
SEC 2/793 Kasama Tour Reports 1952
SEC 2/796 Kasama Tour Reports 1955
SEC 2/797 Kasama Tour Reports 1956
SEC 2/799 Kasama Tour Reports 1958
SEC 2/800 Kasama Tour Reports 1959
SEC 2/801 Kasama Tour Reports 1960

SEC 2/836 Mpika Tour Reports 1933–38
SEC 2/837 Mpika Tour Reports 1938–40
SEC 2/838 Mpika Tour Reports 1940–48
SEC 2/839 Mpika Tour Reports 1948
SEC 2/840 Mpika Tour Reports 1949–50
SEC 2/841 Mpika Tour Reports 1950
SEC 2/842 Mpika Tour Reports 1951
SEC 2/843 Mpika Tour Reports 1952
SEC 2/844 Mpika Tour Reports 1953
SEC 2/847 Mpika Tour Reports 1956
SEC 2/848 Mpika Tour Reports 1957
SEC 2/849 Mpika Tour Reports 1958
SEC 2/850 Mpika Tour Reports 1959
SEC 2/851 Mpika Tour Reports 1960

SEC 2/753 Chinsali Tour Reports 1948
SEC 2/754 Chinsali Tour Reports 1949
SEC 2/766 Chinsali Tour Reports 1961

Other Files

SEC 1/1039 Report of Standing Committee on Human Nutrition, 1937
SEC 1/1042 Survey of Food Conditions, 1937
SEC 2/264 Native Markets, 1939–40
SEC 1/104 Food Supplies, 1947
SEC 5/182 Food Supplies, 1950
SEC 5/183 Food Supplies, 1951
SEC 1/428 Peasant Farms, Chinsali, 1948–52
SEC 5/393 Peasant Farmers' Scheme, 1955–58
SEC 2/227 Vol. 1 African Provincial Council, 1944–5
NR 17/131 Assistance to Peasant Farmers, 1949–51
NR 17/152 Rural Development: Peasant Farming Policy 1948–57
NR 17/106 Peasant Farming, 1956–62
SEC 5/393 Peasant Farmers' Schemes 1955

SEC 2/281 Vols. I and II Native Development: 5-Year Plan, Northern Province, 1943
SEC 2/1300 Annual Report on Native Affairs, Kasama, 1937
SEC 2/94 Annual Report for African Affairs, 1957
SEC 2/111 Annual Report for Native Affairs, Kasama, 1956
SEC 2/1297 Northern Province Annual Reports, 1935–37
SEC 2/406 Vol. I Native Customs, 1936
ZA 1/9/2/3 Native Custom, 1926

White Fathers' Archives: Rome

Chroniques (WFC)

Vicariat Apostolique de Nyassa Vol. 1: 1889–1894
Vicariat Apostolique de Nyassa Vol. 2: 1895–1906
Vicariat Apostolique de Nyassa Vol. 3: 1906–1912

Vicariat Apostolique de Bangweolo Vol. 1: 1912–1922
Vicariat Apostolique de Bangweolo Vol. 2: 1922–1952

Diaries (WFD)

Chilubula (Ituna) Vol. 1: 1899–1907
Chilubula Vol. 2: 1907–1916
Chilubula Vol. 3: 1919–29
Chilubula Vol. 4: 1930–46

Audrey Richards Papers (London School of Economics)

Large File V (Nutrition): Daily Diet Sheets
Large File U (Nutrition): Food Diaries

British South Africa Company Records

B51/31 Native Hut Tax and Labor, 1895
B53/418 Native Labor, Watchtower, 1919

Ann Tweedie Papers (ATW): Private papers

Rhodes House Library: Oxford

Frank Hulme Melland Diaries, 1901–1905

INDEX

Page numbers followed by t and f denote tables and figures, respectively.